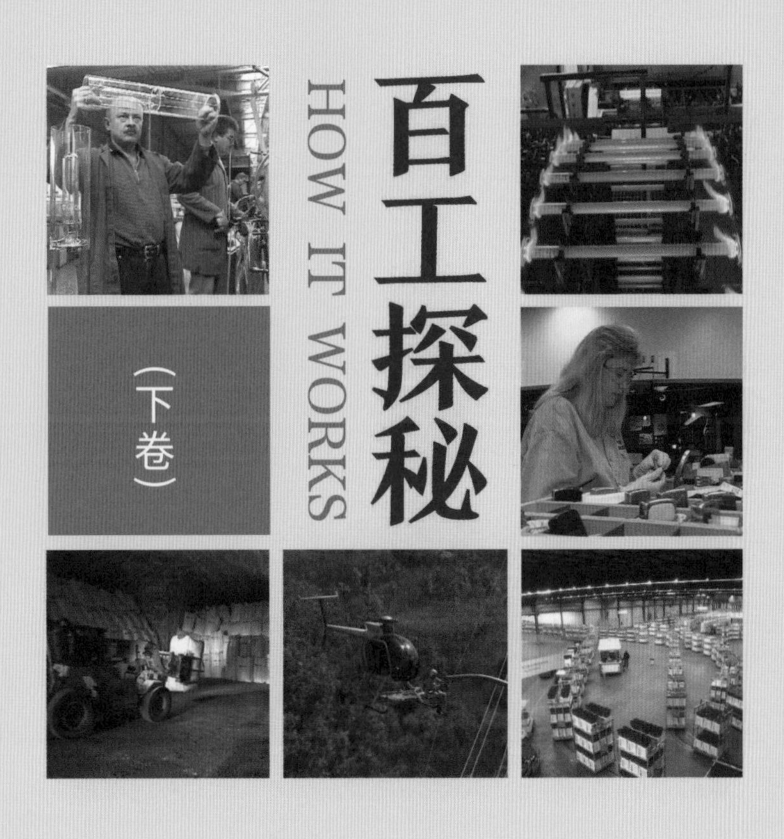

百工探秘

HOW IT WORKS

（下卷）

赵致真 张 戟◯主编 [德]红 箭◯著 高 洁 芒 冰◯译

长江出版传媒 湖北科学技术出版社

图书在版编目（CIP）数据

百工探秘（下卷） / 赵致真，张戟主编. [德]红箭(Red Arrow)著. 高洁，芒冰译-- 武汉 ：湖北科学技术出版社，2021.7
ISBN 978-7-5706-1513-1

Ⅰ. ①百… Ⅱ. ①赵… ②张… ③德… ④高… ⑤芒… Ⅲ. ①产品设计－介绍－世界 Ⅳ. ①TB472

中国版本图书馆 CIP 数据核字(2021)第 081110 号

百工探秘（下卷）
BAIGONG TANMI (XIAJUAN)

特约编辑：于雯雯
责任编辑：彭永东　　封面设计：胡　博

出版发行：湖北科学技术出版社　　电话：027-87679468
地　　址：武汉市雄楚大街 268 号　　邮编：430070
（湖北出版文化城 B 座 13-14 层）
网　　址：http：//www.hbstp.com.cn

印　　刷：武汉精一佳印刷有限公司　　邮编：430034

787×1092　1/16　20 印张　400 千字
2021 年 7 月第 1 版　　2021 年 7 月第 1 次印刷
（上下卷）定价：198.00 元

ABUS
GRANIT

书

斯沃琪手表

信用卡

电池

茶灯蜡烛

芝宝打火机

木炭

日用篇

衣架

剪刀

卷笔刀

镜子

铝合金行李箱

挂锁

铂金戒指

金属丝眼镜框

擦鞋垫

书

扫描二维码，观看中文视频。

沉迷于一部惊悚小说时，你会飞快地看完它。而对于这家德国工厂，印出一本书的速度比你的阅读更快。

小说以光盘的形式来临，在暗室里被加载到计算机上。今天他们将要印刷曾写过《侏罗纪公园》的畅销书作家迈克尔·克莱顿的新书，名叫《猎物》，德文为*Beute*。计算机控制激光，将版面编排刻到一个印刷版上。当激光穿透薄薄的铬层，落在下面的铜层时，文本就被创建出来。

余下的过程就像在工作室洗照片一样。印刷版被浸上显影液，64页书的内容就出现了。他们一次印两本书，所以实际上是一式两份32页书的内容，一册在另一册上面。

印刷版显影后就被送到一个印刷机。墨从印刷机板转到一个橡胶辊，然后压在页面上。他们必须要用一种特殊的纸，让墨在2秒之内干燥。否则印好的页面在快速通过后面迷宫一样的滚轮时，会变成字迹模糊的墨污。

最后的一段传输带，让书页通过一个斜槽，并在那里对折。它们出来后，已是32页的小册子。此后，会和书的其他部分装订到一起。

这本书准备做成硬精装版，所以无须奇怪封皮仅由两片厚硬纸板做成：一片用作封面，一片用作封底。书脊从一大卷硬纸板上切出来。

这种绿纸将用于覆盖硬纸板，用胶粘上去。但不是仅在每个角上涂点胶。一个特殊的机器给绿纸涂上一层薄胶，然后把封皮牢牢压紧。就如书页一样，封皮也是每次印一式两份。

书名用色带打印到书脊上。色带和老式打字机的色带差不多，只不过更厚一些。一个金属印戳压向色带，墨就被转到了硬纸板上。

人们总说，不能仅凭封皮来判断一本书。从技术上说，他们是对的。套在封皮外的护封才是真正吸引注意力的地方。这本书的护封设计由四种不同颜色构成，每次印上一种颜色。他们把印出的成品和原设计相对照，确保颜色组合正确无误。

书的所有部分都完成了。现在进入生产的最棘手部分——书的装订。32页的书帖按顺序码放到传送带上。这本书有14个书帖，当它们依次落到一个接一个传送带上时，一个巧妙的系统按正确顺序将它们汇拢成书。

他们在封面内侧贴进一张黑纸。接着，一台机器把它压紧，如同一块整齐的三明治。

书脊内壁被微微刮毛，这样能帮助书页更好地固定。他们把书页的边沿涂上一层胶水，然后贴上厚纸。以便书页在后面工序中不会脱散，也能保护书脊。

这些书此时仍是头脚相接的双胞胎，但不会太久了。一个刀片把它们切为两半。印出来的大张书页已经被折叠到位，但外边仍有连接缝。这只用通过简单的切边就能解决。

书脊经过机器挤压形成曲线。喷上胶水然

你知道吗？

《圣经》是有史以来最畅销的书，也是世界上被偷走最多的书。

后装上封皮。他们把书脊的边沿压上沟槽，为了能翻阅方便。到此封皮部分就完成了。

在它能登上畅销书榜首之前，还需要护封。一台机器把护封折叠套到封面上。此后书就可以被包装运送到世界各地的书店了。

如果这本书能引起轰动，它将很快脱销，因为第一版只印了 8 万册。但因为有如此快捷的印刷系统，所以脱销不是问题，他们能在一周之内完成第二轮印刷。

1. 激光将版面编排刻到印刷版上
2. 印刷版浸上显影液显出书的内容
3. 印刷版显影后就被送到印刷机
4. 特殊的纸让墨在 2 秒之内干燥
5. 书页通过斜槽并对折
6. 书页成为小册子
7. 硬壳精装书封面封底使用硬纸板
8. 书脊从大卷硬纸板上切下来
9. 书名用色带打印到书脊上

10. 一个金属印戳压向色带
11. 色带的墨被转到了硬纸板上
12. 确保颜色印刷正确无误
13. 书帖依次码放开始装订
14. 封面内侧贴进一张黑纸
15. 刀片把头脚相连的书切开
16. 大张书页连接缝需要切边
17. 开始装封皮
18. 书脊边沿压上沟槽便于阅读

Books

When you get stuck into a good thriller you can tear through it no time and for this German factory it's even quicker to print one.

The novel arrives to the printers on a CD rom which is loaded into a computer in the dark room. Today they are going to print the new book from the best selling author of Jurassic Park, Michael Crichton. It's called "Prey" or "Beute" as it will be known in Germany. The computer controls a laser which engraves the layout of the book onto a printing plate the text is created as the laser scratches through a thin layer of chrome to a layer of copper beneath.

The rest of the process is just like producing a photo in a lab. The plate is dipped in developing fluid and 64 pages of text appear. They print the books two at a time, so it's actually 32 pages in duplicate, one copy above the other.

Once the plate is developed it heads off to a printing machine. The ink is transferred from the printer plate onto a rubber roll and then pressed onto the pages. They have to use a special type of paper which helps the ink to dry in less than 2 seconds. Otherwise the pages whizzing through the maze of rollers would be in illegible smudge.

The last section of the belt takes the sheets through a chute where they are folded in half. They come out as 32 page pamphlets which will be bound together with the rest of the book later on.

This book is going to be a hard back and not surprisingly the cover is made from a couple of pieces of thick cardboard, one for the front cover and one for the back. The spine is cut out from a large roll of cardboard.

This green paper will be used to cover the cardboard and it will be stuck into place with glue but it's not just going to be a blob in each corner a special machine coats the green paper with a thin layer of glue and then stamps the cover firmly into place. Just like the pages the covers are printed two at a time too.

The title gets printed onto the spine by an ink ribbon like you would find in an old fashioned typewriter, just a bit thicker. A metal stamp presses through the ribbon and the ink is transferred onto the cardboard.

People will always say that you shouldn't judge a book by it's cover and technically they are right, it's the dust jacket on top of the cover that's there to attract all of the attention. The design on this jacket is going to be made up of 4 different colour pigments which are applied one at a time. They check the copy against the original design to make sure they've got the right combination of colours.

All of the elements of the book are now finished and it's time for the trickiest part of production, the assembly. The 32 page pamphlets are stacked in order above a conveyer belt. For this novel there are 14. And an ingenious system compiles the books in the correct order as the pamphlets drop onto the belt one by one.

They stick in a sheet of black paper which will go inside the cover then a machine crushes it all together to form a neat sandwich.

The inside of spine is slightly shredded, this will help the pages stay in place. They coat the edge of the pages with a layer of glue and then a thick piece of crate paper is placed on. This helps to keep the pages in position for the rest of the process and also protect the spine.

The books are still twins joined head to foot, but not for long. A blade cuts them in half. The large printed sheets have been folded into position so they still have seams on the outside. This is taken care of in no time by simply chopping off the edges of the pages.

They squeeze the spine to give it a curve. Glue is sprayed on. Then the cover can be attached. They crimp the edge of the spine to allow the pages to fall open easily – and the hard back is finished.

But before it can fly to the top of the best seller list, it needs it's dust jacket. A machine folds it around the covers. Then the books are ready to be packed and delivered to bookshops around the world.

If the book is a hit it will be sold out in no time as they have only made 80,000 copies of this first edition. But with the slick press they've got in place that's not a problem, they can print a second run in just a week.

Did you know?

As well as being the best-selling book of all time. the Bible is also the most shoplifted book in the world.

斯沃琪手表

扫描二维码，观看中文视频。

时间就是金钱，至少对于斯沃琪来说是如此。斯沃琪在全世界已售出超过3亿块的手表。制作手表的第一步，是在底盘装上带动传动齿轮组的螺杆。手表的制造几乎全部由自动化设备来完成。一块手表由51个零件构成。

很快，表身将和底盘结合，但表身首先要装上转动表针的齿轮。总共有5个主要齿轮。3个齿轮对应于时针、分针和秒针，另外2个，则是将三者连接起来。机械手臂拿捏着这些微小的齿轮，将它们精准地放置到位。

接下来是日历环，它和齿轮连到一起显示日期。此后，表身被装到底盘上。一个传送带将它们送到焊接机，在那里两者将合为一体。工作原理是这样的。当这两个部分被紧紧夹到一起的时候，压力使销钉震动产生热量，引起塑料融化。当塑料冷却后，钟表装置也就熔合了。

紧接着要做的，是将表盘从模板上切下来安装到位。当它翻转过来后，就可以装上一次性电池了。

表的指针被做成条带状，绕在卷轴上。它们被切下来后，安装到表盘。这个表有6个指针。3个用于显示时间，3个用于秒表。这些指针必须彼此平行才能正常工作，所以需要工人用镊子对它们进行调整。这是项烦琐的工作，但在大特写镜头的帮助下，每块手表只需花费几秒钟。

现在可以设置时间了。他们用原子钟来做这项工作，所以再也没有迟到的借口。这些手表接下来被装上钢制表壳。

首先，秒表的按钮被装入表壳。接着，是手表本身。它们的结合非常完美，连胶水都不需要使用。这些螺旋表把可以让用户手动设定时间。在边缘加上一层硅胶，使手表具有防水功能。当放置妥当后，用玻璃将硅胶封在里面。

装上外框，剩下的就是封住电池，手表到此完工。 但出厂前还要完成一项耐久性测试。6000块精确校准的手表同时放进一个50摄氏度的箱子。如果在这个转动的热箱子里待上24小时安然无恙，就被允许出厂，得到所有重要的质量担保。安上表带后，已经大功告成。现在唯一需要的，就是找个主人了。

你知道吗？

最小的时间单位之一是阿秒，即10亿分之一秒的10亿分之一。

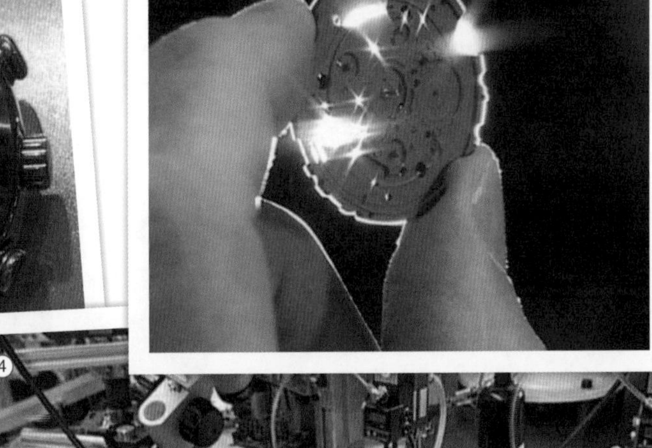

1. 斯沃琪已售出超过 3 亿块手表
2. 一块手表由 51 个零件构成，这是表身
3. 手表制造全部自动化完成

4. 机械把微小的齿轮放置到位
5. 首先装上 3 个表针齿轮和 2 个连接齿轮
6. 日历环和齿轮连接后焊接底盘

7. 准备安装电池

8. 借助计算机调整表盘指针平行

9. 调整好时间准备安装表壳

10. 螺旋表把加一圈硅胶封住

11. 在 50 摄氏度热箱中转动 24 小时

12. 完成耐久性测试后装好表带

Swatch Watch

Time actually does equal money, for Swatch at least. They've sold over three hundred million watches worldwide. One of the first steps in making one is to fit the base with screws that will turn the wheelwork. The watches are made almost entirely by robots and machinery. There are 51 parts in each wrist watch.

Soon the base will be joined by the body but first the body needs to be fitted with the wheels which will turn the hands. There are five main wheels. Three for the hours, minutes and seconds and two that connect these together. Robotic arms pick up the tiny wheels and delicately place them into position. The calendar ring is next, it's connected to the wheels to display the date.

Then the body is fitted onto the base. A conveyor belt takes them to the welding machine where two will become one. Here's how it works. As the two sections are clamped together pressure causes the pins to oscillate and this generates heat. The plastic melts and when it cools down the clockwork is fused.

It's whisked along to the next stop where the face is cut out of a template and lifted into position. Once it's been flipped over a disposable battery can be fitted.

The clock hands have been made in strips and wound around these spools. They're punched out and put onto the faces. There are six on this watch. Three to display the time and three for the stop watch. To work properly they must be parallel so a worker adjusts them with tweezers. It's a fiddly job but with the help of a huge close up it only takes a few seconds for each watch.

Now they can set the time. They use an atomic clock for the job so there'll be no excuses for being late. The watches are off again to get their steel cases fitted.

First the stopwatch pin is inserted. And then the watch itself. It is such a perfect fit that no glue is needed to hold it in place. These screws will allow the user to set the time manually. A layer of silicon is put on the rim to make the watch waterproof. And after it has been set in the right position the glass seals the silicon inside.

Once the outer rim has been slotted into place all that remains is to seal the battery and the watch is finished but it won't be leaving the factory until it has past an endurance test. It's placed in a cabinet with 6 thousand finely tunes time pieces for company and is left to sweat it out at temperatures of up to 50 degrees Celsius. If it survives the rotating hot box for 24 hours it gets the nod and the all importance manufacture's guarantee. With its strap attached the watch is complete all it needs now is an owner.

Did you know?

One of the smallest units of time is an attosecond, a billionth of a billionth of a second.

信用卡

扫描二维码，观看中文视频。

一年时间里，单单是信用卡就在英国消费大约 1240 亿英镑。从香水和礼品到餐厅吃饭都用信用卡支付。这些便利的长方形小卡片是怎么制作出来的，它们又是如何工作的？一张普通信用卡的生命，从类似于斯洛伐克一个这样的工厂开始。它的地址不向公众通告，安全措施严密，以防犯罪分子光顾和觊觎。里面有一个设计中心，新的信用卡诞生于此。

配色方案获得批准后就把颜料混合起来。对于小数量订单，这项工作用手工完成。配方的精确成分也高度保密。

工作人员将塑料片材放入压力印刷机，用来制成信用卡的正面和背面。加进的第一种颜色是白色。在安全摄像机的监视下新压印的卡片做好了。老板们非常警惕，确保没有东西从这个工厂丢失。下一步是实现银行要求的整体色彩设计。将不同的颜料添加到滚筒上，一个特殊的装置用来确保颜料厚薄均匀。

这里就像一个注册的印钞工厂。这台机器每小时能印出大约 7000 张新卡。信用卡公司有巧妙的办法确保生产过程不可被复制。这些真信用卡中隐藏着大量的保密措施，让造假者很难得逞。印刷技术让识别假卡变得相对容易，我们可以向你展示的一个有趣技巧是紫外光印记。当卡完成后这种特殊标记只会在紫外光下显示出来。新的信用卡包含几个部分，有保护膜，打印出来的正面和背面，以及背面的高科技磁条。

它们都是如何工作让你能够花钱呢？是的，用借记卡购物是个非常简单的过程。把你的卡插入读卡器后，信号就被送到你的银行。银行检查卡是否有效，账户是否存在，账号里面有没有钱。如果一切正常，信号被返回商店并要求证明卡是你的。你输入正确的密码后，银行将支付你的账单并且直接记入你的账户。

如果你用的是信用卡，读卡器会联系数据中心，获取信用卡交易请求。公司计算机验证账户有效，而且没有超过限额，在一切正常情况下将支付账单。出现问题则需要进一步核实，但交易通常相当简单。

所有必要的个人信息都存储在信用卡背面的磁条里。这是信用卡如何工作的关键。这些磁条起初绕在卷筒上，然后与一个保护层结合，加入信用卡的构成。

每张卡都要做得完美无缺，否则就不能使用。专家用显微镜检查每张卡。任何有缺陷的卡将被立即剔除。

下一步是让 4 个不同的组件相结合，做出人们熟悉的用来购物的塑料卡。片材经过仔细清洗并叠放入钢板中，装进重型的铜质箱体。然后送入压力机，在 150 摄氏度下产生巨大的 120 吨的压力。这惊人的力量把每层原料压合在一起，使其不再分离。 在压力机的另一端，含有 48 张信用卡的整张塑料片被切成正确形

你知道吗？

如果女王没钱了，她不用担心。白金汉宫有自己的私人取款机。

状。切的过程中还添加了签名条和全息图，这是设计的另一项财务安全措施。

信用卡欺诈是一个世界性问题，它在2006年给英国造成了4.28亿英镑的损失。克隆信用卡是最常见的作假招数之一。安全专家花费数百小时来研究诈骗者是如何做到的。一种窃取个人信息的常用方法，是在取款机上安装一个假卡插槽。通常还伴有一个摄像头，它会在你输入密码时进行记录。

芯片和个人密码技术用来防止盗用信用卡信息。这应该是万无一失的安全措施，但仍有赖于你输入密码时的小心谨慎。

一个特殊的钻头用来在预先制作好的信用卡上钻孔，随后将芯片植入。信用卡会通过严格的检验体系，确保芯片不会在受力过多的情况下弹出来。信用卡做好后，将被打包，计数，安全地存储起来，以备银行需要。

到此为止了吗？不，下一步是你申请一张新卡，或者你的旧卡被盗。发生这种事情时，你打电话给你的信用卡供应商。然后，他们和这家制卡公司联系，为你签发一张新卡。

一盒空白的新卡被打开，并放进这台机器里。它的内存已经有你和许多其他新卡用户的账号信息。

这台机器接下来会把你的名字和其他细节印到正面，同时把账户信息写入磁条。大功告成了。一块不值钱的塑料片，就这样变成了马上可以当钱花的信用卡。

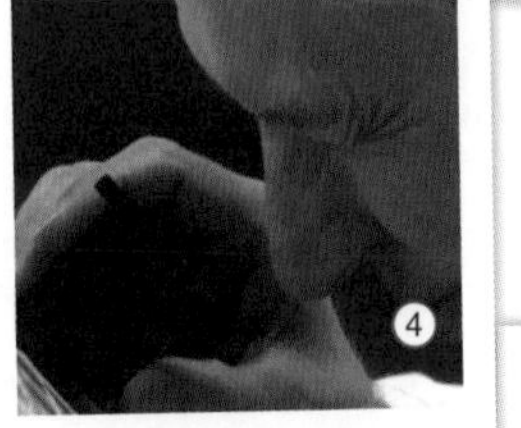

1.用信用卡购物很便利
2.信用卡的生产基地保密
3.用人工调制颜色完成设计
4.信用卡隐藏大量保密措施
5.信用卡含保护膜及磁条
6.检查出有缺陷的卡会当即剔除

7. 加温加压把 4 个组件压在一起
8. 剪切时加签名条和全息图
9. 防诈骗须防密码泄露
10. 在信用卡上植入芯片
11. 办卡时机器把信息写入磁条
12. 做好的信用卡可以使用了

扫描二维码，观看英文视频。

Credit Cards

In one year alone credit cards accounted for almost 124 billion pounds worth of spending in the UK. Everything from perfume and gifts to a meal in a restaurant are bought on credit. But how are these handy little srectangle put together and how do they work? The life of the ordinary credit card starts in a facility like this in Slovakia. Its location isn't advertised to the public and security is high to deter criminals from coming and having a look around. Once inside you'll find the design centre. This is where new credit cards are born.

With the colour scheme approved, the paint must be mixed. For a small order this is done by hand and the precise ingredients are also kept very secret.

The staff will now load the printing presses with these laminated sheets. They form the basis for the fronts and backs of your credit card and the first colour added to them is white. The freshly printed laminates emerge under the watchful eye of another security camera. The bosses are very keen that nothing goes missing from this factory. The next stage is the full color scheme that the bank wants for its design. The colors are added to the rollers, and a special device is used to spread the paint out to an even thickness.

It's like a license to print money. This machine can print around 7,000 sheets of new cards every hour. However the credit card companies have thought of clever ways to make sure the process can't be duplicated. Hidden in these genuine cards are plenty of security measures that make forgers lives very difficult. The printing technique makes spotting a forgery relatively easy, but one of the most interesting tricks we can show you is the UV imprint. When the cards are finished the special mark will only show up under UV light. A new credit card consists of several elements. There's a protective sheet, the printed front and rears and the high tech magnetic strip for the back.

But how does all this work to help you access money? Well, when you go shopping with your debit card a very simple process occurs. You insert your card to the machine and a signal is sent to your bank. The bank checks the card is valid and that the account exists and has money in it. If all's well, a signal is sent back to the shop asking you to prove the card is yours. If you enter the correct Pin number, the bank will pay your bill for you and debit your account directly.

However, if you use a credit card, the machine contacts the data office that collects requests for credit card transactions. The company's computer verifies the account is valid, and that you haven't exceeded your limit and then releases the money if all's well. Any problems would have needed further verification but the transaction is usually quite simple.

All of the necessary personal information is stored on your credit card in the magnetic strip on the back. This is the key to how a credit card works. These strips start out on a long reel but they are combined with one of the protective sheets that make up the credit card.

Each card has to be perfect otherwise the technology just wouldn't work. Experts analyse each sheet microscopically. Any with flawed elements are immediately rejected.

The next stage is to combine the 4 different components to make the familiar plastic cards we all go shopping with. The sheets are carefully cleaned and layered into these steel presses which are placed into heavy duty copper caskets. The caskets are fed into a press that exerts a massive 120 tons of pressure at 150 degrees Celsius. This staggering force bonds the layers together so they are inseparable. At the other side of the press each new sheet of 48 fresh credit cards emerges to be guillotined into shape. The guillotine also adds signature strips and the hologram to the cards, another of the many security measures designed to help keep your finances safe.

Credit card fraud is a worldwide problem and in 2006 it cost Britain almost £428 million pounds. Cloning is one of the most common tricks for making a phony credit card and security experts spend hundreds of hours researching how the fraudsters do it. One popular method to steal your personal details is attaching a phony card slot to the cash point. It's often accompanied by a camera that then records your pin number as you enter it.

Chip and Pin technology has been introduced to combat fraud by using coded information on your credit card. It's supposed to be a foolproof security measure, but it does still rely on you being careful when typing in your Pin number.

A special drill is used to cut a hole in the pre-made credit card and the chip is inserted. The cards are then put through a rigorous exercise regime to make sure the chips won't pop out under too much stress. When they're fit, the cards are packaged up, counted and stored securely until the bank needs them.

So is that it? Well no, the next step is you apply for a new card, or your old one gets stolen. When this happens, you call your card provider. They then call this company, and a new card is issued for you.

A fresh box of blank cards is opened and added to this machine. Its memory has been filled with your account details and those of many others all of whom need new cards.

This machine then imprints your name and details onto the front whilst writing the account information onto the magnetic strip at the back. And it's ready to go. This is how a worthless piece of plastic is transformed into an instant alternative to cash.

Did you know?

If the queen ever runs short of cash, she needn't worry. Buckingham Palace has its very own cash machine.

电池

扫描二维码，观看中文视频。

不起眼的电池。去年英国使用了超过6.8亿个这种“能量小管子”，但电池是如何工作的？

这些外壳是为标准AA电池准备的，做一个电池还需要什么？电池的组成包括外壳、正极、负极、底板，和一根让电在回路中流动的针。

让人难以置信，这种贴着标签的小管子为如此多的家用电器提供能量。从遥控器到MP3播放器都使用电池。但你想过吗，它们是如何工作的？

我们都知道电池有正负极。负极充满了试图跑掉的电子。电力就是电子在电路中从负极移动到正极所发出的能量。

二氧化锰和石墨是制造正极的复合材料。你可能很熟悉石墨，因为它被用来做铅笔。

正极通过吸收电子而工作。复合材料被压成环状，然后插入电池外壳里。加入的材料越多，电池持续的时间就越长。将复合材料送到机器顶部，经过压缩，环状材料在每个电池从下方通过时放置进去。

要使电池工作，正、负两极必须隔开。将一种类似橡胶的材料卷起来，用于建造这层内部屏障。

这台高速机器把每卷隔绝材料插入电池，石墨已经在里面安顿就绪。为了向你展示它如何工作，我们必须让机器停止。内层在这里以红色显示。当电池从机器中出来，速度放慢到刚好能看到每个壳里的白色衬里。

就像阴和阳，电池由两个对立面组成。它既有正极又有负极。我们已经做好了正极，现在需要混合锌粉和胶凝来做负极。这种混合物毒性很大，所以让机器来做，尽量减少与人接触。混合后形成这种黏稠的物质。它是带负电荷的凝胶。一旦混合完成，将被填进电池壳里的内层。

现在我们已经有了电池的两极，要使它们工作，电流必须从一端流向另一端。

如果电荷不能在两极之间运动，电池将无法打开遥控器。要解决这个问题，电池厂聪明的人们有个办法，就是使用底板。虽然沥青通常用于路面，在每个底板上喷点沥青，不失为一种很好的胶合剂。然后加进黄铜针，这就是电子如何跑出来的秘密。电子通过黄铜针向下流出，进入了你的遥控器或当今几百种由电池供电的设备。

现在，你可以使用电池来切换频道，点亮手电或打开MP3播放器了。当电子完成工作后，将通过电线返回电池的正极。底板和黄铜针组装完毕后送入这台机器，液压机把它们压进久等的电池外壳。

至此电池已经完成，但在此阶段，看起来仍不像你能识别的电池。在贴上商标之前，电池需要经过测试。如果电池的衬里被损坏，那

你知道吗？

阿拉斯加州的费尔班克斯镇是全球最大的可充电电池所在地。它重1300吨，可为全城供电7分钟。

么内部就会短路，不能提供电能。这台机器测试每个电池，并清除任何劣质品，这样你就不会因电池失效而烦恼了。

成卷的标签装进机器，并贴到电池上，现在万事大吉了。所以当你下次蜷曲在沙发上按动遥控器时，请想想那些让你无须站起来的电子。

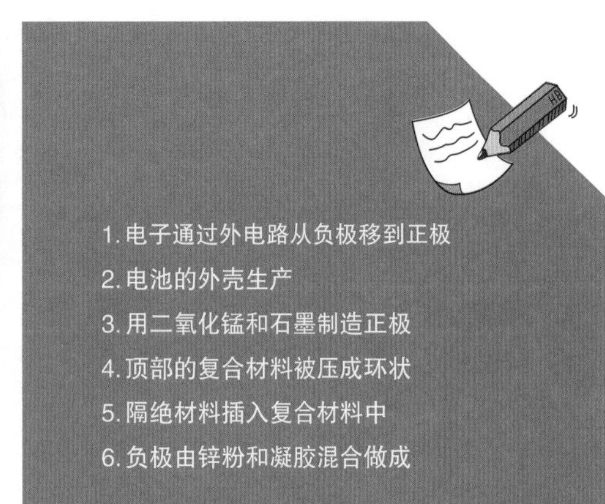

1. 电子通过外电路从负极移到正极
2. 电池的外壳生产
3. 用二氧化锰和石墨制造正极
4. 顶部的复合材料被压成环状
5. 隔绝材料插入复合材料中
6. 负极由锌粉和凝胶混合做成

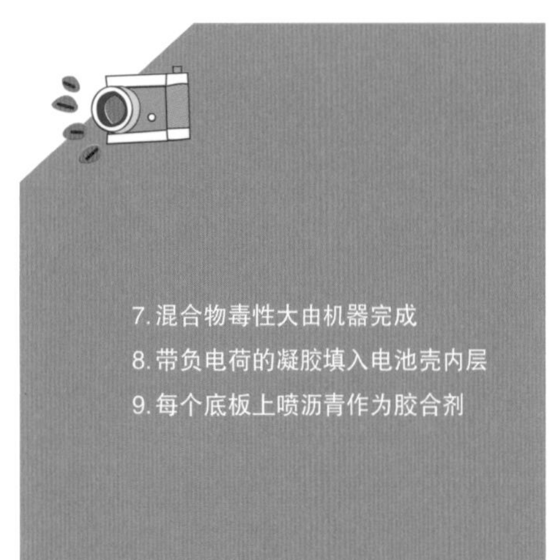

7. 混合物毒性大由机器完成
8. 带负电荷的凝胶填入电池壳内层
9. 每个底板上喷沥青作为胶合剂

⑦

⑧

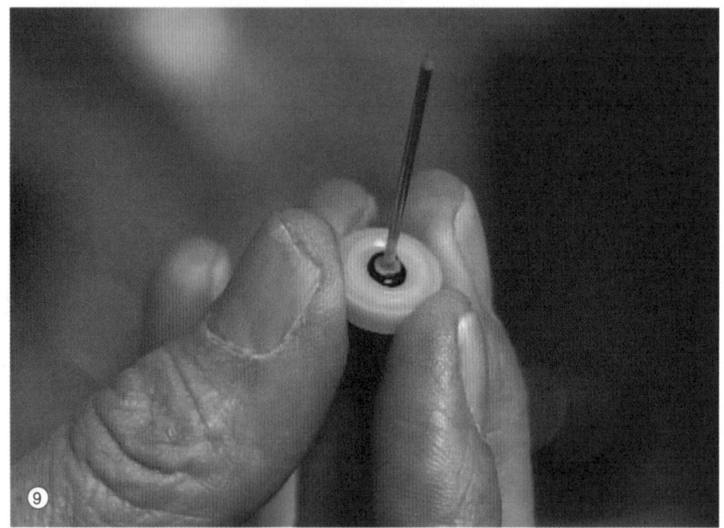
⑨

⑩

10. 出厂检查，剔除问题电池
11. 为电池贴上标签
12. 成品电池出厂

⑪

⑫

Batteries

But first, the humble battery. Last year the UK used more than 680 million of these little tubes of power, but how do they actually work?

These cases are for the standard double A battery, but what else do you need to make a one? A battery consists of a case, a positive pole, a negative pole. A base plate, and a pin to allow the electricity to flow around the circuit.

Its hard to believe that a small tube with a sticker on it, powers so many household devices. They are used in everything from remote controls to MP3 players. But have you ever wondered how they work?

So we've established batteries have a positive and negative pole. The negative pole is full of electrons which are all trying to escape. Electricity is the force used when they move from a negative pole to the positive one through an electrical circuit.

The materials used to make the positive pole are manganese dioxide and graphite. You might be familiar with graphite as it's used to make the pencil.

The positively charged mixture works by absorbing electrons that pass into it. It gets pressed into rings which are then inserted into each battery case. The more that can be put into each battery, the longer that battery will last. The mixture is loaded into the top of the machine here and once it's compressed, the rings are then loaded into each battery as it passes below.

For the battery to work, the positive and negative poles must be kept apart. A rubber-like material is rolled up and used to create the internal barrier that does this. This high speed machine then inserts each individual roll into the batteries which are already lined with Graphite. To show you how it works, we had to have the machines stopped. The lining is highlighted here in red. As the batteries come out of the machine, they're slowed just enough that you can see the white lining in each case.

Like yin and yang, a battery is made up of two opposites. They have both a positive pole and a negative one. We've got the positive, now we need to make its opposite by combining zinc powder and a gelling agent. The mix is very toxic though, so a machine is used to minimalise human contact. The combination creates this gloopy substance which is a negatively charged gel. Once it's fully combined this is added into the lining in the battery cases.

We've now got the two sides too the battery, but to make them work, the electricity needs to get from one side to the other.

扫描二维码，观看英文视频。

If the electrons can't travel between the poles, the battery won't power your remote, so to get round this problem, the clever guys at the battery factory have come up with the solution. That's where these base plates come in. Although it's normally used for the road surface, each plate is loaded with a squirt of bitumen. It makes an excellent glue. Brass pins are then introduced and this is the secret to how the electrons can get out. They travel down through this pin, and into your remote or any of the hundreds of devices that are powered by batteries today.

You can now use them to help you switch channels, light up your torch or power your MP3 player. When their job is done, they'll head back through the wires into the positive pole in the battery. Once they have been assembled the plates and pins are fed into this machine and a hydraulic press squeezes them into the waiting case.

They are now complete, although at this stage they don't look like the batteries you would recognise. Before they receive their trademark stickers they are all tested. If the battery's lining is damaged, then it short circuits and won't give out any power. This machine tests each one and removes any of the faulty one's so you're not left frustrated and powerless.

Rolls of stickers are loaded up and attached to the batteries, and now they are complete. So next time you're crashed out on the couch with the remote control spare a thought for the electrons that are saving you from getting up.

Did you know?

Fairbanks, Alaska is home to the world's largest rechargeable battery. It weighs 1,300 tons and can power the city for 7 minutes.

茶灯蜡烛

扫描二维码，观看中文视频。

蜡烛往往和两个人的浪漫晚餐联系在一起。最常见的烛光之一，是简单的茶灯蜡烛。制作简易，使用方便，它们是怎么做出来的？要做茶灯蜡烛就需要蜡。

做茶灯蜡烛最理想的蜡是石蜡。一种汽油生产的副产品，它是茶灯蜡烛完美的原料，因为从固态液化速度非常快。石蜡的液体性质还使得它便于运输。将它加热到 50 摄氏度后装进这样的罐子中，使得石蜡很容易运载和泵出。

要制造蜡烛，你可能以为会把灯芯悬在一小杯液体石蜡中。但蜡烛在英国的需求量很大，这种办法花费的时间太长。取代由石蜡围绕灯芯凝固的办法，这里先做出蜡烛本身，再从周围向灯芯压紧。

第一步让液态石蜡凝固。液体蜡在非常高的压力下喷向空中，再落到旋转的滚筒上。液态的蜡滴在这里冷却下来并且变硬。当滚筒转动一周后，另外一侧的刮刀就把硬化的蜡刮下来。这种颗粒态的蜡准备放进压力机变成适当的形状。

下一步将粉末状石蜡变为固体块状，在这里形成茶灯蜡烛的主体。粉状的蜡和热的机械放置在高速转盘里，将蜡变成块状。通过慢镜头，可以看到每个空位被预先定量的石蜡粉末所填满。继续旋转，蜡就被压缩成合适的形状。这里演示了机器怎样在蜡块中间留下一个孔。灯芯将在稍后的生产过程中从这里插入。

一个圆柱体从上面压下来并把石蜡紧紧挤在一起，使它再次变为固体块状。直到此时才形成茶灯蜡烛的正确模样，并有一个孔用来插灯芯。

压缩好的石蜡圆片从机器里送出来并加入数百万同伴的大军，准备变成茶灯蜡烛。下一个工序是灯芯，蜡烛没有灯芯就不能使用。普通的线会很快烧光，外面包一层厚厚的蜡，就能延缓燃烧过程，这种结合成为理想的灯芯。

滚筒旋转把一根线拉过融化的热蜡。冷却之后，滚筒把线反向拉过液体石蜡再裹上一层。这个过程反复进行，直到蜡层厚度达到 2 毫米。对于蜡烛孔来说这是完美的尺寸。

将新灯芯绕到这些滚筒上并送入组装机，石蜡块已经在此等候了。灯芯在这里被切割成适合插入的长度。但请稍等，茶灯蜡烛还需要两个很重要的部件。石蜡加热后很快变成液体，如果没有东西接住的话，蜡便会放任自流。成千上万的新座杯放进机器，但其中有些放反了。这并非自身有什么问题，只是需要一个解决办法把它们翻过来。幸运的是这台机器具有一种功能。所有倒置的座杯在流水线上都被高压风机吹走了。他们必须再次地经过系统，直到所有的座杯都放对位置。

正确摆放的圆杯在此处进入组装机。但还有最后问题需要解决。如果你现在把灯芯插进

你知道吗？

要达到太阳光的强度，你需要有足够覆盖 852 亿个地球表面积的茶灯蜡烛……然后来一个打火机。

去，它们就会马上掉出来。至关重要的底座在此处加入。和底座装在一起后，灯芯就能在孔中保持稳定。所有4个元件就绪后，生产就可以开始了。

首先，将灯芯切成合适的尺寸。插入底座封好。石蜡圆盘送进来，灯芯穿过已经打好的孔。组装完毕的蜡烛被整齐地放进尺寸完美座杯里。

成品茶灯蜡烛从组装机里送出来，准备好装袋发送到商店。茶灯蜡烛在英国各地许多商店有售，并且用途各不相同。在家庭中，他们通常拿来营造一个浪漫的氛围，或者用于加热熏香油炉。

可能会让你大吃一惊，当初设计这种蜡烛只为了给茶壶保温，这就是“茶灯蜡烛”名字的来历。

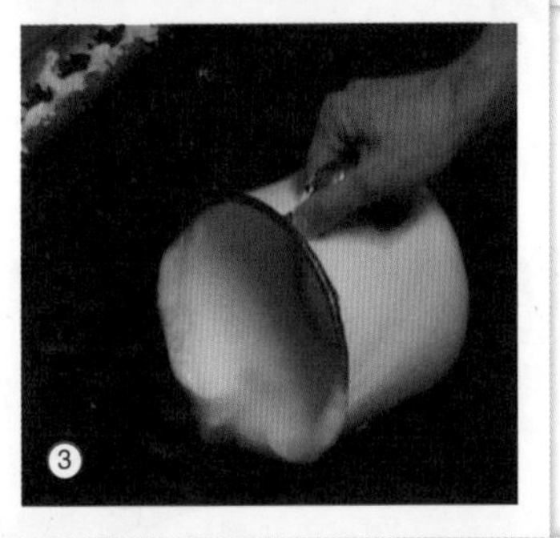

1. 茶灯蜡烛是最常见的烛光之一
2. 茶灯蜡烛的4个主要部件
3. 固态石蜡的液化速度非常快
4. 让石蜡围绕灯芯凝固太耗时
5. 液体石蜡落到滚筒上冷却
6. 滚筒转动一周把石蜡刮下

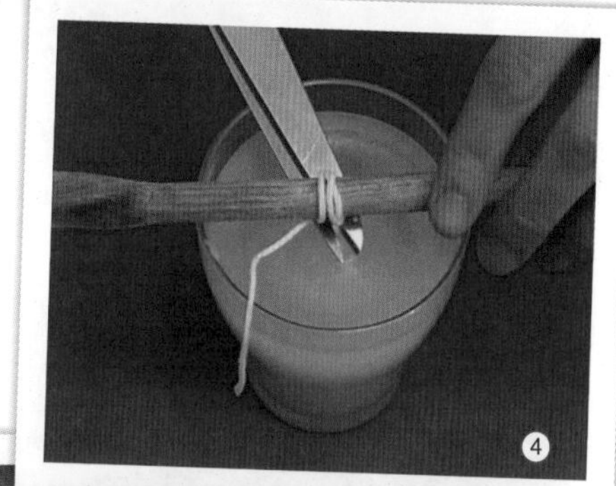

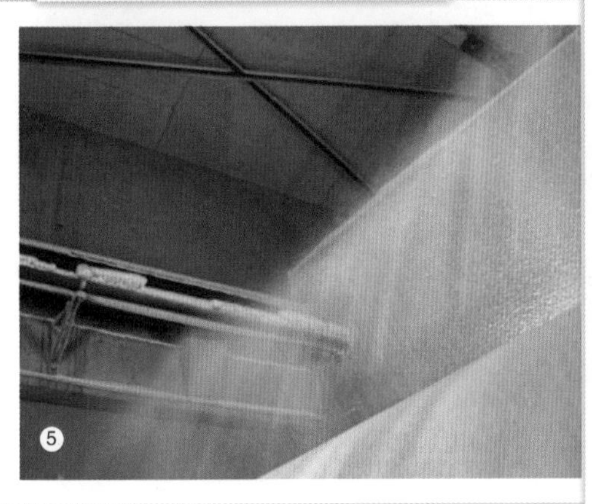

7. 定量的石蜡粉填满转盘

8. 蜡粉压成留有灯芯孔的块状

9. 滚筒把线拉过融化的热蜡

10. 所有杯座都摆放正确

11. 灯芯、底座、蜡块装进杯座

12. 为茶壶保温而得名茶灯蜡烛

Tea Lights

扫描二维码，观看英文视频。

Candles are often associated with romantic dinners for two, but one of the most popular sources of candlelight is the simple tea light. Quick to assemble and easy to use but what goes into making one? To make tea lights, you need wax.

The ideal wax for our tea lights is paraffin. A by-product of petrol production, it's perfect for tea lights as it goes from solid to liquid very quickly. The liquid side to paraffin also makes it easy to transport. It's carried in tankers like this which are heated to 50 degrees Celsius. This makes the paraffin easy to load and easy to pump out.

Now to make a candle you might think it would make sense suspend a wick in a small cup of liquid paraffin. However, demand in the UK is high and this method would take far too long. Instead of waiting for the wax to solidify round the wicks, the body of the candle is made first and then pressed around the wicks.

Step one is to make the liquid paraffin solid. The liquid wax is sprayed at very high pressure into the air and lands on a rotating drum. Here the droplets cool down and go hard. As the drum turns full circle, a scraper on the opposite side removes the hardened wax. In this granulated form it is now ready to be put into the machine that compresses it into shape.

The next step is to turn the powdered paraffin into the solid blocks that form the main part of the tea light and that happens here. Inside this fast moving carousel of powdered wax and hot industrial machinery, the blocks are being formed. Slowed down, you can see each space being filled with a pre-set amount of powdered wax. As it continues round, the wax is then compressed into shape. This demonstration shows how the machines keep a hole through the middle. The wick will be inserted here later in the production process.

A similar cylinder presses down from above and squeezes the paraffin together so hard it becomes a solid block once more. Only now it's the right shape for the tea light cups and has a handy hole ready for the wick.

The compressed discs of paraffin are then ejected from this machine and sent on to join millions of others ready to be turned into tea lights. The next step is the wick, no candle would work without one. Ordinary string burns off too quickly however, a thick coating of wax slows down the burning process and this combination makes the ideal wick.

As the drum rotates it pulls the string through hot liquid wax. Once the wax has cooled, the drum passes the string back into the wax again for another coat. This continues until the coating has built up to a thickness of 2 millimeters. The perfect width for the tea light hole.

The fresh wick is then collected onto these drums which feed into the assembly machine where the paraffin discs are already waiting. Here they will be cut to size to be inserted. But not just yet. Tea lights need two more important items. Paraffin heats quickly and turns to liquid, so without a holder things would get messy very quickly. Thousands of fresh cups are added into the machine but some of them are the wrong way round. There isn't actually a problem with them but a solution is needed. Luckily this machine has one built in. Any cups that are upside down are removed from the line with a high pressure blast of air. They must go round the system again and again until they are the right way up.

Facing the right way round the cups enter the assembly machine here. However, there's one final detail that needs a solution. If you put the wicks in now, they'd just fall straight back out again. That's where the all-important bases come in. Fitted with the wick they will keep it in place in the tea light's hole. With all 4 pieces in place, production can now begin.

First, the wick is cut to the right size. It's inserted into a base and sealed into place. Discs of paraffin are fed in and the wicks are poked through the ready made holes. And the whole completed candle is passed on to be fitted neatly into the perfect sized holder.

The completed tea-light is then ejected from the assembly machine ready to be bagged up and sent on to the shops. Tea light candles are available in many stores around Britain and have a variety of different uses. They're popular in homes where they can create atmosphere like a romantic setting or be used to heat an incense oil burner.

But it may surprise you to learn that they were originally designed to keep a pot of tea warm, which is where the name tea-lights comes from.

Did you know?

To match the intensity of the sun you'd need enough tea light to cover an area as big as 85.2 billion earths... then you'd need a lighter.

芝宝打火机

扫描二维码，观看中文视频。

因为好莱坞明星用它在电影里面吸烟，芝宝牌打火机已经成为世界上最著名的打火机。

芝宝牌打火机在美国生产，它的设计基于一个古老的奥地利打火机型号，由22个零件组成。制造工艺从打火机的外壳开始。

为了生产这种名牌防风打火机，这家位于宾夕法尼亚州布拉德福德市的工厂每天要消耗8千米长的镀镍黄铜金属箔。金属箔首先通过这个100吨的压床，把打火机的底座和盖子冲压出来。成品件被送到压床末端，边角余料收进这个容器里。

新压好的底座和盖子上印有芝宝牌打火机的独特标志，这个传统可以追溯到20世纪50年代。这些标记是一组代码，每个粉丝都知道它们的含义。左边的字母表示生产月份，右边的数字是年份。特别版的芝宝打火机是收藏家的选项，这些识别码能让爱好者了解他们的打火机有多老了。

顶部和底部做完以后，就需要进行组装了。这不是一件令人兴奋的工作而且重复性很强。这些女士们每天需要组装6万多个打火机。是的，每天6万多个！转盘上有一个点焊机，把底座和盖子焊在一起，就可以装进新打火机了。

为了维持良好声誉，芝宝确保每一个新出厂的打火机都毫无瑕疵。每一个外壳都经过多次酸浴，然后电镀，让它闪闪发光。各种各样的成色都有，其中最流行的是镀银类。

把酸洗掉后的下一步是质量检查。每个外壳都要经过20个不同的质检员以确保没有疏漏。这些米老鼠手套不是奇异装束，而是为了保护产品。手套由超柔软材料做成，可以保证每个新外壳纤毫无损。

芝宝打火机的内部结构包括用来产生火花的微小燧石和几个储存燃料的棉球。灯芯从底部插入，并放在浸泡着打火机燃料的棉球之间。燃料从下面被吸进灯芯里后由火花点燃。关键的不同设计是风挡，由一块钢片做成，上面布满了16个洞。能让火焰在强风中也保持燃烧，这一品质让芝宝打火机在全世界深受吸烟者青睐。

接下来是生产线上最奇怪也最重要的一道工序。芝宝打火机以其富有特色的咔嗒声而著名，当内部零件被放进外壳后，工人必须检查咔嗒声恰到好处。

尽管说来奇怪，特色的咔嗒声却是非常重要的。即使在维修部，修理工也必须确保品牌的咔嗒声得以保留。工人们生产这些经典的打火机时感到非常自豪。精细的工艺，让公司承诺对每个打火机终身维修。打火机修复完毕回到主人身边时，总会仍然带着品牌的咔嗒声。有些返修的打火机实在坏的太厉害，完全无法修复了。这种情况下，工厂会奉送一个新打火

你知道吗？

有史以来最昂贵的芝宝打火机之一是1933年的型号。它在公司成立75周年纪念会上，以3.7万英镑（约为人民币32.33万元）的价格售出。

机来代替。

每个星期有数量惊人的2500个打火机被能工巧匠们修复。

不只是咔嗒声需要修理。有时整个外壳都必须从头开始制作，这仍然属于质保的范围。一旦修复后，打火机被免费寄回原主。听起来好像花费不少，但如果你知道，自1933年以来卖出了超过4亿个打火机，他们应该还是付得起邮费的。全球对芝宝打火机有巨大需求，工厂里每周要生产约20万个新打火机。

尽管目前的趋势是，吸烟越来越不流行，但人们对这种标志性打火机的需求依然火爆。

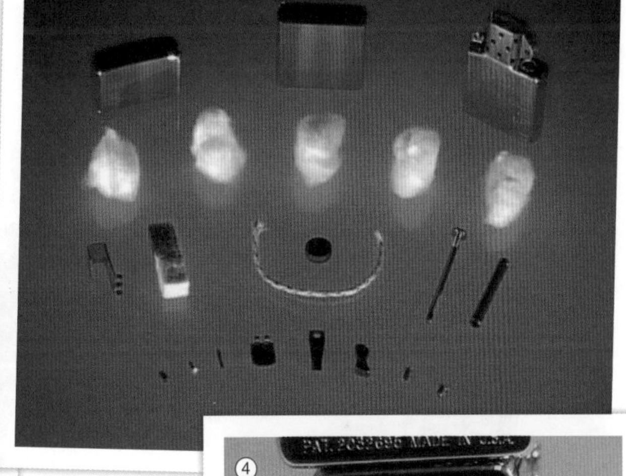

1. 芝宝牌打火机是世界著名品牌
2. 芝宝牌打火机由22个零件组成
3. 底座和盖用镀镍黄铜冲压出来
4. 特别版芝宝打火机受收藏家青睐
5. 手工把底座和盖子组装在一起
6. 经多次酸浴后电镀

7. 外壳要经过 20 个不同的质检员
8. 芝宝打火机的内部结构
9. 风挡由钢片做成布满 16 个洞
10. 零件装入外壳并检查咔嗒声
11. 修理工必须确保咔嗒声正确
12. 如无法修复将奉送新打火机

Zippos

Made famous by Hollywood stars smoking in the movies, the Zippo has become one of the world's most famous lighters.

Made in the USA, the design is based on an old Austrian lighter model and is made up of just 22 pieces. Construction starts with the famous lighter's outer shell.

To produce these wind-proof icons, this factory in Bradford, Pennsylvania uses a massive 8 kilometers of nickel plated brass every single day. It's first passed through this 100 tonne press which stamps out the bases and the lids for each new lighter. The finished pieces emerge from the far end of the press whilst any left over waste is collected in this bin.

The freshly pressed lids and bases now have Zippo's unique markings imprinted on them, a tradition which started in the 1950's. There's a code to those markings, which any dedicated fan would know about. The letter on the left signifies the month of its production and the number on the right is the year. Special edition Zippo's are collector's items and the identification helps enthusiasts know how old their lighter is. Once the tops and bottoms have been made, they need to be assembled. It's not a very exciting job and it's also quite repetitive. These ladies have to produce over 60,000 every single day. That's right; 60 thousand a day! The carousel includes a spot welding device which joins the bases to the lids ready to house a new lighter.

With a reputation to uphold, Zippo makes sure that every new lighter that leaves their factory is absolutely spotless. Every case is sent through several acid baths and then on to the plating baths where they get their characteristic shine. A wide variety of textures are available and one popular variety is the silver-plated model.

Once the acid's washed off the next step is quality control. Each case is checked by 20 different controllers to ensure nothing is missed. The Mickey Mouse-style gloves aren't for fancy dress, but product protection. Made of ultra-soft material, the gloves keep brand new cases stay spotless.

The Zippo's inner workings include a tiny flint stone to create the spark to a selection of several cotton wool balls which make up the fuel reservoir. The wick is inserted from the bottom and sits in the compartment with the cotton wool which is soaked in lighter fluid. The fuel is drawn up through the wick to be ignited by the sparks, but the key design difference is the windshield. Made of a piece of steel peppered with 16 holes, it keeps the flame alight in strong winds. A quality that's made Zippos popular with smokers the world over.

The next step is one of the strangest, but also most important jobs on the production line. Zippo's are famous for their characteristic click, and whilst the inner workings are put into the case, this worker has to check that click sounds just right.

Strange as it sounds that characteristic noise is very important. Even here in the repair department the repairers must make sure that trademark click is retained. The workers take great pride in making these classic lighters, and the precision engineering allows the company to offer a life-time guarantee on each one. Lighters returned to be repaired always go back to their owners with the trademark click intact. Some of the lighters that are sent back to Zippo for repairs are just too far gone for anything to be done to resurrect them. If this is the case, the factory will send out a replacement instead.

A staggering 2,500 lighters are repaired each week in the capable hands of these repair shop workers.

And it's not just the click that needs repairing. Sometimes the whole case has to be built from scratch, but that's all part of the guarantee. Once it's been rebuilt, it's sent back to the owner free of charge. Now that might sound expensive, but if you think that over 400 Million lighters have been sold since 1933, they can probably afford the cost of the stamp. Worldwide demand is enormous and every week the factory produces 200,000 new lighters to be sent out.

And despite current trends that are seeing smoking becoming less popular, the demand for this iconic lighter remains hot.

Did you know?

One of the most expensive Zippos ever sold was a 1933 model. It sold at the company's 75th anniversary party for £ 37,000.

木炭

没有烧烤的夏天是不完整的，而没有木炭的烧烤也是不完整的。可木炭仅仅只是焦木头吗？还是其中有更多名堂？

在这家工厂，他们要制造出上乘的木炭，所用的只是廉价的木材边角余料。木料被一个巨型破碎机切成小块。木头中仍然含有水分，在变成木炭之前，水分必须被完全去除。所以将木头送进一个筒仓，在那里以100摄氏度的温度把它们烘干。

但要让木头干透，还必须放进蒸馏器中加热。这是一种特殊设备，可以提取出每一滴水分。倒进25吨的木头后，舱口关闭使蒸馏器保持气密。等木材重见天日的时候，就已经变为成品了。

在实验室，科学家演示了它的工作原理。这个气密的容器代表蒸馏器。木材在容器内被加热。因为没有氧气，所以木头不会燃烧。容器内越来越热，木材中的水分被完全烤出，通过这些管子送走。剩下的便是炭化的木头，或我们日常所说的木炭。从蒸馏器中排出的液体也会被出售，用来调味和烟熏食品，比如意大利香肠。

当他们打开蒸馏器盖子后，需要冷却木炭，并且要快。木炭的温度达500摄氏度，当它和氧气接触时，可能会毫无预警地着火。木炭会通过地面活门，跌落到下面凉爽的地堡里。他们用喷水管消除刺鼻的尘烟。一旦木炭被锁入地堡，危险就结束了。待到木炭冷却，工人们按大小将它们分类。大块的木炭会作为烧烤燃料，在全国各地销售。

木炭细末与小麦淀粉混合，以便木炭碎屑结合到一起。此后混合物被压缩成煤块。在这个阶段，它们仍然是软的，需要经过烘烤，变成更坚固的块状。

最后，大量成品被装袋。这家工厂每天发售8万袋。它们被送到加油站和其他经销点。到了夏季，就会出现在你附近的某个后院里烤炙香肠。

你知道吗？

木炭是世界上最古老的艺术用品之一。它在史前时期用于洞穴壁画。

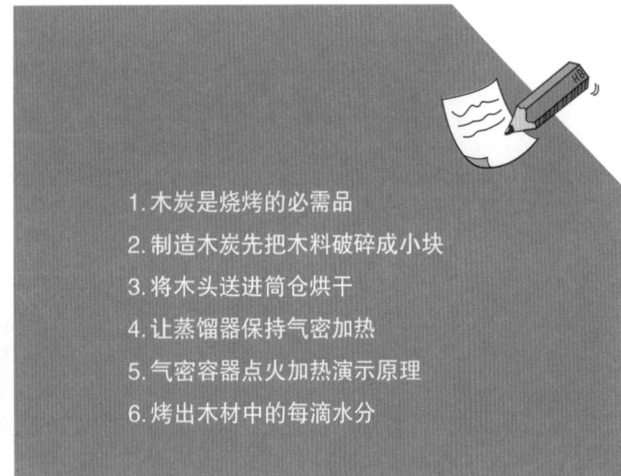

1. 木炭是烧烤的必需品
2. 制造木炭先把木料破碎成小块
3. 将木头送进筒仓烘干
4. 让蒸馏器保持气密加热
5. 气密容器点火加热演示原理
6. 烤出木材中的每滴水分

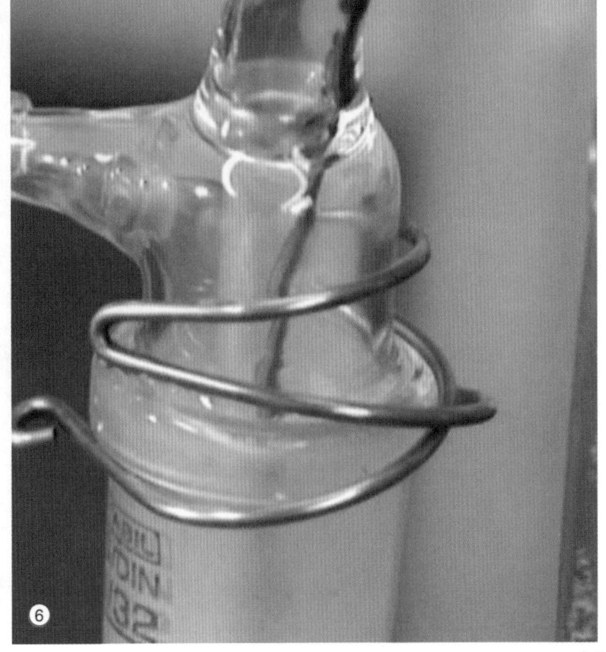

7. 剩下的炭化木头就是木炭
8. 蒸馏木材排出的液体用来做调料
9. 蒸馏器打开木炭快速跌入地堡

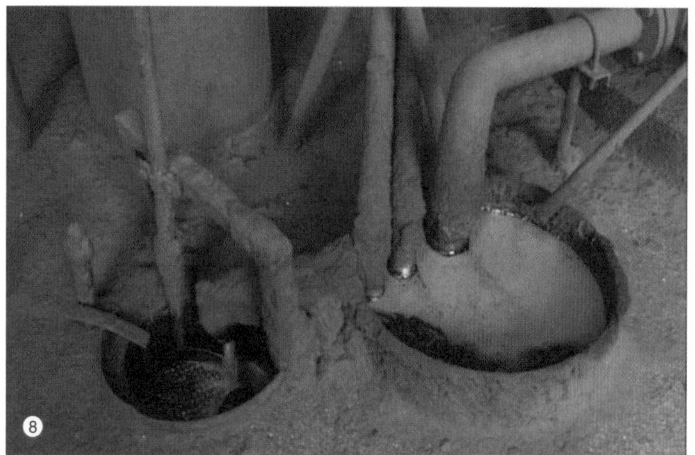

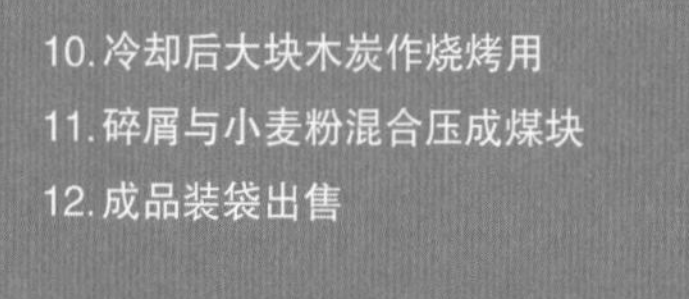

10. 冷却后大块木炭作烧烤用
11. 碎屑与小麦粉混合压成煤块
12. 成品装袋出售

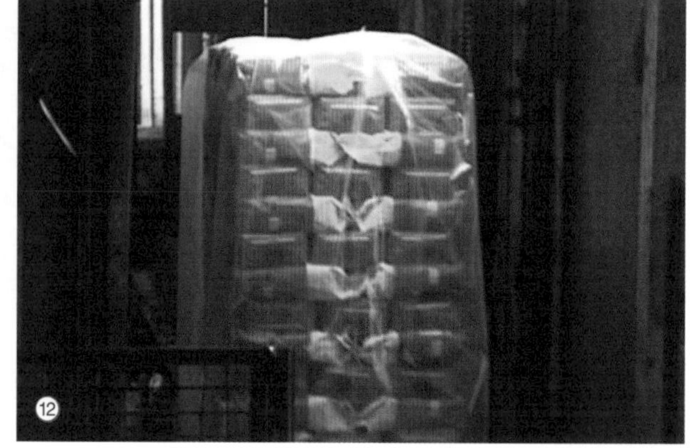

Charcoal

No Summer would be complete without a barbecue, and no barbecue would be complete without charcoal . But is charcoal just burnt wood? Or is there more to it?

At this factory, all they need to make the fiery coals are cheap offcuts of timber. It's brought down to size by a giant cutter. The wood still contains moisture that needs to be completely removed before it's turned into charcoal. So it goes into a silo where it'll dry out at 100 degrees Celsius.

But to make it bone dry, they have to heat it in a retort, it's a special apparatus which will extract every drop of moisture. After 25 tons of wood is piled in, the hatch is closed to make the retort airtight. When the wood next sees the light of day, it will be the finish product.

In the lab a scientist demonstrates how it works. this air tight container represents the retort. the wood is heated inside, and because there's no oxygen it can't catch fire. As it gets hotter it sweats out the last drops of moisture which is then taken away through these tubes. What's left is carbonized wood or charcoal to you and me. The liquid which is drained off the retort gets sold to. It'll be used to flavour smoky foods like salami.

When they open the lid they've got to cool the charcoal down, and fast! It's 500 degrees Celsius and as it comes into contact with oxygen it could burst into flames without warning. It plummets down into cool bunkers through this trap door. They have to spray water hoses to settle the cloud of acrid dust. Once it's locked in the bunkers the danger's over. After it's cooled down they sort it by size. Large lumps will be sold as barbecue fuel around the country.

The fine dust is mixed with wheat starch to help it bind. Then it's compressed into briquettes. At this stage, they are still soft, so they get baked off and turned into solid lumps.

Finally the large lumps are bagged up. this factory sents out 80,000 bags everyday. They are delivered to the petrol stations and other outlets. And come at summer months, they will be blackening sausages in a back garden near you.

Did you know?

Charcoal is one of the oldest art materials in the world. It was used in prehistoric times for cave paintings.

衣架

扫描二维码，观看中文视频。

你冬天出门，穿走外套后总会留下一样东西。尽管是件简单的物品，制作这种普通木衣架也需要很多工夫。

方便的木衣架让衣服不用放在地上，有助于让衣服或多或少避免折皱。

生产从锯木厂开始。这样尺寸的木头能制作出约2000个衣架。使用山毛榉有两个原因。首先它是短纤维硬木，因此有足够强度。其次，它不会产生过多树液，使得木材更容易加工。木板切割出来后，木匠把它们逐个放到另一个桌子上，切成更容易处理的尺寸。

做一个好衣架，树皮绝对要除去。下一步是让这些方块原木成形。木匠让木板来回通过弧线锯。就像在肉铺切火腿一样，每次都切下来一块新的曲形木材。这仅仅是开始。

接着将香蕉形木片放进这台机器。在这里，它们被赋予更为人熟悉的衣架特征。首先去掉粗糙的边缘。接下来在上部刻出凹痕，以便悬挂背带连衣裙。

你可能已经注意到，这些衣架看起来有点偏大。是的，你是对的。下一个机器把每个木块切成几片。

快速摇掉木屑，一块厚厚的衣架变成了5个正常大小的衣架。然后按这种特殊结构把它们搭起来。刚锯好的木头需要干燥，但又不能太快，否则木头会变形。这个缓慢的干燥过程大约需要2周。

人们所熟悉的传统衣架形状慢慢出现了，但在这个阶段，材料仍然相当粗糙。机器把它打磨平滑。每个新衣架通过砂磨机时，把前面完工的衣架顶开，落到下面等待的盒子里。

到此，衣架厂已经生产出平滑而具有流畅曲线的木架，但如果拿来挂衣服，还需要一个钩子。在木料上打孔，每个工人每小时可以加工约400个衣架。

钻孔完成，衣架就准备挂起来了，但还要等等。首先木匠把它们拿到这些架子上。放稳后用锤子固定。目的何在？上清漆。衣架作为日常用品，会在任何新衣柜里安家。这层清漆能让它们更加结实。

从清漆里出来后，清漆滴被刷掉，衣架被放置12小时进行干燥。用这套设备，木匠每天能给3000个新衣架上漆。

那么现在可以上挂钩了吗？还没有。首先必须加一个支撑杆。杆子沿着底部钉进去。这条杆最重要的任务是挂裤子，比如和西装配套的长裤。

衣架终于可以安装钩子了。工人取一个钩子和一个新衣架。他们被放到这里，剩下的活由机器来干。

这个工厂用一根山毛榉的木材，可以生产出2000多个方便的衣架，用来放置你的衣服。

你知道吗？

谢菲尔德的前售货员韦恩·利发明了一种圆珠笔，可以变成功能齐全的衣架。

1. 制作普通木衣架需要很多工夫
2. 山毛榉更易加工并有足够强度
3. 大木板锯成方块形状
4. 让方块原木来回通过弧线锯
5. 弧线锯让方木块变成曲形木材
6. 去掉粗糙边缘刻出凹痕

7. 曲形木片切成正常衣架大小
8. 砂磨机把衣架打磨光滑
9. 在衣架上钻出安挂钩的孔
10. 把光滑的衣架固定在架子上
11. 衣架浸过清漆会更结实
12. 榉木原木变成了漂亮的衣架

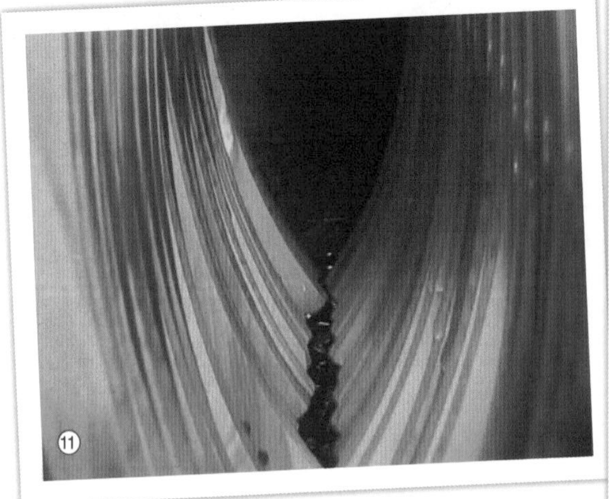

Coat Hangers

When you pop out in the winter you always take your jacket, but you also always leave something behind.

It's a simple device but a lot of work goes into the ordinary wooden coat hanger.

The handy wooden hanger keeps clothes off the floor and helps keep them more or less crease free.

Their life starts out here at a saw mill. A log this size could probably produce about 2,000. Beech wood is useful for two reasons. Firstly, the wood is a hardwood timber with short fibres which gives it strength. And secondly, it doesn't produce excessive sap and that makes the wood easier to work with. Once the planks have been cut, the carpenter swings each length around to another table where he can cut it down to a more manageable size.

For a good hanger, the bark definitely has to go. And the next step is to shape these square panels of the raw wood. The carpenter passes the block back and forth across this curved saw. Like slicing a ham at the butcher's, he slices off a fresh curve of wood each time. But this is only the beginning.

Next the banana shaped wood slices are fed into this machine. Here they will be given some of the more familiar features of the coat hanger. First some of the rough edges are removed. Next the indentations which will help hang spaghetti strap dresses are carved into the tops.

Now you may have noticed that these hangers are looking a little on the large side. Well, you'd be right. This next machine produces several slices from each chunk.

A quick shake to remove the sawdust and one thick hanger has now become 5 normal sized ones. These are then stacked up in this special formation. The freshly cut wood needs to dry, but not too quickly or it would warp. This slow drying process takes about 2 weeks.

The traditional shape of a familiar hanger is slowly emerging, but at this stage the material is still quite rough. This machine sands them down. As each new one is passed along the sanders, it knocks the finished one in front into the waiting bin below.

So far the hanger factory has produced nice smooth curves of wood, but if they are to hang on to your coat on they are going to need a hook. Holes are drilled into the wood and each worker can prepare about 400 an hour.

With the holes complete they're ready to be hooked up, but not yet. First this carpenter is adding them to one of these frames. They need to sit securely so he hammers them into place. Their destination? A varnish bath. The hangers will be handled on a daily basis in whichever cupboard becomes their new home. This varnish coat will make them more robust.

When they're removed, any drips are brushed off and the hangers are left for 12 hours to dry. Using this system, the carpenter can coat over 3,000 new hangers every day.

So are they ready for their new hook now? Well actually no. First a support bar must be added. This will be nailed into place along the bottom. The most important task for this bar is to hold up the trousers that go with a suit for example.

And finally the hangers are ready to get their hook. The worker will select 1 hook and one fresh hanger. They are placed in here and the machine does the rest.

So from a single beech log, this production plant can turn all that wood into over 2,000 handy hangers to store your clothes.

Did you know?

Wayne Leigh, an ex–salesman from Sheffeld, has invented a ballpoint pen that can be turned into a fully functional coat hanger.

剪刀

打开一箱果汁，一瓶啤酒，甚至一个罐子。现代的剪刀除了剪东西还有很多用处。为了达到足够强度，它们用钢做成。加热到1100摄氏度后，粗钢变得可塑。工匠可以用工业锤子打出一对新剪刀的样子，用一吨半的压力把金属锻造成形。

接着压床以100吨的巨大压力，把剪刀的两半从模板里切出来。

为了把两部分结合到一起，有些公司使用铆钉，但它们在这里开一个螺丝孔。如果剪刀用久了会变松，可以很容易地拧紧螺丝。

孔内套好螺纹后，把剪刀的两半送去硬化。不经过加热或回火，钢会相当软。加热到1085摄氏度并冷却，使得剪刃变硬，不易弯曲或偶然损坏。

每一半都研磨出一个平面。它还并不锋利，但剪刃最终会从这里做出来。接下来打磨剪刀的手指孔。光滑很重要，因为那是用户手指接触的地方。剪刃的其余部分也要打磨以确保光滑，没有尖锐的毛刺或金属的细齿，但这只是最初的打磨过程。这个盆子盛满了磨石和半只剪子。 8个小时后，任何凸起都被石头磨掉了。完成之后，磁铁传送带下降到盆中。吸走金属，留下石头。

现在给剪刀开刃，这种刃有额外的长处。除了非常锋利，剪刃呈锯齿状。意味着剪刀用来剪布时，金属锯齿会抓住布料避免打滑。

最后把两半剪子组装起来。加一滴油让新剪刀润滑。通过前面做好的螺丝孔，两半剪刀拧在一起。螺丝的背面打平，这样就不会滑脱。接着用锤子把剪刀尖敲成正确形状，并进行测试。

我们在这里看到高品质剪刀的秘密。制作精良的剪刀刃轻微的向内弯曲，因此，它们闭合时总保持一个点接触。这能最大限度发挥剪切的潜力。剪刃之间的空隙能让材料完好无损地通过。为确保这个接触点不会损坏剪刀，要对样品进行常规测试。剪刀启闭8万次，经过这般磨损后，下一剪仍然应该和第一剪同样锋利。

最后，新装好的剪刀被抛光，准备发送到商店。这就是剪刀的前沿技术故事，它让剪刀外形美观并保持锋利。

你知道吗？

世界上最小的剪刀只有几十亿分之一米长。它们可以抓住单个分子，用光打开和关闭。

1. 粗钢加热后锤打出剪刀形状
2. 压床把剪刀的两半从模板中切出
3. 在两部分结合处开螺丝孔
4. 剪刀的两半送去回火硬化
5. 每一半都研磨出一个平面
6. 打磨剪刀的手指孔

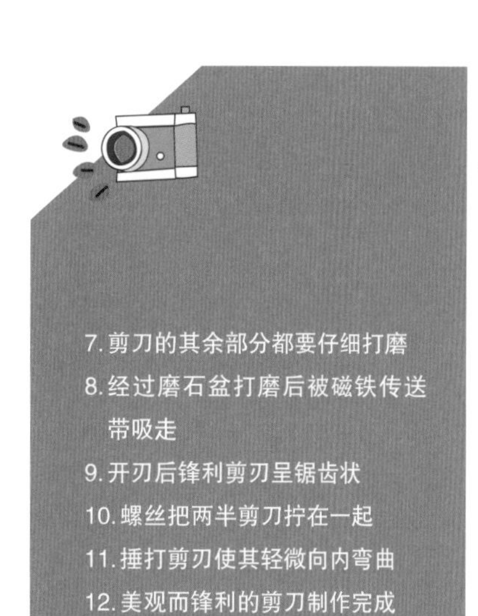

7. 剪刀的其余部分都要仔细打磨

8. 经过磨石盆打磨后被磁铁传送带吸走

9. 开刃后锋利剪刃呈锯齿状

10. 螺丝把两半剪刀拧在一起

11. 捶打剪刃使其轻微向内弯曲

12. 美观而锋利的剪刀制作完成

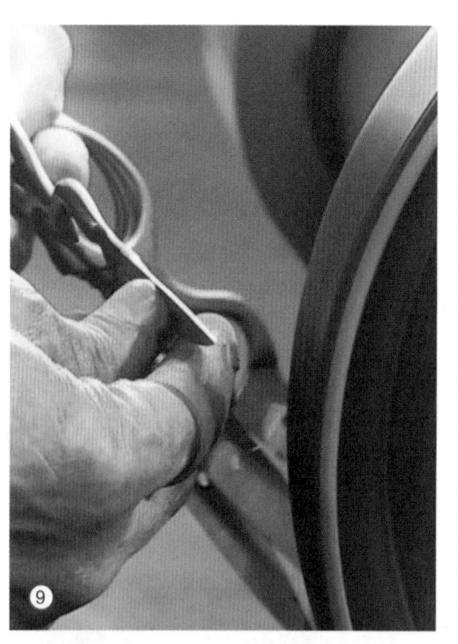

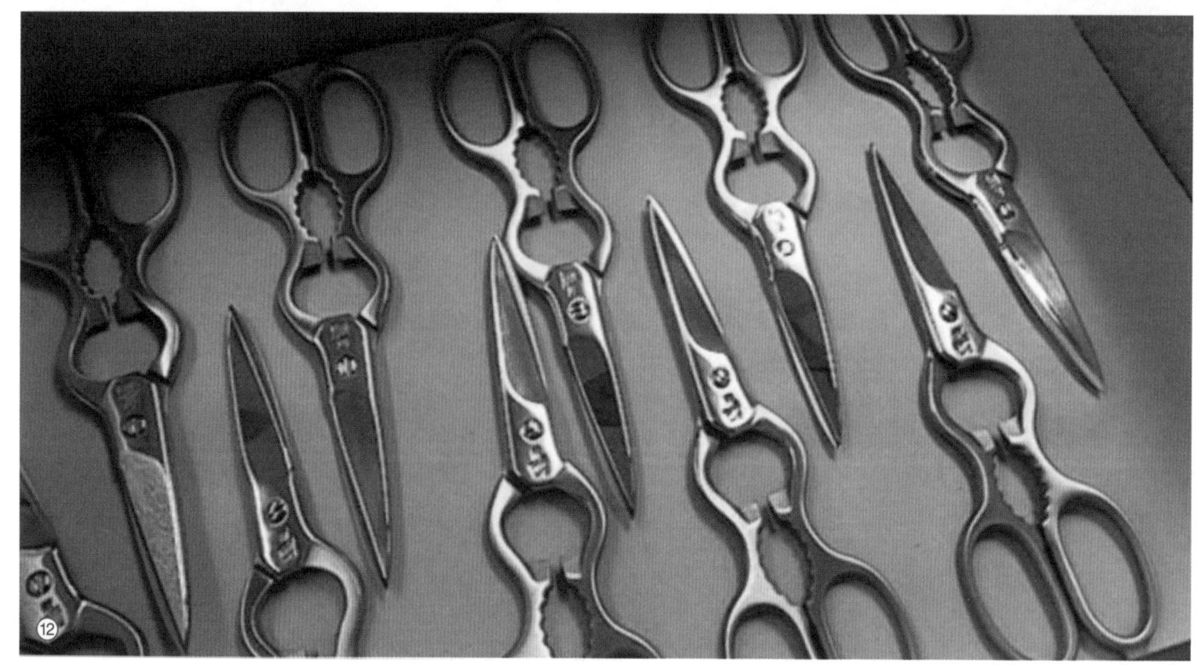

Scissors

Opening a carton of juice, a bottle of beer or even a jar. A modern pair of scissors is used for a lot more than just cutting stuff. And to make them strong enough, they're made out of steel. Heated to over 1,100 degrees Celsius, the raw steel becomes malleable. The engineer can then shape each new pair of scissors using an industrial hammer. One and a half tons of pressure mash the metal into shape.

Next the two sides of the scissors are cut out of the template by a press that exerts a staggering 100 tons of pressure.

To stick the two sides together, some companies use a rivet, but here they cut a hole for a bolt. Then if the scissors loosen over time the bolt can easily be tightened.

After the hole is threaded each half of the scissor is sent to be hardened. Without heating or tempering, steel is quite soft. Heating the parts to 1,085 degrees Celsius before cooling toughens the blades so they won't bend or break by accident.

A face is ground into one side of each half. This isn't sharp yet, but it's where the blade will eventually be carved out. Next the eyes of the scissors are ground. It's important they're smooth because that's where the user's fingers will sit. The rest of the blade is ground down to ensure it's smooth and free of sharp burs or jagged bits of metal, but this if only the first smoothing stage. This tub is full of abrasive stones and fresh scissor halves. After 8 hours any bumps are worn away by hours of rubbing up against the stones. When they're done, a magnetic conveyor is lowered into the vat. This extracts the metal parts while leaving the stones behind.

And now the scissors receive their edge - an edge that in this case comes with an added bonus. As well as being very sharp, the blade is serrated. This means that when the scissors are being used to cut cloth, the metal teeth of the serration grip it so it doesn't slip.

扫描二维码，观看英文视频。

And finally the two sides can be assembled. A drop of oil is added which helps lubricate the new scissors. They're then screwed together through the hole that was threaded earlier. Here the back of the bolt is flattened so it won't slip out. Next the tips are hammered into shape and the scissors are tested.

And here's where we learn the secret of high quality scissors. Well made blades have a slight inward curve, so they always touch at a specific point when they close. This maxmizes their cutting potential. A gap between the blades would allow material to pass through uncut. To ensure this point of contact doesn't ever damage the scissors, tests are regularly conducted on samples. They are opened and closed 80,000 times. Even after all that wear the next cut should be just as good as the first.

Finally the newly assembled scissors are given a good polishing, ready to be sent to the stores. So that's the story behind this cutting edge technology that ensures that these scissors look and stay sharp.

Did you know?

The world smallest scissors are just a few billionths of a metre long. They can grip single molecules and are opened and closed using light.

卷笔刀

扫描二维码，观看中文视频。

HB 铅笔是靠不住的。当你打算写一份购物清单或者一篇日记时，铅笔芯常常不是钝了就是断了，解决的办法是用卷笔刀。

要做一个卷笔刀，你需要预先成形的狭长镁条。将它送入工业切刀里，变成大小合适的镁块。镁是一种廉价的轻质金属，用来做卷笔刀最为理想。从下方把刚切好的镁块收集起来，然后送进下一台机器，真正的制作过程在这里进行。

曾对数学和历史感到无聊的人，都肯定会认出慢慢显现的卷笔刀形状。

在每块材料通过时，机器做出插铅笔的主要孔洞，还有安装刀片的螺丝孔。为螺孔套上丝扣，这样螺钉就可以拧紧。再雕刻出刀片安放的位置，把可能伤害使用者的锋利边缘锉平。镁块看上去像卷笔刀了，但前面的流程使它们变得肮脏，与一个闪亮的新铅笔盒很不相配。将镁块装入一个特殊的筐子里，使它们能经受住下一个生产步骤——强腐蚀性的硝酸洗浴。如果时间过长，卷笔刀就会溶解，而快速的酸浴让它们熠熠生辉。然后将坯料漂洗并旋转甩干。这时你可以插进铅笔了，但拿出来时笔尖还会和先前一样钝。

下面的工序是刀片，它从长卷钢板上切下来。工业压力机冲压出成千个刀片准备安装在刀架上。在冲压刀片过程中，压力机使用了润滑油，所以刚做好的刀片都粘在一起。它们需要好好洗个澡除去油污。现在刀片前往烘炉了。在这里它们被加热至 850 摄氏度，停留 3 分钟。出来时通过另一次油浴来降温，快速冷却使金属硬化。但刀片再次沾满了油，需要重新洗个澡。然而这一次不是用水清洗，而是把刀片放到沙子里吸去油渍。这是一个相当肮脏的过程，工人最终得到装满了油污、沙子和刀片的大桶。为了收拾干净，让它们通过一系列的刷子和筛子留下干净的硬刀片。经过所有洗涤和清理过程，终于准备磨刀了，但首先需要排好方向。要使卷笔刀工作，必须把刀片正确的一边磨利。用人工排列刀片太麻烦，所以请这台机器代劳。只有方向正确的刀片可以通过，其他的被打回去重来一遍。

一旦刀片通过了，最终将来到这里。一个工人此时能够收集刀片，并放置到打磨机上，让它们变得锋利。每个刀片耐心等待着获得锋利刀刃的时刻。传送带上每个插槽只能容下一个刀片，所标识的一侧被打磨，使它能削得动最顽固的 HB 铅笔。剩下要做的是把刀片和刀架安装到一起，工作就完成了。和前面检验刀片一样，只有朝向正确的刀片才能到达最后一关。用一个小螺丝连接刀片和刀架，到此收官。每年要生产成千上万的卷笔刀，让铅笔和英国孩子的头脑变得锐利。

你知道吗?

瑞典发明家彼得·斯文森使用 670 马力的百夫长坦克引擎，制造了世界上最强大的卷笔刀。

1.卷笔刀用于削铅笔

2.把狭长的镁条放进工业切刀

3.把长镁条切成合适的镁块

4.依次做出插笔孔及刀片螺孔

5.镁块进行快速硝酸洗浴

6.洗干净的刀片走向烘炉

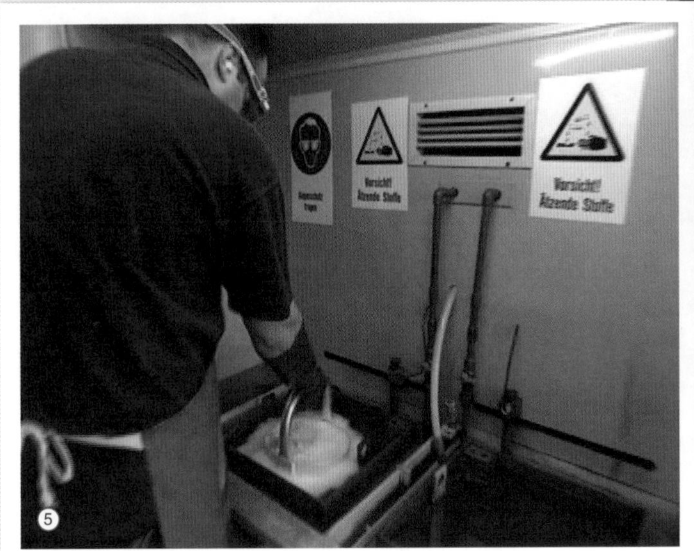

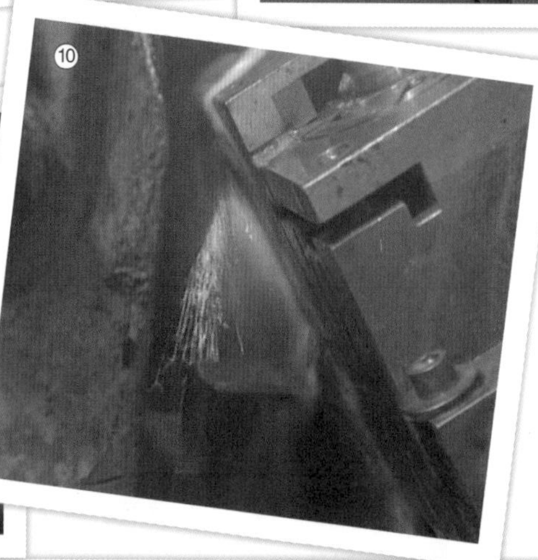

7. 刀片倒进搅动的沙子里去除油污
8. 机器让排列方向正确的刀片通过
9. 每个插槽只容下一个刀片
10. 打磨机让刀片一侧变得锋利
11. 摆放正确的刀架才能来到这里
12. 螺丝连接刀片刀架完成制作

Pencil Sharpeners

HB pencil's are unreliable. The lead is always blunt or breaking just when you need to write a shopping list or an entry in the diary, but the solution is the pencil sharpener.

To make one, you'll need a long bar of pre-formed magnesium. It's fed into this industrial cutter, which then slices it into the right sized blocks. Magnesium is a cheap, light metal ideal for the job. The freshly cut blocks are collected below. They're then loaded into the next machine which is where the real work of making pencil sharpeners takes place.

Anyone who was ever bored in maths or history will certainly recognize the shape that's beginning to emerge.

As each blank passes around the machine, it creates the main hole for the pencil. The screw hole to attach the sharpener's blade. It threads the hole so the screw will hold tight. Carves the space where the blade will sit on the sharpener. And files it down to remove any sharp edges that may injure the user. The blocks look like sharpeners' but all the work has made them rather dirty and they wouldn't go well with a shiny new pencil-case. They're loaded into special baskets which will help them survive the next stage of the process, a highly corrosive bath in nitric acid. Too long and the sharpeners will dissolve, but a short bath will shine them up nicely. The blanks are then rinsed and spun to dry them. At this point you could put your pencil in but it would come out just as blunt as it was before.

The next stage is the blade. This gets cut from a long roll of sheet steel. An industrial press stamps out blades by the thousand ready to be attached to the sharpener bodies. However, in order to punch them out, the press uses oil to lubricate the process, so now the fresh blades are all stuck together. They'll need a good bath to rinse away the excess oil. They are now heading for the ovens. Here they will be superheated to 850 degrees Celsius for 3 minutes. When they emerge they're cooled with another oil bath and the rapid cooling is what hardens the metal. But, yet again the blades are coated with oil, so another bath is necessary. However this time, they're not washed, but instead the blades are immersed in sand which absorbs the oil. However, this is rather a messy process and the workers end up with a gigantic drum full of oil, sand, and blades. To clean this off, they're passed through sets of brush heads and over big sieves, leaving clean hard steel. After all the bathing and cleaning they're finally ready to be sharpened, but first they need to be sorted. For pencil sharpeners to work the blade must be sharp on the correct side. It's far too much hassle for a person to sort each blade by hand, so this machine does it for them. Blades facing the right way can pass through, the others are dumped back into the system to go round again.

Once they have passed through, they end up here. A worker can now collect them and load them into the grinder for sharpening. Each blade now waits patiently for its turn to receive a razor sharp edge. Each slot on the carousel takes one blade only and the highlighted side here, is ground down so it'll be ready to take on even the most stubborn HB pencil. All that remains is for the sharpener bodies to be joined with the blades and the job is done. As with the blades earlier only the bodies facing the right way are allowed through to the final stage. A tiny screw is used to connect the blades to the bodies and that's it. Thousands of sharpeners are made every year, which help keep the pencils and the young minds of Britain sharp.

Did you know?

Swedish inventor Peter Svensson built the world's most powerful pencil sharpener using a 670 horse power Centurion tank engine.

镜子

扫描二维码，观看中文视频。

早在中世纪，如果一个女人想在出席宴会前检查她的头发，她能找到的最好映像是一碗水。如今千万人在使用普通的镜子。但制造镜子要比你想象的更复杂。

镜子是由覆盖着金属和化学涂层的高品质玻璃制成。在工厂里，玻璃从头开始生产。这个过程从很多沙子起步。炉子把沙子变成液体玻璃，再形成平板玻璃。

炉火温度在 1600 ～ 1700 摄氏度。在这样灼热的温度下沙子融化只需几分钟。操作员密切关注着永不歇息的炉子。炉火每年 365 天，每天 24 小时都无条件地燃烧，可以持续 12 年。

这是一种灼热而有潜在危险的工作，所以工人炉前操作时要穿上重型防护服。

超长的铲形工具把固体锡锭放入炉中，经过一种巧妙的处理，能生产出完美的平板玻璃。锡熔化后形成一层液态金属。从右侧添加的沙子融化后，变成我们前面看到的液体玻璃。它被迫上升到熔融态锡层的顶部。然而这两种物质不会结合。它们形成一种状态，一层均匀的玻璃浮在金属上。这层玻璃向前滚动，冷却后就成为完美的玻璃，用于制作镜子。

当玻璃板做出时要不断监视防止缺陷。反射到一块白板上的强光能暴露出玻璃中的任何杂质。关键要知道什么现象出现时寻找什么，在未经训练的眼睛看来，它只像一团漩涡灰雾。

通过质量控制的玻璃，准备切成合适的形状。轻轻在表面划线后通过辊上，易碎的玻璃板就不会破裂。当辊子抬起时，玻璃板的重量使它恰好沿着划线断裂。粗糙的边缘以同样的方式除去。

这样巨大而易碎的玻璃板过于沉重粗笨，很难用手移动，所以用特殊的传输工具在工厂搬运玻璃。

新状态下的玻璃薄板很适合做镜子。它各处一样光滑无瑕，但却沾有灰尘和污垢。玻璃表面进行清洗并轻轻擦净，让成品镜子产生最好的光反射。现在制作最重要的反射涂层。玻璃被镀上一层薄薄的液态银。喷嘴来回移动，实现多层喷涂。在完成的时候，你已经能看到它们被银层反照出来。然后镜子通过加热辊。这能让银层干燥到位并送去进行另一次质量控制检查。

闭路电视和先进的扫描设备能检测到所有异常。如演示所见，银层上的划痕被立即识别出来。扫描仪检测到的任何问题，都在白板的反射中清晰可见。这部分将被除掉，其余部分通过。

下面将一种合成液体涂在背面。它能保护薄薄的银层，不在日常磨损中被刮掉。镜子被翻转过来，送到楼下进一步处理。在这里，它们接受一层带静电的粉末。这个厂每天生产 5 万多平方米新镜子，它们被大批量存储和运输。

你知道吗？

专家认为，大象可以在镜子中认识自己，像人类、猿猴和海豚一样表现出自我意识。

带电粉末能防止镜子粘在一起，导致破碎。

当它们进入最后工序，巨大的平板被切成家用镜子大小。这种工作不适合迷信的人。这个工人一整天都在折断镜子。玻璃再次划线，然后轻轻地使它沿划线断开。虽然会有事故，但对于映像而言，家用镜子制造仍是一个惊人的成功。

1. 这里是生产高品质玻璃的车间
2. 炉子把沙子变成液体玻璃
3. 这是灼热而有潜在危险的工作

4. 超长铲形工具把锡锑锭放入炉中
5. 液态玻璃升到熔融态锡层顶部
6. 玻璃自身重量使它沿划线断开

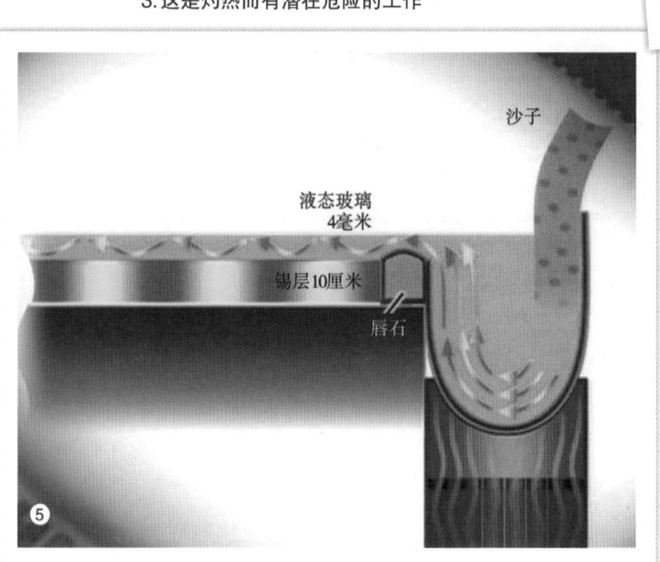

7. 沉重易碎的玻璃用特殊工具搬运
8. 干净的玻璃镀上薄薄的液态银
9. 镜子通过加热辊让银层干燥
10. 合成液体涂在背面保护银层
11. 喷上带静电的粉末防止镜子粘连
12. 镜子沿划痕断成合适大小

Mirrors

But first, back in the middle ages if a woman wanted to check her hair before heading out for a banquet, the best reflection she could find would have been a bowl of water. Today, millions of people can use the ordinary mirror. But it's more complicated to make one than you might think.

A mirror is made from a layer of high quality glass coated with metals and chemicals. The glass is made from scratch at the factory. The process begins with sand and plenty of it. This furnace is used to turn the sand into liquid glass ready to be formed into flat sheets.

The fires burn between 1600 and 1700 degrees Celsius. At these searing temperatures the sand melts in just a few minutes. Operators keep a close eye on the furnace which never takes a break. The fires burn 24 hours a day 365 days a year for anything up to 12 years.

This is hot and potentially dangerous work, so workers have to wear heavy duty protective clothing when operating the furnace.

Extra long spatula-like tools are used to drop solid tin ingots into the furnace, in an ingenious process that makes perfect sheet glass. The tin melts to form a layer of liquid metal. The sand, which is being added from the right, melts and becomes liquid glass as we saw earlier. It's forced up on top of the layer of molten tin. However, the two don't combine. Instead they create conditions where a uniform layer of glass can float over the metal. This layer rolls forward to cool into perfect glass ready to be turned into a mirror.

As the glass sheet emerges a constant watch is kept for flaws. A powerful light reflected onto a white board will expose any impurities in the glass. Knowing what to look for is key in what appears as a swirling grey mist to the untrained eye.

Having passed quality control the glass is ready to be cut into shape. So the fragile sheet won't shatter, the surface is gently scored and passed over rollers. As the rollers lift, the weight of the glass sheet causes it to break exactly along the lines scored. The rough edges are removed in the same way.

Such large, fragile glass sheeting is too heavy and unwieldy to be moved by hand, so special transporters are used to shift the glass around the factory.

In its new state, the glass sheeting is ideal for making mirrors. It's flawless and uniformly smooth, but it's also picked up dust and dirt. The surfaces are washed and gently scrubbed clean, so the finished mirror will give the best reflection possible. And now for the all-important reflective coating. The glass is plated with thin layers of liquid silver. Several layers are built up by spray nozzles as they pass back and forth. And by the time they're finished, you can already see them reflected in the silver coating. The mirrors are then passed over heated rollers. This dries the silver into place and they can now be sent on for another quality control check.

This time CCTV and state of the art scanning equipment detect any anomalies. As this demonstration shows, a scratch in the silver is identified immediately, any problems picked up by the scanner show up clearly in the reflection of a white board. This section will be removed while the rest passes on.

Next a synthetic liquid is painted onto the back. This protects the thin silver layer so it isn't scratched off through daily wear and tear. The mirrors will now be flipped over and sent downstairs for further processing. Here they receive a statically charged layer of powder. Everyday, this plant produces over 50,000 square meters of new mirror so they're stored and transported in large quantities. The electrically charged powder stops them from sticking together which might cause them to shatter.

When they are ready for final processing the enormous sheets are cut down to domestic mirror size. It's not a job for the superstitious. This technician has to break mirrors all day. Once again the glass is scored and then gently forced to break along the score line. Accidents will happen, but on reflection, the manufacture of the household mirror has been a shattering success.

Did you know?

Experts believe elephants can recognise themselves in a mirror, showing self-awareness just like humans, apes and dolphins.

铝合金行李箱

扫描二维码，观看中文视频。

自从低廉的航空公司蓬勃兴起，全球旅行已经成为巨大产业。但这里有个问题。航空公司希望行李尽量轻些以节省燃料，你希望行李箱足够结实来保护物品，该如何解决呢？一个办法便是铝合金行李箱。

这家专门的德国工厂50年来一直制造轻便的铝质行李箱。铝的强度大、重量轻，成为制作行李箱的理想材料。这种金属在铝中掺进镁，既提高强度又增强柔韧。

制造行李箱的第一步，是让金属带上特殊的波纹。每张材料送进滚转机，变成适当的形状。波纹有两个作用：第一个是赋予行李箱传统的外观。第二个功能性更强一些，凹槽就像工棚屋顶的铁皮一样，使材料变强。通过平面和曲面的组合，能够更好抵御破坏，比如搬运工不小心把行李箱掉落。一旦铝板被压上波纹，还必须重新变平，否则下一个机器将无法加工。这样，我们就有了一个平的波纹板。

为了做出行李箱，板材需要成形。使用工业切割机，夹紧板材，然后冲压出合适的形状。制造出半个箱体的模板。为使两半箱子能安全闭合，还需要两条铝带，每半个箱子边缘各有一条。工人把铝带固定到位，用重型压力机弯曲成最佳角度，使它能包住行李箱的每一半。然后把铝带用铆钉连在一起形成封闭环，以便安装到行李箱上。接下来的工作非常简单。只要把铝框放在一起，确保配合完美。如果闭合不好，行李箱就不会严丝合缝并确保安全。

除了铝带之外，每个箱子的侧面也需要使用专门设备，扭转和弯曲成恰好的形状。现在可以连接所有的手柄、搭扣和螺丝帽了。铝板的尖锐边缘要求所有工人必须戴防护手套。安好把手后，现在可以将铝带安装到箱子的一侧，把面板边缘牵拉到位。然后用更多铆钉密封起来。接下来安装手提箱的塑料角，它们能加强铝制箱体的强度。最后安装滚轮。虽然行李箱很轻，但你不会愿意总提着它！

使用了这么多不同的零件，难免哪里会出问题。但工厂有一个专门部门整天测试各种模型，还从事修理和恢复损坏的行李箱。同时在厂房里，工人给行李箱添加更多有用的附件，包括关紧行李箱的锁扣和铰链。该把两半箱子装在一起了。机器用重铆钉连接，箱子可以关闭了。

我们做好了耐磨的外表，但箱子里面仍需要一些工作。首先涂一层胶水，其次加入纺织层，这能够保护你的衣服和纪念品安全。行李箱再装上高科技的伸缩手柄。使用铝合金箱旅行时，你的行李可能会被送到距离遥远的机场，但你至少相信它不会散架。

你知道吗？

你需要回收180个软饮料罐，才能得到足够的金属来制造一个铝制手提箱。

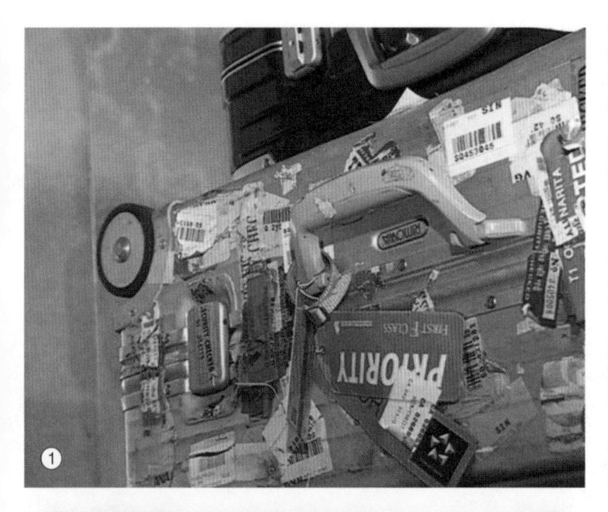

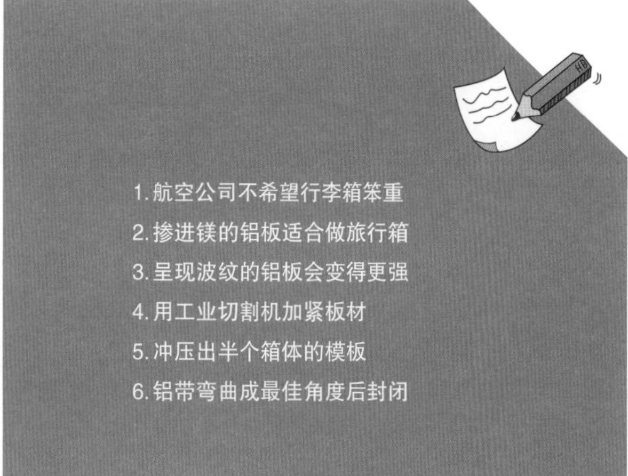

1. 航空公司不希望行李箱笨重
2. 掺进镁的铝板适合做旅行箱
3. 呈现波纹的铝板会变得更强
4. 用工业切割机加紧板材
5. 冲压出半个箱体的模板
6. 铝带弯曲成最佳角度后封闭

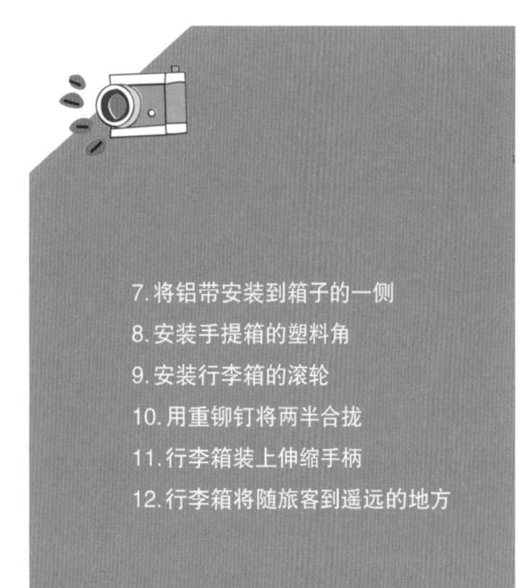

7. 将铝带安装到箱子的一侧
8. 安装手提箱的塑料角
9. 安装行李箱的滚轮
10. 用重铆钉将两半合拢
11. 行李箱装上伸缩手柄
12. 行李箱将随旅客到遥远的地方

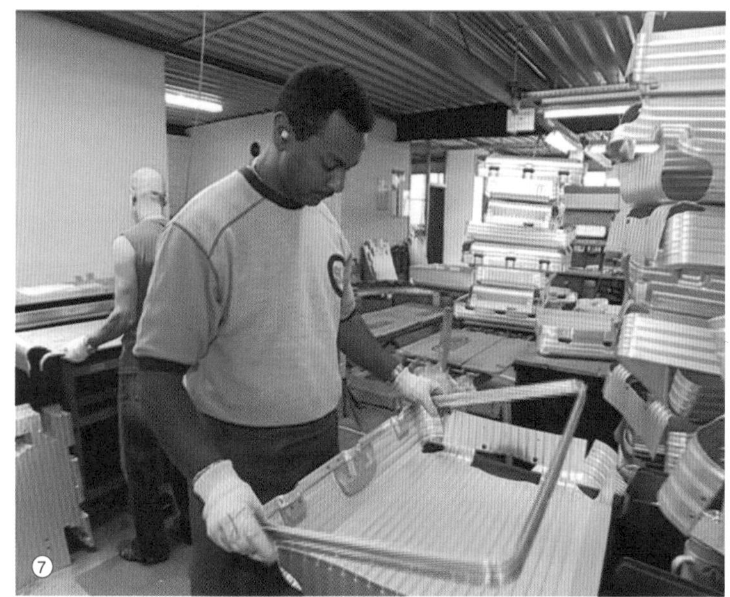

Aluminium Suitcases

Ever since the low cost airline boom world travel has become big business. But here's the problem. Airlines want light luggage to save their fuel bill and you want it strong to protect your souvenirs, so what's the answer? Well one solution is the aluminium suitcase.

At this dedicated factory in Germany they have been making lightweight aluminium cases for 50 years. The metal is strong, yet light, making it perfect for a suitcase. The metal is mostly aluminium mixed with magnesium. This improves the strength, but keeps it tough as well as flexible.

The first stage of making a suitcase like is to give the metal its characteristic corrugation. Each sheet is fed into a roller which bends it into shape. The corrugation serves two useful purposes. The first is to give the case it's traditional look. The second reason is far more useful as the grooves strengthen each sheet like corrugated iron on a shed roof. By folding the metal through a combination of flats and curves, it can now resist damage for example when baggage handlers accidentally drop it. Once it's been corrugated, it must be re-flattened otherwise the next machine wouldn't be able to work with it.

So, now we've got a flat corrugated sheet. To make an individual case, this now needs to be shaped. Using an industrial cutter, the sheet is clamped into the machines grasp, which then stamps out the right pieces. This makes the template for one half of a case. For the two separate lids to close securely, the case must have two bands, one for each lip. This worker clamps the band into place and a heavy duty press bends it with the perfect angle so it will encircle one side of each half. He'll then rivet the band together to form the closed loop to fit the case. The next worker has quite an easy job. He just has to bang on the loops to make sure they fit together. Without a perfect match, the case won't be watertight or secure.

As well as the bands, the sides for each case also need to be bent using specialist equipment. Here the panels for each half of the case are twisted and contorted into the right shape. A worker can now attach all the handles, latches, and screw covers. The sharp edges of the aluminium means all the workers must wear safety gloves. With the handles in place, he can now fit one of the bands to the side which will help draw all the panel edges into place. This is then sealed using more rivets. The plastic corners of the suitcase are next and they add to the strength of the aluminium body. Finally the wheels are fitted into place. Well it may be light, but you wouldn't want to have to carry it!

Now when you're using so many different parts, something is bound to go wrong. However, the factory has a special department that spends all day testing each model out. They also works to repair and restore any luggage that may have taken too much of a beating in the past. Meanwhile back on the factory floor, this worker is adding some of the more useful elements to the case, including the fasteners that will hold it shut as well as the hinges. It's time to put the two sides of the case together. Using heavy duty rivets, this machine seals the deal and the case can now be closed.

So we've got our hardwearing exterior the inside could still do with a little work. First a layer of glue was added followed by a layer of fabric. This will protect your clothes and keep your souvenirs safe. It's also fitted with the high tech extendable handle So with an aluminium suitcase you can travel knowing that your bag may end up in a different airport from you, but at least it will be in one piece.

You would need to recycle 180 soft drink cans to get enough metal for an aluminium suitcase.

挂锁

挂锁。它们在世界各地广泛使用。从家庭到企业，保护着各种财产。但既然它们如此坚不可摧，那么最初是怎么做出来的呢？

在这个挂锁制造公司里，生产从锁体开始。这些重型钢条每根达25千克，有着令人难以置信的高强度。它们被强大的锯片切割成8厘米长度一段。切下的小段用筐子运到工厂的另一侧，为后面工序做准备。

所有的零件必须完美配合，否则挂锁将不会安全。它们被送去钻孔，以便安装锁闭构件。这项工作由机器完成，确保孔的尺寸绝对准确。

为了制作挂锁臂，钢筋被送入这台机器，强有力的压床把它们弯曲成合适的形状。钢筋被转移到另一台机器，在锁臂的侧面钻出缺口，用来把锁臂固定在锁体内，直到转动钥匙把锁打开。

为了增加强度，挂锁被烧到炽热。然后被迅速冷却。这个过程使金属更柔韧，更不容易被扭断或打碎。

用肥皂水快速洗涤后，下一个步骤是确保挂锁在雨中不会生锈。它们和盛满镍粉的绿色麻袋一起浸泡在溶液里。镍被连接电路的正极，当挂锁连接负极时，便吸引镍粉形成镀镍层。然后，镀镍的锁体经过再次冲洗。下面要电镀另一层材料铬，使用和前面相同的方法。有着镀铬层的锁是防水的。

该制作复杂的锁定构件了。这里钥匙被切割出来，每把有一个数码编号。这些7位数字的编码，分配给每个挂锁特定的锁紧机构，总共有1000万种组合。工人们使用编号与钥匙代码相对应的锁环制作机构。

锁紧机构包括两个组成部分。首先是一对球轴承。这两个球轴承紧靠锁芯放在锁臂的缺口里，当钥匙转动时，它们滑入锁芯凹槽，锁臂被放开。

第二个零件是锁芯一侧的销子。它确保只有正确的钥匙能将锁打开。当钥匙转动时，锁环旋转，排列成能够容纳销子的凹槽。如果销子不落入槽里，锁芯就不能进一步转动，锁将保持锁定状态。完成后整个挂锁被封闭起来。然后，安装锁臂和外壳。

由于挂锁经常用于户外，它们必须经受得住最严酷的天气条件，所以在低于零下40摄氏度的环境下进行测试。它出色地通过了测试，是一把非常安全的挂锁。

你知道吗？

大约4000年前，埃及便使用锁来保护贵重物品。

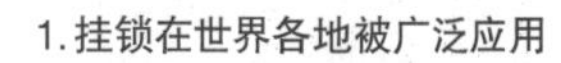

1. 挂锁在世界各地被广泛应用
2. 高强度的重型钢条被送进切割机
3. 钢条变成了 8 厘米长的小段
4. 在锁体上钻孔以安装锁闭结构
5. 压床把钢筋弯曲成挂锁臂
6. 烧红的锁体和锁臂要迅速冷却

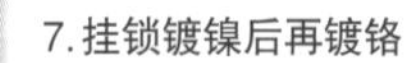

7. 挂锁镀镍后再镀铬
8. 每把钥匙有一个数码编号
9. 锁环和钥匙代码相对应
10. 球轴承在锁定状和打开状的演示
11. 销子确保配套钥匙才能打开
12. 冷冻后的强力破坏测试挂锁

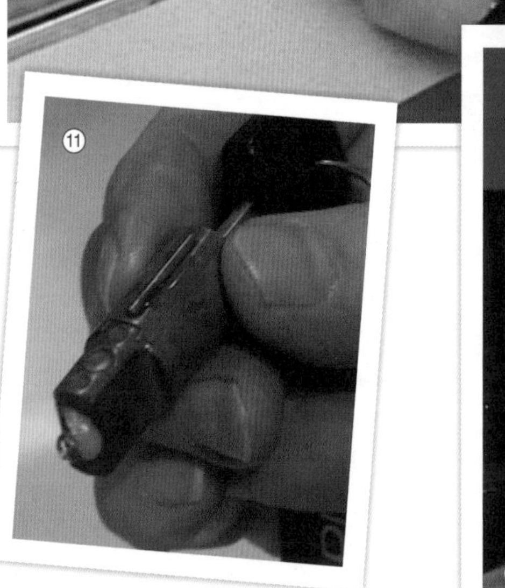

Padlocks

Padlocks. They're used all over the world to protect everything from homes to businesses. But if they're so hard to break, how do they put one together in the first place?

Here at this padlock manufacturer they start with the body. These heavy duty steel bars weigh twenty five kilos each and are incredibly strong. They have to be cut into eight centimetre lengths by a saw with a powerful blade. Crates of the cut bodies are then taken off to the other side of the factory for the next step of the process.

All the elements have to fit perfectly of the padlock won't be secure. They're going to be drilled so the locking mechanism can be fitted. It's done by a machine to make sure that the holes are exactly the right size. To make the padlock arms, steel bars are fed into this machine where a powerful press bends them into shape. The steel bars are moved along to another machine which drills grooves into the side of the arm, later on these will anchor the arm in the body until the key is turned.

To strengthen the padlock its baked until it's red hot. Then they're rapidly cooled. This process makes the metal more flexible, therefore less likely to be snapped or shattered.

After a quick wash in soapy water the next step is to make sure the padlocks wont rust in the rain. They're submerged in a bath along side green sacks full of nickel dust. The nickel has been given a positive polarity so when the padlocks are set to a negative polarity they attract the nickel dust which forms a coating. Then the nickel coated logs are given another rinse. To galvanise the steel another layer has to be added, this time of chrome. They use the same method as before and when they have a chrome coating the locks are waterproof.

Its time to build the complex locking mechanism. Here keys are cut and a code is assigned to each one. This 7 digital number assigns each padlock each specific locking mechanism, of which there are 10million combinations. The lock builder constructs each mechanism using the corresponding numbered rings that match the key code.

The locking mechanism consists of two elements. The first is a pair of ball bearings. These two ball bearings sit next to the barrel and in the grooves of the arm, when they key turns the barrel they slide into it and the arm is released. The next element is a pin on the side of the barrel. This ensures only the correct key will open the lock. When it's turned the disks rotate and line up to form a slot for the pin. If the pin doesn't fall into the slot then the barrel can't be turned any further and the lock stays locked. Once completed the whole lock is sealed shut. Then each lock receives it's arm and an outer covering.

As padlocks are often used out doors there have to be able to stand the harshest weather conditions so they're tested at extremely low temperatures below minus 40 degrees Celsius. And it passes the test with flying colours and is one very secure padlock.

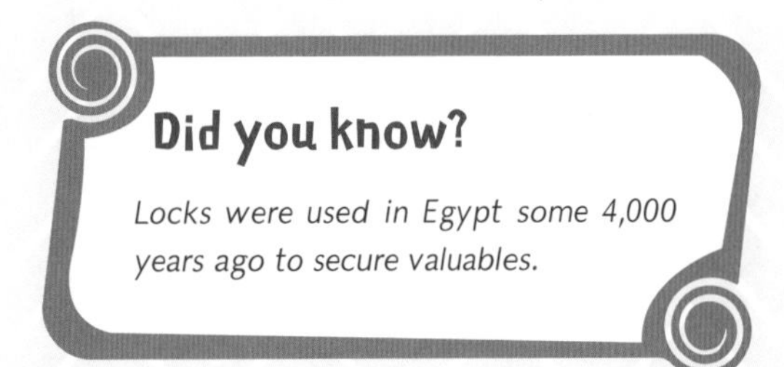

Did you know?

Locks were used in Egypt some 4,000 years ago to secure valuables.

铂金戒指

扫描二维码，观看中文视频。

如果你想表示爱她，订婚戒指是一个很好的方式，但不要被廉价仿制品，譬如普通黄金所糊弄。如果你真想让她芳心大悦，只有铂金才行。

铂金是昂贵的东西，每千克价格是黄金的两倍以上，所以要锁起来保管。由于它非常稀少，工人便尽量从一切可能的地方回收和再利用，这个盒子中的工业废料由此登场。

这种废料含有纯净的铂，化工厂的聪明人需要把它从其他金属中分离出来。所有的金属碎片都被放进这个巨大的玻璃烧瓶。使用玻璃很重要，因为它经得住将要使用的化学物品腐蚀。

不错，这就是化学的有趣之处。硝酸和盐酸倒进去后将烧瓶密封。在此后的 2 天里，金属会溶解，直到你得到这种酸性的黑色羹汤。

铂金回收就像一个大的化学实验，只有科学家知道结局会怎样。下一步是把铂与其他金属分离，要通过这种粉末实现。粉末添加到烧瓶顶部，当它沉降下去时，铂的微粒粘到上面，把其他便宜的金属留在后头。粉末慢慢沉降到底部，当科学家满意的时候，所有液体会被排出，留下一大桶沙子般的粗砾。

你大概很难说服别人这是世上最有价值的金属。砂砾随后被放置到托盘中进行焚烧。在 950 摄氏度下，除了金属之外的其他物质都烧掉了。当技术专家打开炉子时，剩下的东西看上去就像……沥青吗?

它实际上是铂金，但充满了炉子中的气泡。要把它变成制作首饰的珍稀璀璨金属，需要放进密闭炉内再加热。

这个强大的炉子能达到 2000 摄氏度，并保持气密性。加入铂炉渣，温度调高，含有气泡的金属熔化了。然后关上盖子并抽出全部空气。即使魔鬼终结者也无法在其中生存。

出来的是世上最昂贵的砖块。1 千克这种贵金属的价值超过 20000 英镑。

这种形态的铂金不适合佩戴，所以工厂的专家把铂金块做成铂金管。珠宝商可以用它制作戒指，或者你的女朋友喜欢的其他首饰。在一个密封的环境中，切片从管子上切下来，以确保没有任何丢失掉。

这就是你最终得到的东西。黄金和白银比较软，所以要与其他金属混合以增加强度，但这会降低价值。铂金的强度足以单独使用，这是它格外宝贵的另一个原因。

拿捏尺寸是一个简单过程，但如果顾客喜欢更纤细的风格，就要切掉多余的金属。使用车床和高档润滑油，纤细的薄片从铂金环上雕刻下来。同样没有任何浪费，珠宝商和计算机都密切盯着整个过程。

因为铂金如此稀有，你的女朋友也不想要旧戒指，所以铂金首饰往往是定做的。只有专

你知道吗?

在英国，一张白金唱片销售量为 30 万张。迈克尔·杰克逊的专辑《颤栗者》170 次打破了白金唱片的纪录。

家才能加工这种贵金属，并且必须知道该如何做。加热是最常用的手段。熔化的金属使它变得柔韧，但如果火焰烧得太久，铂金就会坍塌成极其昂贵而十分难看的一坨。

这种锯是用来制作宝石底座的另一个工具。微量的铂金被刻掉，加工出安放宝石的位置。你可以用百得工具试着干这种活，但我们真的不建议你这么做。

选出的较小宝石，可以放置到这种备好的底座里。电锤轻轻将周围的铂金闭合，把宝石固定到位。

最后戒指抛光准备佩戴。黄金和白银会随着岁月流逝而改变色泽，但铂金因为高纯度而永葆容颜。

就像你一直渴望的信用卡，这是你梦寐以求的铂金，它的价值超过同等重量黄金的两倍。

1. 铂非常稀少，回收含有纯净铂的工业废料可再利用
2. 含铂废料在硝酸和盐酸中溶化
3. 加进一种粉末把铂分离出来
4. 粉末沉降时铂的微粒粘到上面沉下来
5. 沙子般的粗砾放进托盘进行焚烧
6. 焚烧后变成了沥青样的铂金渣

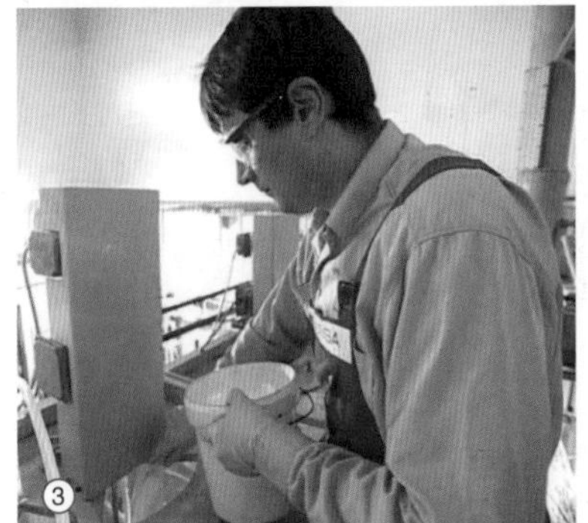

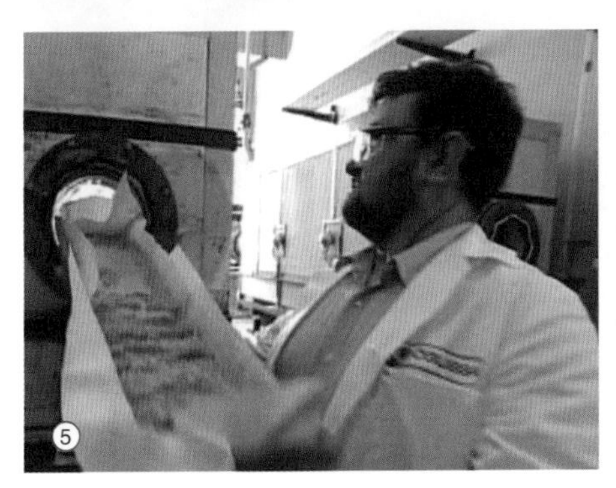

7. 铂金渣放进 2000 摄氏度的炉内熔化
8. 关上炉盖抽出全部空气
9. 从炉内出来的是世界最昂贵的砖块

10. 工厂的技术人员把铂金块做成铂金管
11. 微量的铂金从圆环上刻掉，加工出安放宝石的位置
12. 昂贵的铂金戒指

Platinum Rings

If you want to show her you love her, an engagement ring is a nice touch, but don't be fooled by cheap imitations like ordinary gold. If you really want to sweep someone off their feet, only platinum will do.

It's expensive stuff, worth twice as much per kilo as gold is, so it's kept under lock and key. But because its also quite rare, workers recycle and reuse it wherever possible and that's where this box of industrial waste comes in.

Contained within this waste is pure platinum, and the brainiacs at this chemical plant are going to separate it out from the other metals. All the debris is put into this enormous glass flask. The glass is important, because it resists the chemicals they are about to use.

Yes this is the fun bit of chemistry. Nitric acid and hydrochloric acid are poured in then the flask is sealed. Over the next 2 days, the metals will dissolve until all you're left with is this acidic black soup.

Platinum recovery is like a big chemistry experiment, only the scientists know what the outcome is going to be. The next stage is to separate the platinum from the other metals and that's done with this powder. The powder is added at the top of the flask and as it filters down, the platinum particles stick to it leaving the other cheaper metals behind. Slowly it settles to the bottom and once the scientists are happy, all the liquid will be drained off, leaving them with a big bucket of sand like grit.

Now you might have a hard time convincing anyone that this is the most valuable metal in the world. So the grit is then placed into trays and incinerated. At 950 degrees centigrade everything but the metal is burned off and when the boffins open the furnace up what's left looks like ... asphalt?

It's actually platinum, but it's been filled with air bubbles from the furnace. To turn it into the shiny metal that is so sought after for making jewelry, it needs to be reheated in an airtight furnace.

This powerful oven can reach 2,000 degrees Celsius and can be kept airtight while it does this. The platinum slag is added in, the heat is turned up and the aerated metal is melted down. The lid is then closed and all the air extracted. Even the Terminator wouldn't survive in here.

What emerges has to be the most expensive brick in the world. Just 1 kilo of this precious metal is worth over £20,000.

In this form it would be a bit awkward to wear, so specialists at the factory turn the blocks into tubes. Jewelers can use them to make rings or any other jewelry your girlfriend might want. Slices are cut off the tube in a sealed environment to ensure nothing is lost.

And this is what you end up with. Gold and silver are weak so they're mixed with other metals to strengthen them, but this reduces their value. Platinum is strong enough to be worked alone, which is another reason it's so valuable.

Sizing is quite a simple process, but if the customer prefers a thinner style, more metal must be carved away. Using a lathe lubricated with high quality oil, micro-fine slivers of the ring are carved away. Again, nothing is wasted and the jeweler and a computer both keep a very close eye on what's going on.

Because it's so rare and your girlfriend isn't going to want any old ring, platinum jewelry is often made to order. Only experts get to work with this precious metal and they have to know what they're doing. Heat is the most common tool. Melting the metal makes it pliable but if he keeps the flame on too long, it'll collapse into an incredibly expensive, but very ugly blob.

Another tool that can be used to create a mount for jewels is this saw. Tiny quantities of the ring are carved away to create the setting for the gemstone to fit into. Now you could try this with your Black and Decker workmate but we really wouldn't advise it.

Alternatively smaller stones can be placed into ready made settings like this one. An electric hammer gently closes the metal around the stone which will hold it in place.

And finally the ring is polished ready to be worn. Gold and silver change colour as they age, but because of it's purity platinum doesn't.

So like the credit card you've always dreamed of, this is platinum and its worth twice its weight in gold.

Did you know?

In the UK, a platinum album is one that has sold 300,000 copies. Michael Jackson's album "Thriller" has gone platinum a record-breaking 170 times.

金属丝眼镜框

扫描二维码，观看中文视频。

金属丝眼镜框一定要结实。它们的使命艰巨，考虑到只是用细金属丝制成，从开始就需要认真打造。做一副金属丝镜框，你要从设计开始。为了牢固，必须制作精良，正确的规格至关重要。设计完成后，就可以按形状制作镜框了。这台机器装有大卷的优质钢丝。镜片尺寸和钢丝被输入后，机器就开始工作。

为了确保形状绝对精准，一块空白的玻璃镜片用来做引导。机器根据玻璃片的弧度来弯曲钢丝。每隔几秒钟就有一个新镜框弹出来汇入其他做好的产品，但这些镜框还没有完成。任何镜片放入其中将立即掉出来。下一个零件——铰链提供了双重的解决方案。添加铰链后能把钢线框连在一起，并为此后的眼镜腿提供了连接的地方。现在，我们已经有了一个带铰链的框架，但是这只是个单片眼镜。为了做一副眼镜，我们需要将两个带铰链的框架从中间连接。这个连接的零件叫镜桥，它用焊丝焊接到位。在625摄氏度，焊丝熔化，把两部分连在一起。

这些镜框会这样坐在你的鼻子上却不能久留。因此下一步是添加眼镜腿。眼镜腿由更多的钢丝压制，但这样会造成边缘粗糙。那么用什么办法解决呢？接下来它们被放到一个大桶里与锥形物和肥皂混合。锥形物由石头做成，与肥皂搭配后能磨掉粗糙或锐利的边缘。于是你得到了一个光滑的眼镜腿，不会划伤头部两边。这些眼镜腿现在可以从先前的铰链部位连接镜框了。但每个戴眼镜的人都会告诉你，这些细小的螺丝非常容易丢失。

制造商在不断寻找各种方法来防止螺丝脱落。在这个部门，选来的新的眼镜架放在测试机上重复打开和闭合。在眼镜的一生中平均每副眼镜要开合10万次以上。它们还将经历很多的磨损，比如被一个顽皮孩子抓弄，或匆忙中掉在地上。它们需要有足够的强度来应对。为了改进耐受力，框架还会获得一些有用的附加层。

首先被电镀。这意味着覆盖一层坚固的金属外壳，保护它们免于划伤和锈蚀。

接着是上清漆。这一工序可用手工完成，如果有大量镜框，机器能够为许多副镜框同时上漆。上好清漆的镜框随后在烤箱烘焙。14分钟以后这些镜框就有了一层坚实的外衣，帮助他们抵御损害。

因为人的脸和耳朵都相当敏感，所以厂家用特殊橡胶包裹镜架，包括眼镜停在鼻子上的两个小支点和眼镜腿部位。接下来是折弯。这台机器将眼镜腿折成曲线，让佩戴者放在耳朵上感觉舒适。几乎所有人在生活中的某个时刻需要一副眼镜，所以这个行业对未来有着清晰的视野。

你知道吗？

埃尔顿·约翰曾说他拥有两万多副眼镜。其中著名的一款带有闪光灯和雨刷器。

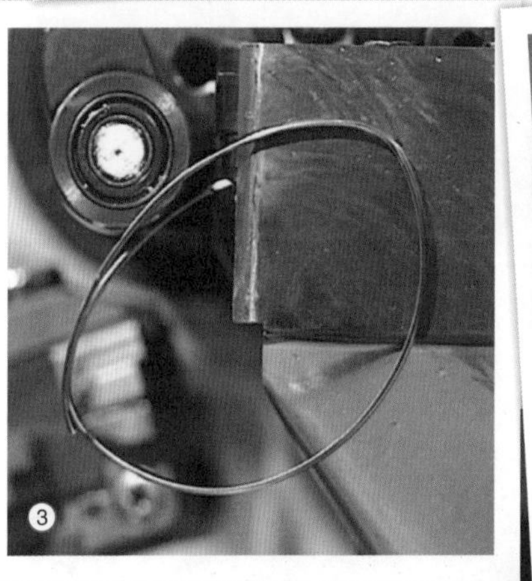

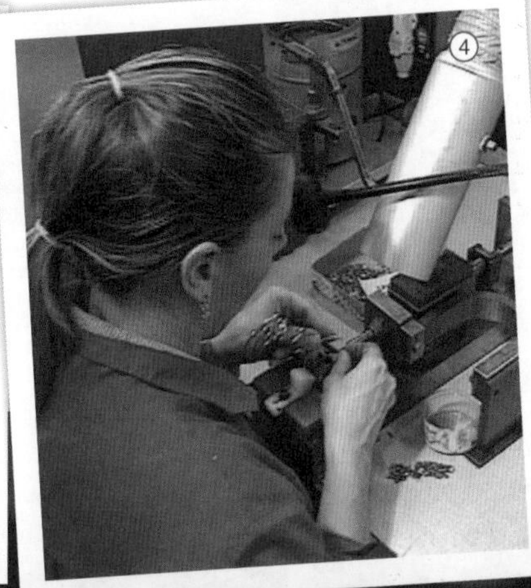

1. 一副金属丝眼镜框从设计开始
2. 一块空白的玻璃镜片用来做引导
3. 机器隔几秒就有新镜框弹出
4. 铰链把钢线框连在一起
5. 镜桥将两个带铰链的框架连接
6. 眼镜腿由更多的钢丝压制而成

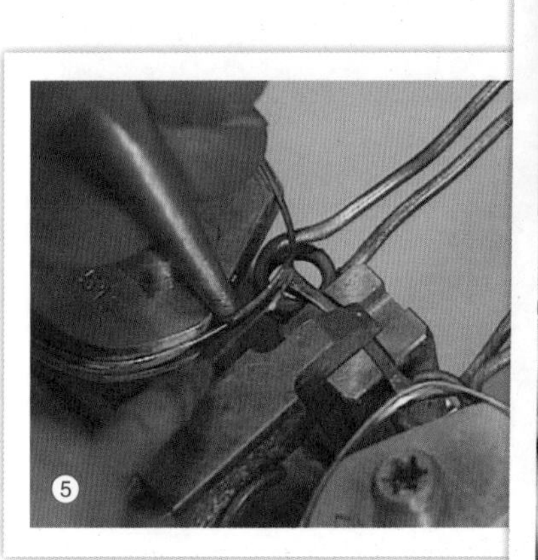

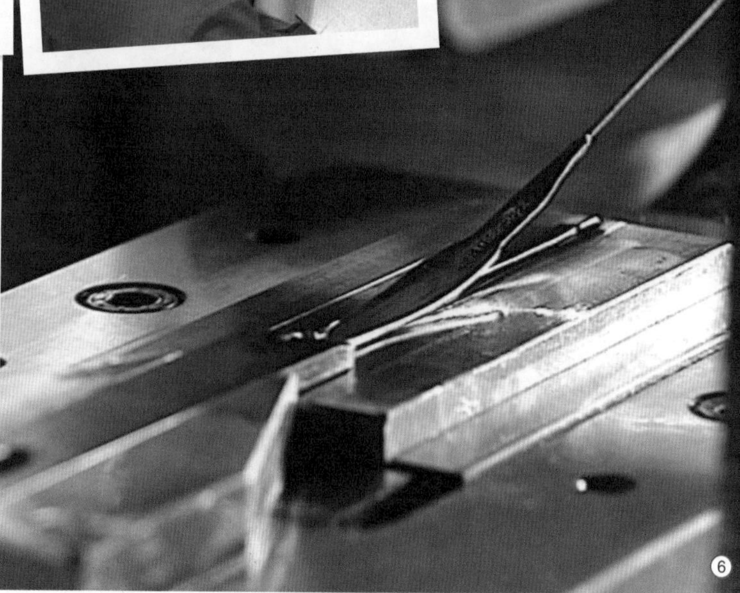

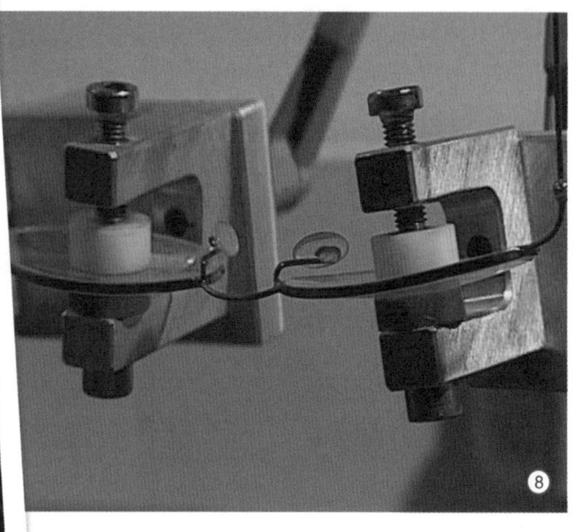

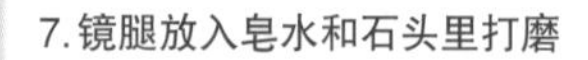

7. 镜腿放入皂水和石头里打磨
8. 检测镜桥的承受强度
9. 电镀为镜框覆盖一层金属外衣
10. 上好清漆的镜框在烤箱烘焙
11. 用橡胶包裹眼镜腿并折弯
12. 精心制作的金属丝眼镜框

Wire Glass Frames

Wired framed glasses have to be tough. They've got a tough job and considering they're only made from thin wire, they need to be well-built to start with. When you're making a pair of wire frames, you start with a design. To be strong, they have to well-made so getting the specifications right is paramount. With the plans complete, the frames can begin to take shape. This machine is loaded up with huge reels of high grade steel wire. The lens dimensions and the wire are fed in and the machine gets to work.

To make sure the shape is absolutely perfect a glass lens blank is use as a guide. The machine bends the wire according to the curves of this blank. Every few seconds a new frame pops out to join the others, but they're not finished yet. Any lens put into one of these would immediately fall out, but there's a double solution provided by the next piece, which is the hinge. When this is added, it holds the wire frame together, and provides a place for the arms to be attached to later. Now we've got a frame with a hinge, but that's a monocle. To make glasses, we need two hinged frames that are attached in the middle. This attachment is the bridge and it's soldered into place using lead soldering wire. At six hundred and twenty five degrees Celsius the solder melts and bonds the two sides together.

These frames would sit on your nose like this, but they wouldn't stay there very long. So the next step is to add some arms. These are pressed from more steel wire, but pressing them like this gives them rough edges. So what's the solution? Well next, they're now dropped into an enormous vat with these cones and some soap. The cones are made of stone and in combination with the soap, they rub away the rough or sharp edges. What you're left with is a smooth arms that won't scratch the side of your head. These arms can now be attached to the lens holders with the hinge from earlier. But, as any glasses wearer will tell you, these tiny screws can go missing all too quickly.

The manufacturers are continually looking for ways to stop losing their screws. Here in this department, a selection of new frames have been connected to a test machine which opens and closes them repeatedly. In their lifetime, the average pair of glasses will open and close over 100,000 times. They will also experience a lot of wear and tear in that time like being grabbed by a playful child to being dropped when you're in a rush. They need to be tough enough to cope. To help improve their durability, the frames will also receive a couple of useful additional layers.

First, they're galvanized. This means they are coated in a robust outer layer of metal which will protect them from scratches and rust.

Next they're varnished. This can either be applied by hand or if there are a large number of frames this machine can varnish many pairs at a time. The freshly coated frames are then baked in this oven. 14 minutes in here and they've got a solid coat to help them resist any damage.

Now, your face and ears are quite sensitive so the manufacturers add special rubber covers to the frames. There are two nubs where the glasses rest against the nose and the arms get covers as well. The bending stage comes next. This machine will curve the arms so they sit comfortably behind the wearers' ears.

Almost everyone will need a pair of glasses at some time during their lives, so this is one industry with a clear vision of the future.

Did you know?

Elton John once said he owned more than 20,000 pairs of glasses. Famous examples had flashing lights and windscreen wipers.

擦鞋垫

扫描二维码，观看中文视频。

一年之中，秋天是逛公园的最佳季节。但雨水和泥泞意味着你回家时，粘在鞋上的污垢和树叶会弄脏你的地毯。有个极好的解决方法来自椰子，或更准确地说，椰子的壳。

椰子壳里的纤维，称为椰壳纤维，能制造擦鞋垫。它的抗菌特性有助于杀死细菌。它吸水性强，最重要的是坚韧，便于擦掉泥污。

天然椰子纤维擦鞋垫从这里开始制造。数英里长的椰壳纤维纺成长线，绕在大卷轴上。这些线卷把椰壳纤维送进制垫机。

做一个擦鞋垫要用 6 个椰子壳，所以工厂必须供应充足。当一个线卷快用完时，将连接另一个线卷持续送料。工人必须把纤维编织在一起，如果打结将会阻断整个系统。机器像一个大蜘蛛坐在巨型椰子纤维之网中央。400 条线同时经过这些孔，做成擦鞋垫的基础。用椰子壳制作垫子的秘诀是胶水。椰壳纤维线不是纺在一起的，而是垂直地立在垫子上。胶水是线的根基。首先这把铡刀将纤维切成适当长度。然后落入下面的胶水里。当铡刀切过纤维时，漏斗形状使切出的小段垂直落下，在迎候它们的胶水上着地。

如果做这种擦鞋垫，切好的纤维大约伸出 16 毫米高，非常适合清理运动鞋或泥泞的靴子。按压纤维以确保它们牢牢植入胶水中，然后胶水和纤维结合的整体通过加热板。这样让胶水变硬，并将刷毛固定到位。

为保证质量，工人对每卷擦鞋垫进行常规抽样检查。必须经历的第一项检测是强度试验。工人使用高科技拉伸装置和一把钳子，分析将纤维从胶基中拔出来需要多大的力。如果刷毛能承受 6 千克以上的拉力，就可以通过。

下一个测试是垫子的弹力。这个机器人设备，看来就像从科幻电影中拿来的，但它并不复杂。它的工作是不断行走。模拟长期使用，并测试擦鞋垫的耐久性。

当今擦鞋垫最重要的品性之一是吸水。为了防止泥和水跑进家里，有效吸水性很重要。没有湿脚印意味着椰子纤维发挥了作用。

质量过关后，擦鞋垫可以切成合适的形状了。首先将粗糙的边缘裁掉，然后“理发”。接着使用切刀，像割草机一样，这台机器将纤维修剪成统一长度。

下面是形状。每个卷轴可以切成许多个垫子。工人把垫子收集起来送往打印部门。流行的设计包括“欢迎”标志和人们最喜欢的足球队颜色，这些标志将在此处印上去。首先对应所需形状铺一层胶水。随后是几层带静电和颜料的薄片。薄片粘到胶水上，最终现出图片。静电能吸附鞋上的灰尘，所以，即使图片在垫子上，也有助于你的房子保持清洁。

最后一步是把垫子清理干净。生产过程中有大量灰尘和纤维粘在垫子上，没人愿意买一

你知道吗？

在泰国，猕猴被训练从高大的树上采摘椰子。

个脏垫子。这些机器把附着的脏东西抖掉。

垫子终于做好了。最初保护椰子肉的椰壳纤维，现在变成了保护你家免受脏鞋玷污的擦鞋垫。

7. 步行机测试垫子的持久性
8. 检测椰子鞋垫的吸水性能
9. 机器将纤维修剪成统一长度
10. 切割好的垫子送去印制
11. 贴上带静电和颜料的薄片
12. 抖掉垫子上的碎屑杂物

⑦

⑧

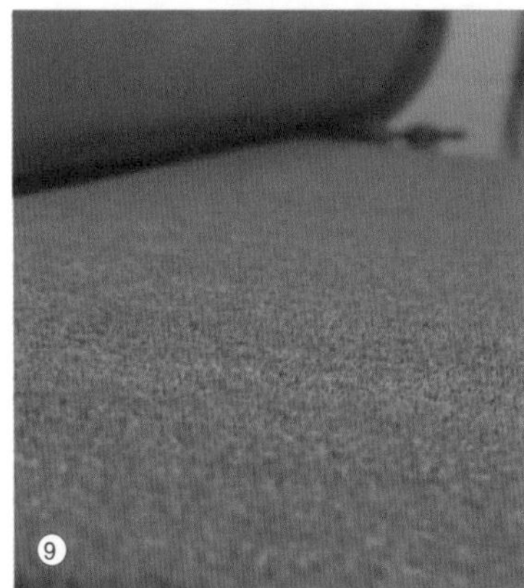

⑨

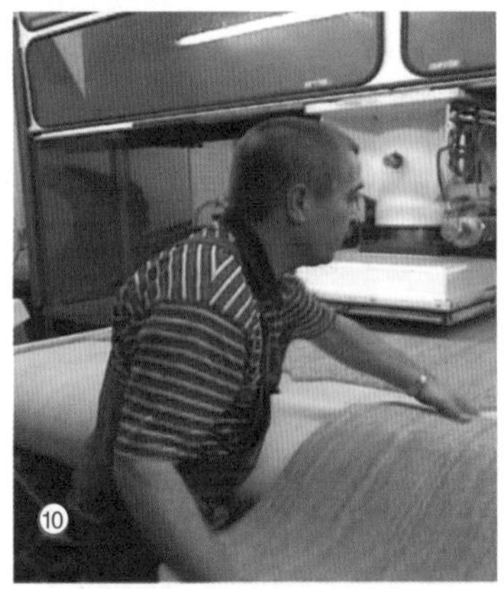

⑩

⑪

⑫

Doormats

Autumn. It's the perfect time of the year for a nice walk in the park. But rain and mud mean that when you get home, the dirt and leaves stuck to your shoes may ruin your carpet. One brilliant solution comes from coconuts or the coconut's husk to be more precise.

Coconut husk fibre, known as coir, is used to make doormats. Its antibacterial properties help to kill germs. It's absorbent, which traps water and most importantly it's tough, which helps remove unwanted dirt.

Natural coconut fibre doormats start their lives here. Miles of the coir are woven into long threads which are kept on big spools.

These reels feed the mat making machine.

It takes the fibre from 6 coconuts to make just one doormat so the factory needs a plentiful supply. When a reel runs low, it's connected to another one to keep a continual feed to the machine. The worker must weave the reels together because any knots would jam the whole system. The mat machine sits like a big spider at the centre of an enormous coconut thread web. 400 lines are fed simultaneously through these holes to form the basis for the doormats. Now, to make a mat using this coconut husk the secret ingredient is glue. The fibrous threads aren't woven together, they all sit vertically on the mat. The glue is used as a foundation for the threads. First, this guillotine slices the fibres into the right lengths. These lengths then fall into the glue-base waiting below. As the knife slices through the fibre, the shape of the funnel forces the pieces to fall vertically where they land in the waiting glue.

In the case of this doormat, the fibre lengths are cut so they will stick about 16 millimetres high, perfect for cleaning a pair of trainers or muddy hiking boots. The fibres are pressed to make sure they're seated firmly and the whole glue-fibre combination is then passed over a large heating plate. This hardens the glue and seals the bristles into place.

To keep quality up the workers take regular samples from each roll of new doormat. The first test they must undergo is the strength test. Using a high tech tension device and a pair of pliers, the worker will analyse how much pressure is needed to wrench the bristles free from the glue base. If the bristles survive more than 6 kilos of pressure, they've passed.

扫描二维码，观看英文视频。

The next test measures the mat's resilience. This robotic device looks like it's from a sci fi film set, but it's not that sophisticated. Its job is to walk up and down continually. This simulates many hours of use and tests how durable the doormat is.

And one of the most important qualities of the modern doormat is absorbency. To help stop mud and water coming into your home, it's important they can trap liquid effectively. No damp footprints mean the coconut fibre is doing its job well.

With the quality assured, the doormats can now be cut into shape. First the rough edges are cut off and a hair cut follows. Using a guillotine which looks like a lawnmower, this machine will trim the fibres down to a uniform length.

Next, comes the shape. Many mats can be cut from each new roll. The workers collect them up and they're sent on to the print department. Popular designs include a big Welcome sign, to the colours of your favourite football team, and those logos will be applied here. First a layer of glue is laid down corresponding to the desired shape. This is then followed by several different layers of statically-charged, coloured flakes. The flakes stick to the glue and a picture eventually emerges. The static helps attract dirt from the user's shoes, so even the picture on the mat helps to keep your house clean.

The final step is to clean the mats off. During the production process plenty of dirt and fibres may have been caught and no one wants to buy a dirty mat. These machines dislodge anything that may be stuck in place.

And finally the mat is finished. So having started life protecting the flesh of the coconut, the coir now protects your home from dirty shoes.

Did you know?

In Thailand, macaque monkeys are trained to collect the coconut harvest from tall trees.

水龙头

储物架

柴火壁炉

荧光灯

马桶

器物篇

豪华皮椅

膨胀螺丝

斧头

割草机

园林工具

购物车

暖气片

水龙头

有些东西我们非常熟悉，但却从来没有真正想过，它们是怎样制造的，或者是如何工作的。单杆冷热水混合水龙头的生产，从一桶石英细砂开始。

石英砂被压入加热的模具中，然后在260摄氏度下烘烤。变硬后的石英砂本身就成了一个模具。经过边缘平滑处理，就可以用来铸造水龙头了。与此同时，回收的黄铜杂件投入两吨半的巨大炉子中，在1050摄氏度下熔化。工人把模具放进低压铸造机，并用气枪吹走脱落的颗粒物。当液态黄铜开始倾倒时，喷灯被用来加热管道以保持铜汁流动顺畅。两分钟后，一对水龙头就成型了。

工人需要除去水龙头内的固体砂子，为此他们用超声波把砂模打碎成颗粒。铸造的水龙头是连在一起的双胞胎，但很容易用金属锯切开。现在可以给每个水龙头钻孔和切割了。这个大机器包含有17种不同的工具，一手包办所有的工序。每个开口和螺纹都经过仔细检查。如果一个水龙头没有达到标准将会回炉和重塑。

那些通过检查的水龙头需要进行擦洗。首先一个机器人用砂纸打磨让所有粗糙的边缘变得平滑。然后用纯棉抛光水龙头的表面。黄铜的水龙头的主体完成了。

另一个机器人在照相机镜头前展示水龙头的每个角度，检查它是否完美。这个水龙头有问题，所以放在一边等待处理。由工人做最后的决定，质检员确认机器人的发现。这个水龙头被拒收并送去回炉。对于幸运通过检查的，该洗个澡了。一个计算机控制系统引导它们通过几个水槽。首先洗掉污垢或油脂，然后把它们泡进镍溶液里。这会形成一个银色的涂层，有助于保护它们免受腐蚀。

最后让它们在铬溶液中洗浴4分钟。冒泡的溶液表面下，水龙头接收到最终的涂层，这将有助于它们经久耐用。将水龙头连接电流的负极，铬溶液连接正极。能使水龙头从溶液中吸附微小的铬颗粒，并在表面镀上一层比头发还细的铬层。

当水龙头的主体进行干燥时，冷热水混合的中心部分也被组装起来。一种简单的机制使得水龙头高效又环保。旋转手柄会带动水龙头内部一个有椭圆形开口的部件。向左转动打开通往热水的管道；处在中间位置，则同时连通冷热水管；移动到右侧会关闭热水，只有冷水流出了。和标准的立柱式水龙头相比，具有混合喷洒功能的水龙头节省多达80%的水，因为它可以在较低水压下维持稳定的水流。

在内部机关安装完毕后，需要进行最终测试以确保水能顺畅流出。水龙头已经由机器人抛光、镀镍和镀铬，其内部结构有助于节约世界的水资源。也许这个看起来不起眼的水龙头，并非那样乏善可陈。

你知道吗？

一个水龙头每秒漏一滴水，每年会浪费10000升水——足够装满50个浴缸。

1. 水龙头从石英砂模具开始
2. 黄铜杂件在高温下融化
3. 石英砂模具放进低压铸造机
4. 倒入铜汁后一对水龙头成型
5. 水龙头开口和螺纹在这里完成
6. 逐个检查所有开口和螺纹

7. 机器人用砂纸打磨水龙头
8. 人工确认检测出的问题水龙头
9. 龙头依次在镍及铬溶液中浸泡
10. 组装冷热水混合中心部分
11. 椭圆形开口转动管控冷热水
12. 冷热混合水龙头备受欢迎

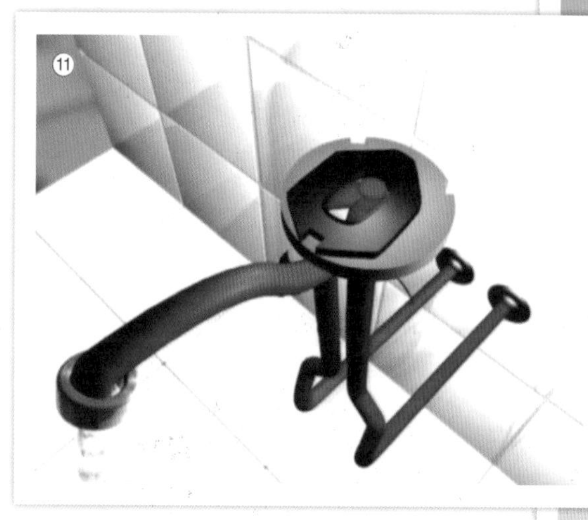

Water Tap

Some objects are so familiar to us that we never actually consider how they are made or how they work. A single-lever mixer tap begins life as a bucket of very fine quartz sand.

It's forced into heated moulds and then baked at two hundred and sixty degrees Celsius. When it's hardened the sand has been transformed into a mould itself. Once they've smoothed off the edges it can be used to cast the taps. Meanwhile recycled brass parts have been melted into a huge two and a half ton furnace at one thousand and fifty degrees Celsius. They place the moulds into a low pressure casting machine and blow away any loose grains with an air gun. As the liquid brass is poured in, the duct is heated with a blow torch to keep it flowing smoothly. Two minutes later a couple of pairs of tapes drop out.

They need to get rid of the solid sand inside the tapes so they blast them with ultrasonic sound waves to break it down into grains. The taps have been created is co-joined twins but they are easily separated by a metal saw. Each tap can now be drilled, countersunk and cut. This large machine contains seventeen different tools and does all of the jobs single-handedly. Each opening and thread is carefully checked. If any of the taps aren't up to scratch they get melted down and re-moulded.

For those that pass, it's time to get scrubbed up. First a robot smoothes out any rough edges by rubbing them on sand paper. Then it polishes the surface with pure cotton. The brass body is finished.

Another robot displays every angle to a camera lens which checks it's perfect. This one is faulty so it's put to one side for inspection. A human has the final say and the quality controller confirms the robot's findings. It's rejected and will be recycled. For the lucky survivors it's time to take a dip. A computerized system guides them on a route through several troughs. First a rinse removes any dirt or grease. Then they get dunked in nickel. It forms a silvery coating which helps to protect them from corrosion.

Finally they're immersed in a chrome bath for four minutes. Beneath the bubbling surface the taps are receiving a final coating which will help them last a lifetime. An electrical current sets the taps to a negative polarity and the Chrome bath to positive. These causes the taps to attract tiny chrome particles from the water and creates a fine surface layer of chrome that is thinner then a human hair.

While the bodies are drying the heart of the mixer tap is put together. The simple mechanism is what makes the tap so efficient and environmentally friendly. Turning the handle moves an element with an oval opening inside the tape. To the left opens the waterway to the hot water pipe; in the centre to both hot and cold; and moving it to the right closes the hot water flow so just cold water flows out. Mixer taps with spray mechanisms like this can save up to eighty percent of the water used by standard pillar tapes as they produce a steadier flow with less water pressure.

Once the mechanism is put inside the tap a final test ensures the water flows out smoothly. They've been polished by robots, coated with nickel and chrome and their inner workings will help to conserve the world's water supply. Maybe this humble tap is not so humble after all.

Did you know?

A tap that leaks one drop per second wastes 10,000 litres of water per year—enough to fill 50 bath tubes.

储物架

自行组装储物架，价廉，物美，高效，而且据说装起来很容易。装储物架是个复杂过程，你需要确定拥有全套部件，才能把它们搭建起来。虽然瑞典被认为是这种自装家具的发源地，现在生产厂家已经遍布全世界。

在德国的这家工厂，卡车运来刚刚生产的碎木胶合板。这是制作储物架单元的基本材料。每个货盘都超过1吨重，需要铲车从卡车上卸下来再运进工厂。

无须惊讶的是，胶合板是由碎木片制成。胶水将碎木粘到一起，然后被压成这样的板材。这个工厂一天用掉多达3万块。

碎木胶合板便宜而实用，但看上去却不够品位。为了让做成的储物架显得更光鲜，碎木胶合板外表会装上木质面板。这个栎木面板厚度正好，半毫米厚，还需要切割成适当的长度，镶在每块胶合板的表面。有些面板还需要再次用机器切割成正确的宽度。有些面板需要被拼接起来，用于镶饰更宽的碎木胶合板。

每片面板被放在强光之下检验，有小洞的地方会被贴上胶条进行修补。

现在是给碎木胶合板装上这些面板的时候了。成垛的胶合板在传送滚轮上穿过工厂，接着机器把胶合板逐一吸起，放到镶面机器上。它们先被快速去除灰尘，然后两个滚筒把胶合板上下双面都涂上胶。因为黏胶表面会粘住传送带，所以涂完胶的碎木胶合板，由薄金属盘运走。

胶合板的上下各放一张面板，经过机器加压，粘接成一个胶合板三明治。

现在该打上全部重要的孔洞了。20个钻头固定装在同一个框架上，打孔的时候同时抬起，确保所有孔的间距以及深度准确无误。每块木板不到一秒钟就能打孔完毕。为确保家具的各个部分能完美地拼装到一起，质检员会从每个批次抽出一个样品进行检验。样品会拿去和标准模板进行对照，甚至会给样品安上销子和木钉。

储物架是设计用来装书和CD光盘的，但它们还需要能够经受住饮料泼洒一类的考验，譬如在家庭宴会上。所以，这些板子被整齐地摞起来，掸掉浮灰后，边缘被喷上一层清漆。喷过清漆的这堆木板后是一个水墙，能吸收散在空气中的清漆。在80摄氏度下，清漆10分钟就可以干燥。接着，木板边缘会快速抛光，以去掉沾在上面的木头纤维。现在要修饰一下表面了。在一个100米长的自动化系统里，它们也被上好清漆。这些板子先上清漆，继而干燥，最后被抛光。

当板子被做好后，它们会通过传送带，和其他所有部件堆放在一起。一组工人往每个包装里放入组装所需的全部零件，原则上应如此。

你知道吗？

4000年前埃及人就用薄木板装饰家具。

1. 自装储物架需要全套部件才能成功
2. 碎木胶合板是储物架基本材料
3. 镶饰更宽胶合板的面板需要拼接
4. 在强光下检验面板并进行修补
5. 两个滚筒把胶合板两面涂胶
6. 胶合板送去压合

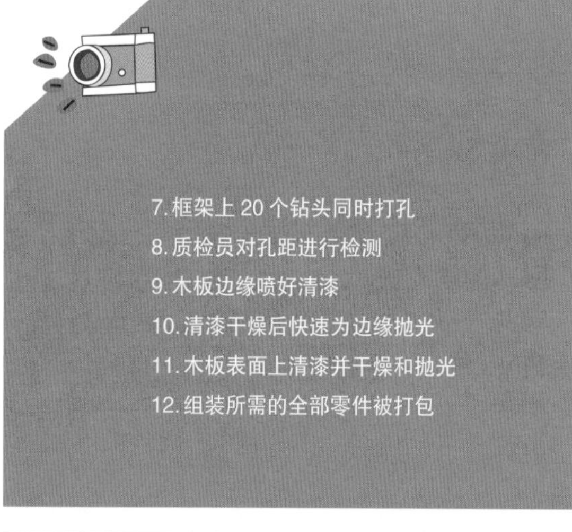

7. 框架上 20 个钻头同时打孔

8. 质检员对孔距进行检测

9. 木板边缘喷好清漆

10. 清漆干燥后快速为边缘抛光

11. 木板表面上清漆并干燥和抛光

12. 组装所需的全部零件被打包

⑦

⑧

⑨

⑩

⑪

⑫

Flat Pack Shelf

The flat pack shelf; cheap, effective and allegedly easy to put up. This is the intricate process that should make sure you get all the pieces you need to put up those shelves Although it's thought that this type of shelving was originally invented in Sweden, it's now made all over the world.

At this factory in Germany trucks arrive with freshly manufactured chip board which is the basic building material of the shelving unit. Each pallet weighs over a tonne so a forklift helps unload the trucks cargo and take it into the factory.

You won't be surprised to learn that chip board is made out of chips of wood. They are glued together and compressed to make boards like these. This factory uses up to 30,000 of them in a single day.

Chip board is cheap and practical, but it doesn't exactly exude quality. So to smarten up the soon to be shelves the chip board is going to be covered with wooden veneer. This oak veneer is the right thickness, half a millimetre, but it has to be cut down to the right length for each piece of chip board. Then some of the pieces are chopped again this time by a machine to make them the right width. Other pieces need to be sown together so they can cover wider boards.

Each veneer panel is then checked over a bright light and weak spots are repaired by gluing on small strips.

Now it's time for the chip boards to get kitted out. Rollers carry the stacks through the factory, then suction cups lift the boards to the veneering machine piece by piece. They are given a quick dusting down then two rollers cover both sides of the boards with an adhesive. The tacky surfaces would stick to a conveyor belt so instead they are moved on by thin metal disks.

A sheet of veneer is placed underneath and another one on top. Then it's glued into a chip board sandwich by an industrial press.

Time for the all important holes. Twenty drills fixed in a frame are all raised simultaneously to ensure that all the holes are the right distance apart and the correct depth. Each panel is drilled in less than a second. To make doubly sure that every piece of furniture will fit together perfectly they check a sample from the batch against a template and even fit the pegs and dowels.

The shelving units might be intended for storing books and CD's but they've also got to be able to withstand the odd spilled drink at a house warming party. So after they have been brought into line, dusted off, the edges are sprayed with a coat of vanish. Behind the stack is a water wall which absorbs any stray varnish from the air. At eighty degrees Celsius the varnish sets in just ten minutes. Any wood fibres that are sticking to the edges are smoothed down by a quick polish. Then it's time for the main surfaces. They are varnished in a one hundred metre long automotive system. First the varnish is supplied then the boards are dried off. And finally they are polished.

Once the boards are finished they are stacked by a conveyor belt next to all the other parts and a team of workers add every single item to each pack, in theory.

Did you know?

Egyptians used wood veneer 4,000 years ago to decorate their furniture.

柴火壁炉

没有什么比传统的开放式壁炉更浪漫了，但如果你想有一个现代化的替代品，可以试一试这种柴火壁炉。

炉子由炉室和外壳组成。它主要用钢制成，大块钢板被剪切成炉壁，用来制作20个炉室。

钢板被装入激光切割机，激光就像热刀划过黄油一样把钢板切开。激光切割令人难以置信地准确，以超过1600摄氏度的高温把金属烧穿。剩下的边角料被起重机的巨型钩子吊走，它们将被熔化并重新使用。

这是用来制作炉室的46块钢板之一。显然，炉壁需要完美地安装在一起，所以质量控制员把它们全都检查一遍。一个3吨的辊子把钢片压成形，用来做炉子弯曲的侧面。

钢板还不能给你提供温暖的亮光，但这将很快改变。用模板检查曲率。另一个操作员把每边弯两次，这样炉室壁就可以安装在一起。同样也使用模板检查连接处是否完美配合。炉室壁在这里被组装起来。用一系列的夹子把它们固定。

炉室初具规模了。四个角用手工焊接到位。机器人把所有的接缝都焊起来，形成一个坚固的结构。只需要12分钟即可完成工作。

虽然炉子是为了加热客厅，但炉子外壁不宜太热，否则将存在火灾隐患。隔壁的车间里，正在进行外壳组装。只有接缝需要处理，所以用手工焊接，并不需要很长时间。

这是炉室放置在外壳里的位置。每个炉子重量为80千克，所以为了爱惜脊背，这个人用起重机把它抬起来。

现在该做一些外观工作了。在这个房间里，数千个微小的钢珠将会围绕壁炉击打15分钟，产生出带纹理的表面。升降机把壁炉从房间里面提起来，然后倒出钢珠。接着放炉灰的托盘和铸铁的炉箅被放置到位。

一个工人连接上喷涂装置，给整个炉子加上一层保护性的清漆。它能经受得住600摄氏度的高温。下一步，绝缘的耐热层被插入到炉中。它可能看起来像软木塞，但实际上是一种叫作蛭石的矿物。

为了让你可以凝视浪漫的炉火，而又把它安全锁定，需要安一个装甲玻璃门。它必须足够结实，能经受得住偶然的撞击，所以用这个10千克钢球进行测试。

壁炉已经完成，但还有事情要做。安装壁炉时必须严格遵守安全规则，所以派有经验的工人来认定正确无误。他们需要确保烟雾被安全地抽走。在这里，把壁炉接到一个弃置不用的烟囱上。检查到有气流存在，确定烟囱的通风道没有被堵塞。壁炉移动到位后，烟道被随之固定。

把一个底板滑到壁炉下面以保护地板，安

你知道吗？

火焰的颜色由火的温度和燃烧的材料所决定。

装到此完成了。

工人点燃第一把火，以确保所有工作完美。柴火壁炉不仅给你带来爱的感觉，也非常有效率，用环保方式让你的家充满温暖。

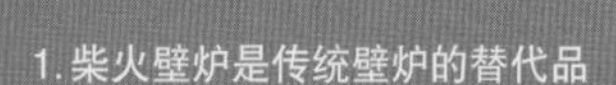

1. 柴火壁炉是传统壁炉的替代品
2. 激光切割机把钢板切开
3. 3 吨的辊子把钢片压成形
4. 每片钢板弯两次后检查连接处
5. 炉室壁在这里被组装起来
6. 机器人把炉室接缝焊接起来

7. 将炉室放入外壳中

8. 壁炉经小钢珠击打表面产生纹理

9. 放入炉灰托盘和铁铸的炉箅

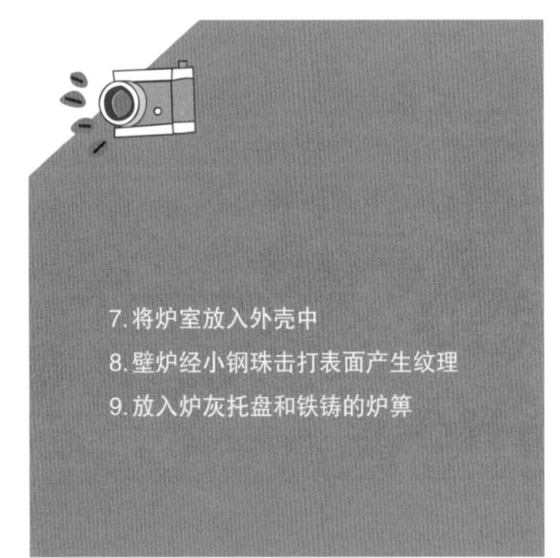

10. 壁炉安上装甲玻璃门

11. 壁炉连接烟囱确保烟雾抽走

12. 底板放到壁炉下保护地板

Wood Burning Stoves

There's nothing more romantic than a traditional open fire but if you want a modern alternative you can go for one of these wood-burning stoves.

The stove consists of an inner furnace and an outer shell. It's mostly made out of steel and here, a large sheet is cut into furnace walls for 20 stoves.

It's loaded into a laser cutter that slices through the steel like a hot knife through butter. It's incredibly accurate and burns through the metal with a temperature of over 1,600 (sixteen hundred) degrees Celsius. All the left over steel is hauled away with a heavy duty hook attached to a hoist and will be melted down and re-used.

This is one of 46 steel pieces that will make each stove. Obviously the walls need to fit together perfectly so they're all checked over by a quality controller. A 3-ton barrel rolls the sheets into shape to make the curved sides of the furnace.

The steel sheet might not be giving you that warm glow right now, but all that will change. The curvature is checked on a template. Another operator bends each edge twice so the chamber walls can be slotted together. Again the template is used to check the joints will fit together perfectly. Here the walls of the furnace are assembled. They're held together using a series of clamps.

And the chamber begins to take shape. The corners are welded manually to hold the pieces in place. A robot then welds all the seams to form a solid structure. It only takes 12 minutes to finish the job.

Although the stove is meant to heat the living room the exterior shouldn't get too hot or it would be a fire risk. So in the workshop next door, an outer shell is being assembled. There are only seems so it doesn't take long to weld by hand.

This is where the furnace gets put in the shell. Each unit weighs eighty kilos, so to save his back, this man uses a hoist to carry it for him. It's time for some cosmetic work. In this chamber thousands of tiny steel balls are blasted around the stove for fifteen minutes to give it a textured finish. The hoist lifts the stove from the chamber and then the steel balls are tipped out. Then a tray for the ash and a cast iron grate are put into place.

One of the workers hooks up an air-brush system and gives the whole unit a fine coat of protective varnish. It will resist temperatures of up to 600 degrees Celsius. Next a heat-resistant layer of insulation is inserted into the furnace. It may look like cork but it's actually a mineral called vermiculite.

So you can gaze at your romantic fire, but keep it safely locked up an armoured glass door is installed. It's got to be tough enough to take the odd knock so it's tested with this 10 kilo steel ball.

The stove is complete but there is still work to be done. Strict safety guidelines must be followed when installing one, so experienced fitters make sure it's done properly. They need to ensure the smoke will be extracted safely. Here they're going to channel it through a disused chimney shaft. They check there's a draft and the shaft hasn't been blocked. After the stove has been moved into place the flue is fixed into position.

A base plate is slid under the furnace to protect the flooring and the job's done.

The workers start the first fire to make sure everything is working perfectly. Not only do wood burning stoves give you that loving feeling they're also so very efficient and an ecologically sound way to heat your home.

Did you know?

The temperature of a fire and the material being burned determine the colour of the flames.

荧光灯

扫描二维码，观看中文视频。

来到工作地点，开门，开灯。

英国各地的办公室都使用荧光灯照明，因为它便宜，从长远看更有效率，一个充满气体的管子是怎样点亮你的生活呢？

荧光灯制造从一大团热玻璃开始。在液体状态下，玻璃可以被拉伸并吹成一个很长的管子。金刚石刀具把长玻璃管切成2米长。偶尔玻璃管破裂，将被抛弃和回收。

下一步是修整玻璃管两头。700摄氏度的火焰切出一条整齐的断裂线，小刀具敲掉不整齐的末端。锋利的边缘被修整圆滑后，玻璃管就可以装进各种发光部件了。

荧光灯比普通灯泡更高效，使用起来更便宜，因此对企业更有吸引力。要让这种经济的光源亮起来，灯管内部必须涂磷光体，一种让灯管发出荧光的化学物质。

下面安装灯管的内部元件。像普通灯泡一样，荧光管也有通过电流的灯丝。灯丝必须装配在一起，以便安放到涂有荧光粉的灯管里。荧光管用电流激励气体粒子。使这些粒子放出能量并转化为光线。灯丝做好后密封到灯管里。

你可能已经注意到每个大玻璃管的末端伸出一个小玻璃管：发射能量的气体从这里注入。这种气体是由汞制成。汞是一种有毒的液态金属，从电池到医院温度计，许多日常物品中都用到它。

从这些物品中回收汞，在这样的工厂汇集起来。这是个危险的过程，科学家们必须小心。处理汞是一种有危害的工作，需要佩戴防毒面具，这种液体金属可以用普通的桶收集，一个简单的窍门是用水进行安全存放。

汞蒸气是荧光灯管中的发光气体，如果大量吸入会很危险。把汞储存在水层下面可以防止蒸发。如果你担心灯管内的汞蒸气，也大可不必。在荧光灯中使用的剂量非常之小，危险性可以忽略不计。

灯管现在准备填充汞蒸气。玻璃管和汞被装入灌装机。汞被加热，产生汞蒸气注入管内。热使气体释放能量，让灯管发光，虽然还没有接通电源。

那么电、汞蒸气和荧光粉涂层是如何产生荧光的？我们已经知道充气的玻璃管内有灯丝。灯丝从一端发送电子到另一端，撞击其间漂浮的气体粒子。气体受激发，进而发出辐射。但辐射是看不见的。

这正是荧光粉涂层的重要之处。它并不是给灯管上色的。荧光粉能在辐射的轰击下发出荧光，这就是照亮你办公室的光芒。

灯管内壁涂好荧光层，安装好灯丝，填充好气体，剩下的就是安装外部金属帽，以便安装到灯座上。金属帽被小心盖在灯丝上，密封到位，于是安装成功。

你知道吗？

如果每个英国家庭都用节能灯泡换掉3个普通灯泡，以此节省的电力足够照亮我们的街道整整一年。

最后灯管通过测试，在送往到商店之前及时发现不合格的灯管。

最初的路灯是气体点亮的，它们大概已经不符合现代标准。但我们今天仍然能看到更清洁、更环保的气体照明技术——荧光灯。

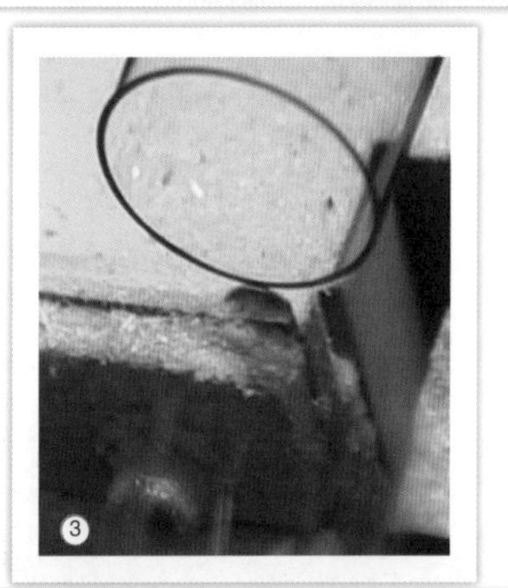

1. 制造荧光灯从一团液态玻璃开始
2. 液态玻璃吹成一根很长的管子
3. 敲掉玻璃管不齐的末端
4. 灯管锋利的边缘被修整圆滑
5. 灯管内壁必须涂上磷光材料
6. 灯管的内部元件在这里装配

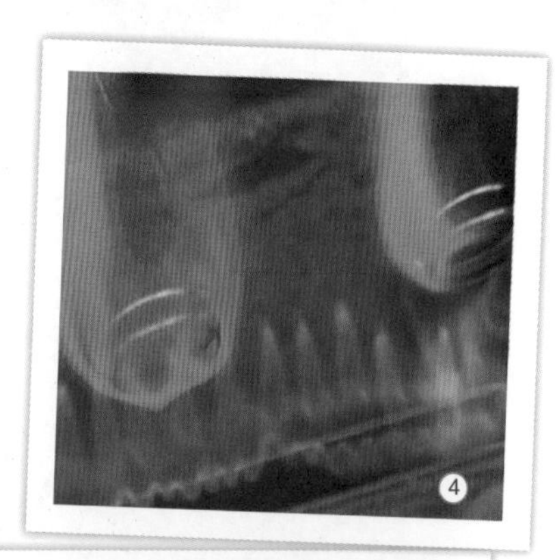

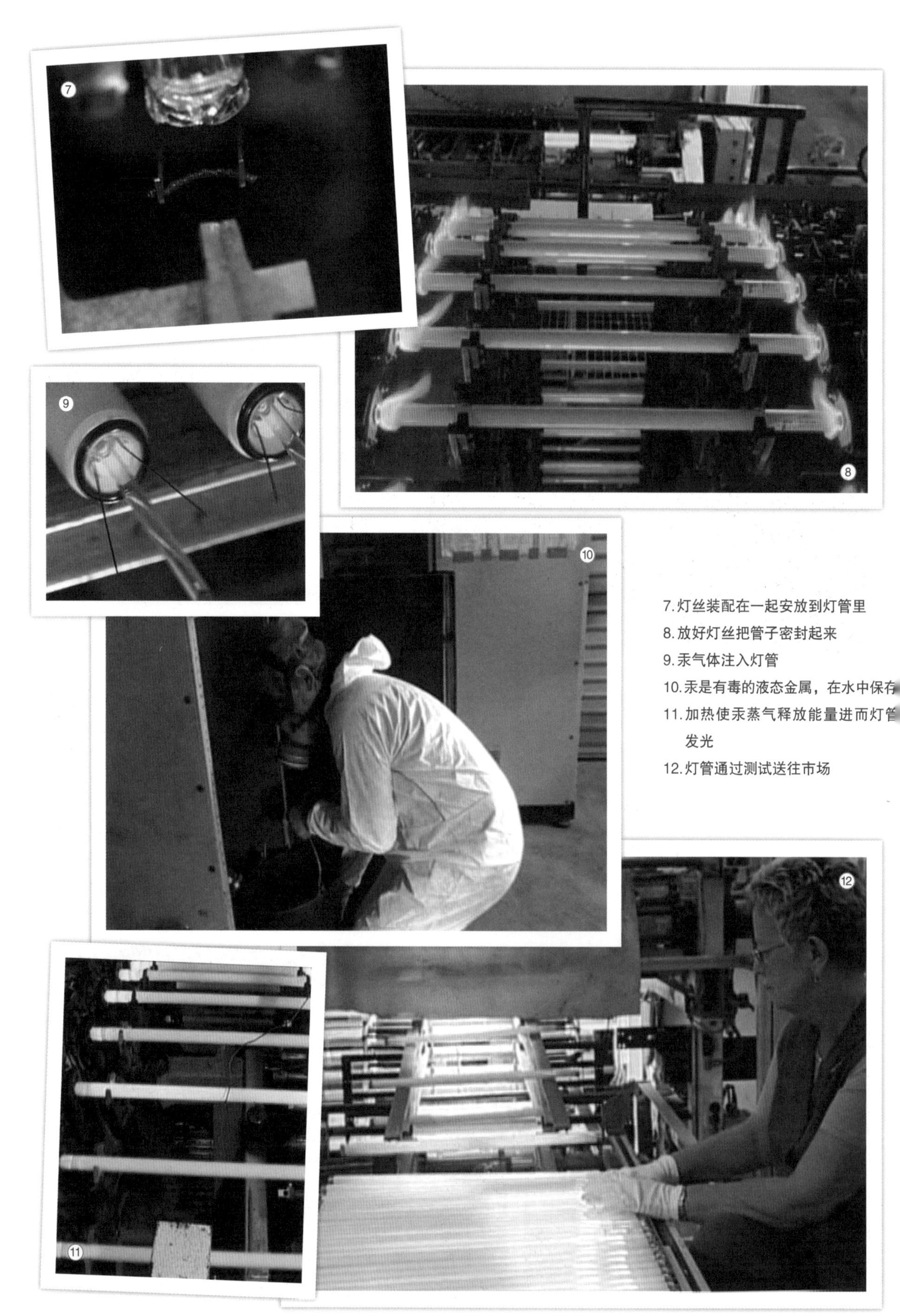

7.灯丝装配在一起安放到灯管里

8.放好灯丝把管子密封起来

9.汞气体注入灯管

10.汞是有毒的液态金属，在水中保存

11.加热使汞蒸气释放能量进而灯管发光

12.灯管通过测试送往市场

Fluorescent Lights

But first, arrive at work, unlock the door, turn on the lights.

Offices all over Britain use fluorescent lighting because it's cheaper and more efficient in the long run, but how does a tube full of gas light up your life?

Fluorescent light bulbs start off as a large lump of hot glass. In this fluid state, the glass can be stretched and blown into a very long tube . Diamond tipped cutters then trim the long tube down to 2 metre lengths. Occasionally the tubes break but they're discarded and recycled.

The next step is to trim the ends. Flames burning at 700 degrees Celsius create a neat fracture line, and a tiny cutter knocks the uneven ends away. Sharp edges are rounded off and the tubes are now ready to hold the various elements that will generate light.

Fluorescent bulbs are more efficient than ordinary light bulbs making them cheaper to use and therefore more attractive to businesses, but for that economical light to glow, the inside of the tubes must be coated with phosphor, a chemical that will cause the bulb fluoresce.

Next the internal workings of the bulb are constructed. Like an ordinary light bulb the fluorescent tube also contains a filament which carries an electrical charge. This filament must be put together so it can be fitted inside the phosphor-coated tube. Fluorescent bulbs work by exciting particles of gas with electricity. These particles then emit energy which can be turned into light. Once constructed the filaments are sealed into the tubes.

Now you may have noticed a glass tube sticking out of the end of each larger tubes; this is where the energy-emitting gas will be injected. This gas is made out of mercury. A toxic, liquid metal used in a range of everyday objects from batteries to hospital thermometers.

Mercury is recycled from these items and collected at a facility like this one. It's a dangerous process and the scientists have to be careful. While handling mercury is a hazardous job that requires a gas mask, the liquid metal can be collected in an ordinary bucket and a simple trick using water keeps it safe.

Mercury vapour is the gas that creates light in the florescent tubes...but it's dangerous if inhaled in large quantities. Storing mercury beneath a layer of water stops it vaporising. Now, if you're worried about the gas in the tubes, don't panic. The amounts used in fluorescent lighting are so small they are not considered dangerous.

The bulbs are now ready to be filled with the mercury vapour. The tubes and the mercury are loaded into the filling machine. The mercury is heated, which vapourises it and this is injected into the tubes. The heat causes the gas to release energy and this makes the bulbs glow even though they're not connected to a power supply.

So how do electricity, mercury gas and a phosphor coating create fluorescent light? We already know there are filaments inside the gas-filled tube. They send out electrons from one end to the other which strike the gas particles floating in between. This excites the gas, which in turn emits radiation. But radiation is invisible.

And that's where the phosphor coating becomes important. It wasn't to colour the tube. The phosphor fluoresces or glows when bombarded with radiation and this glow is what lights your office.

With the insides now phosphor-coated, filament-fitted and gas-filled, all the tubes need are the external caps so they can be fitted into lamp-holders. The caps are carefully fed over the filaments, sealed into place and the job is done.

Finally the bulbs are passed through a tester where any that don't work can be spotted before they're sent out to the stores.

Although streetlights were originally powered with gas, they probably wouldn't pass modern standards. But today we can still see, thanks to cleaner, greener gas powered technology. The fluorescent light.

Did you know?

If every UK home replace 3 normal light bulbs with energy saving bulbs, we'd save enough power to light our streets for a whole year.

马桶

平均每个人一生会用去 3 年时间上厕所。这里要讲到的，就是工厂如何生产你房间里最不可或缺的这件家具。

生产马桶所用的模具在这个箱体中制作出来。工人向内注入含石膏的液体混合物。石膏是一种能吸收水分的天然矿物。它可以从黏土马桶吸走水分，使它更快干燥。

模具由四个部分组成，仅用 20 分钟就可完成制作。石膏只能吸收一定分量的水分，此后就达到饱和，所以每个模具只能做出大约 100 个马桶。

陶瓷马桶主要由黏土制成，但也加入了一些其他天然矿物增加强度。这些原料都在一个大锅里加水混合。震动的机器能消除任何结块。经过快速检查，确定液体的黏稠度已经合适，就可以倒入石膏模具了。

1 小时后，压力软管中的气流很快把模具冲开。多余的边角料被切掉，并重新回收利用。马桶的排水口由手工安装，然后用湿海绵让它和其余部分融为一体。

成品是一件曲线优美的雕塑，足以让任何卫生间更加优雅。趁着材料还是软的，工人们切出排水孔。接着马桶被快速擦拭一遍，然后放在架子上过夜晾干。

第二天早上马桶已经晾干，但吸水性仍然很强，所以工人为它涂一层玻璃粉、白垩和水混合而成的釉，它经过烧制将形成防水层。

马桶外部的釉是喷上的。当工厂生产繁忙时，额外的工作量由人工承担，但大部分时间，是靠自动化机器。无论哪种方式，最终的成品都一样。

马桶仍然是易碎的，所以最后阶段它将在一个窑里焙烧。这是一个很大的窑。它有 120 米长，分为 3 个独立的区域。在第一个区域里先将它预热，因为短时间内温度升高过快会让黏土变形。第二个区域里，它在 1200 摄氏度的高温下烧制，从而变得石头般坚硬。在通过最后一个区域时马桶被冷却，此后便大功告成了。

这些马桶现在就可以装上坐垫，安装到全国各地的卫生间里。平均每人每年会上 2500 次厕所。尽管事实上不起眼的厕所如此常用，它却很少受到感激。

你知道吗？

根据世界厕所组织数据，女性上厕所用去的时间是男性的 3 倍。

1. 生产马桶石膏模具用的箱体
2. 向箱体注入含石膏的混合物
3. 石膏模具 4 个部分拼接在一起
4. 每个模具能做约 100 个马桶
5. 检查黏土和天然矿物的黏稠度
6. 黏稠度合适后注入石膏模具中

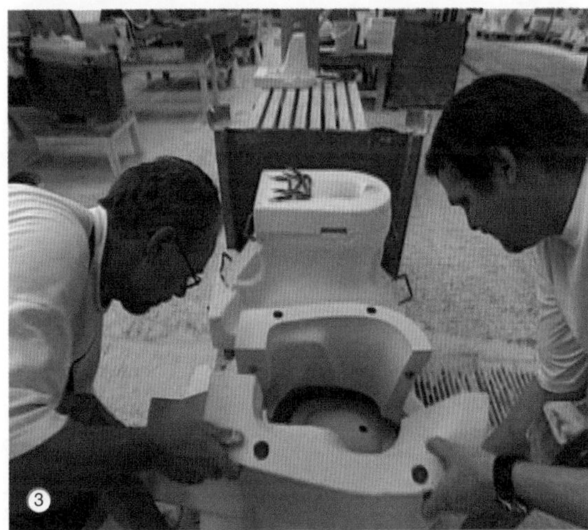

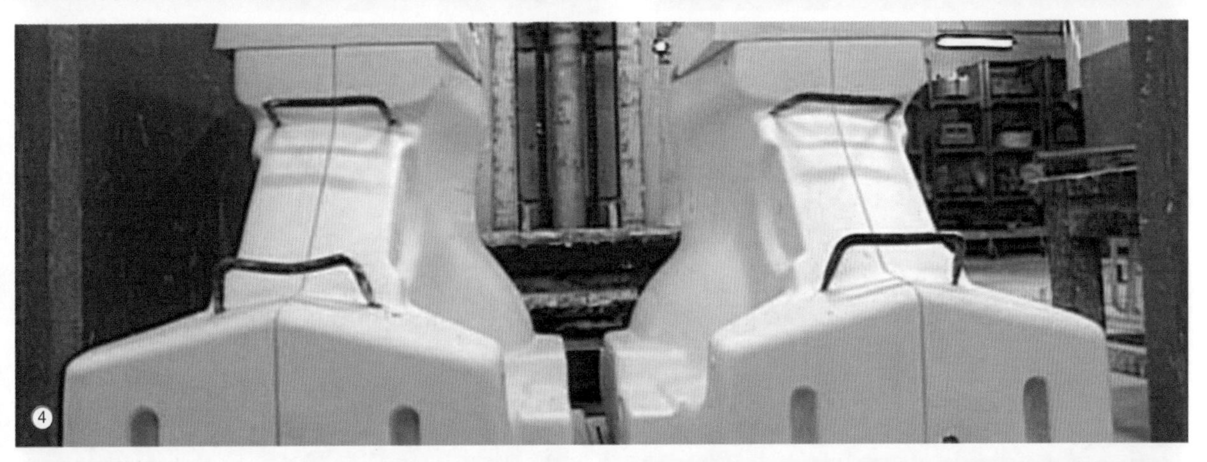

7. 气流冲开模具切掉多余的料
8. 马桶的排水口由手工粘贴
9. 人工切开黏土马桶的排水孔
10. 黏土马桶里外都要涂一层釉
11. 马桶进窑经过预热—烧制—冷却
12. 马桶像石头般坚硬，制作完成

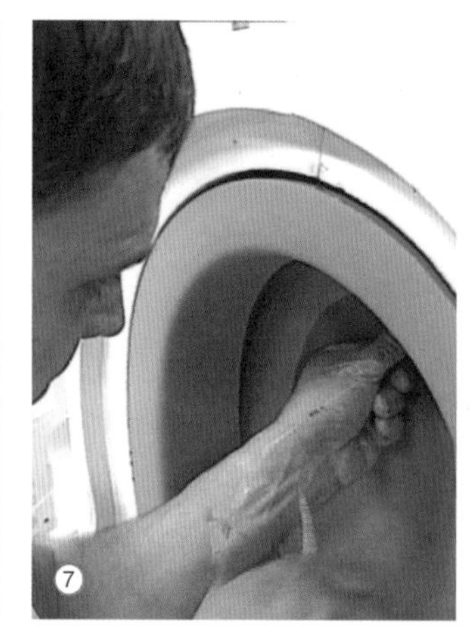
⑦

⑧

⑨

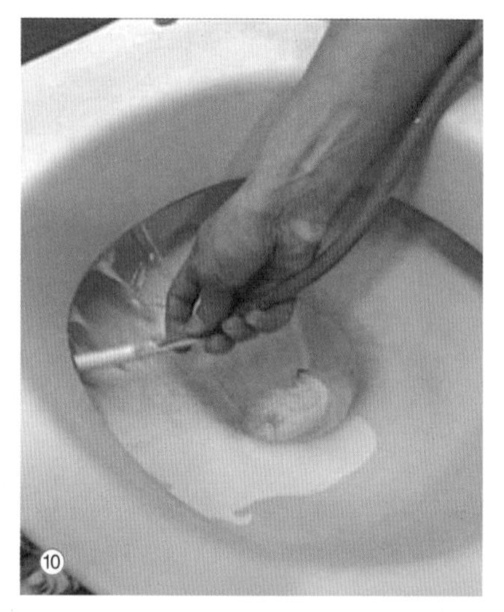
⑩

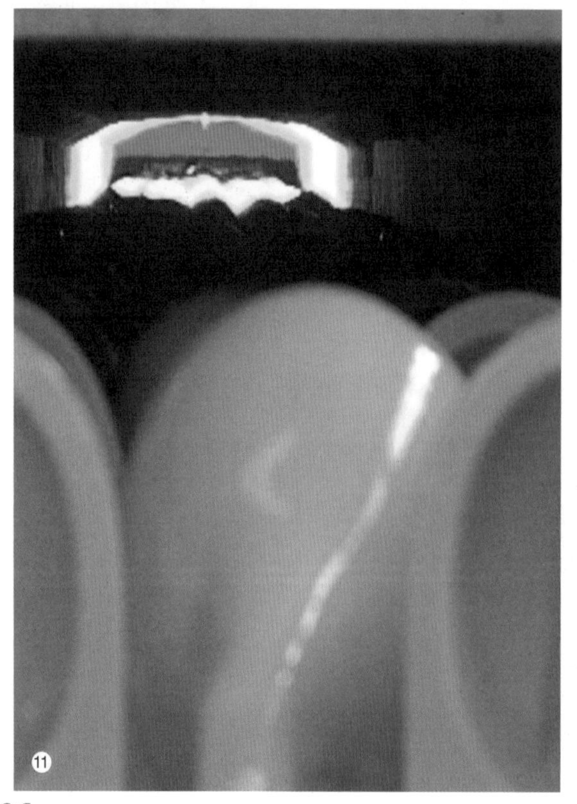
⑪

⑫

Toilet

The average person will spend three whole years of their life on the loo. This is the story of how a factory makes the most essential piece of furniture in your house.

The moulds for the toilets is made inside this casing. They pour in a liquid compound that contains gypsum. It's a natural mineral that soaks up moisture. This will draw water out of the clay bowls so they will dry quicker later on.

The mould is made in four parts which set in just twenty minutes. The gypsum can only draw out a certain amount of water before it's saturated, so these moulds can only be used to make about a hundred toilets each.

The ceramic for the toilet is mostly made out of clay but a mixture of other natural minerals have been added to provide added strength. It's all mixed together with water in a cauldron. These vibrating machines get rid of any lumps And then after a quick check to make sure the liquid is at the right consistency it's ready to be poured into the gypsum moulds.

One hour later a quick blast from a pressure hose opens them up. The excess is cut away and will go off to be recycled. The toilet shoot is added by hand then blended in using a wet sponge.

The result is a fine sculpture with sweeping curves that would grace any bathroom. While it's still soft they cut out the drainage hole. After that it's given a quick wipe and then left to dry on a rack over night.

The next morning it's dry but it's still very absorbent so they coat it with a glaze of powered glass, chalk and water. This will form a waterproof layer when it's fired.

The glazing on the outside of the toilet is sprayed on. When the factory is busy the extra work load is done by hand but most of the time it's all done by robots. Either way the end result is the same.

The toilet is still brittle. So in the final stage it's baked in a kiln and this is one seriously big kiln. It's 120 metres long with three separate zones. In the first zone it's gently warmed up, too much heat too soon will distort the clay. Then In the second, it can be blasted with 1,200 degree Celsius which turns it rock solid. As they pass through the final zone of the kiln, the toilets are cooled down and then they're finished.

扫描二维码，观看英文视频。

The bowls are ready to be fitted with seats and plumbed into bathrooms across the country. On average we pay two and a half thousand visits to one every year. Despite the fact the humble loo is used so often it's seldom appreciated.

Did you know?

According to the World Toilet Organisation, women take three times as long as men when visiting the toilet.

豪华皮椅

你已经抵达了，在重要日子里去专属俱乐部度过良宵，这里的服务和氛围让你感到奢华与成功。但还有件东西。是的，一个豪华的真皮座椅。它们可能是英国专属会员俱乐部的传统设施之一，这些高品质的真皮座椅从南非开始它的生涯。

为了制造这些舒适的座椅，你需要一个设计。这正是工人们入手的地方。所有椅子的一个共同要素，是结构。工人使用模板在原材料上标记，然后将木料切割成合适的形状。他在这里把两块木板钉在一起，这样切割出来的部件会完全相同，给椅子带来精密的匀称。

加工木材是一项非常危险的工作，他使用电锯时必须特别小心。工人非要经过训练才能操作这种设备，如果不知道怎么用，电锯很容易锯掉手指。

所有部件备齐后，他开始组装框架。需要用钉子、螺丝，乃至木胶水把零件安装到位。他会不断量度做出的活，确保尺寸正确，否则坐垫将装不进去。传统的俱乐部皮椅深度来自高扶手和靠背，这使得人们更加舒适地享受白兰地和雪茄。

不，这不是大腹便便的商人。这是印度水牛，我们感兴趣的是它的皮革。一个高质量的俱乐部皮椅给人一种老旧和使用经年的感觉。如果你看过一张染过颜色而质量低劣的皮革，你会发现它平淡乏味，难以引起兴趣。

对于一个豪华的座椅，上乘的天然皮革是首选条件。它突出仿古外观，营造会员俱乐部专属的特权氛围。理想的状况是每个座椅都使用同一头动物的皮革。这有助于自然颜色保持一致。每个座椅需要用 17 片皮革。然而，裁缝必须非常小心，因为如果她犯一个简单错误弄坏其中一块，整张皮革都变得没用了。

为了能让大腹便便的商人舒适就座，俱乐部皮椅需要有良好坚固的支撑。为椅子装潢的工人用坚韧的纤维带来回编织，这给下一阶段安装弹簧打下坚实基础。弹簧用钢丝制成，它们显得很不稳定，但能改善对顶部座垫的支撑。当工人用复杂的绳结体系把弹簧捆绑在一起，弹簧变得越发牢固。

这位工人忙于捆绑弹簧时，坐垫工人也在加紧干活。豪华座椅需要一个良好的座垫，每个装有多达 5 千克的填充物，以确保它尽可能舒适。

这里我们看到正做好的真皮表面。缝线能补充皮革的颜色，但要非常小心，不留下任何痕迹。虽然每个人都小心翼翼不损坏材料，但为了安装到位，工人实际上还要击打皮革。

皮革的内表面和泡沫填充物粘在一起。这很有好处，因为意味着皮革不会打滑。但这也首先需要工人有相当力气把皮革放置到位。整

你知道吗？

2007 年，马克·纽森以 74.85 万英镑的价格卖出了他的玻璃纤维铝合金洛克希德休闲躺椅，创下了在世设计师的纪录。

个椅子的皮革下都有一层垫子。这样可以保护它不被木架损伤，也使座椅更加舒适。定制的外皮套在垫子上面，然后被捶打到位。假设每个工人的尺寸量度都正确无误，一切安装就应该恰到好处，但在椅子完成之前总有个别小地方需要修整和调理。

手工俱乐部座椅完美制作的最后一个诀窍是无缝密封。这些金属条像一个拉链，不再需要通过缝合让皮革到位。工人把卷边掖到接缝下，使它隐藏不见。用锤子敲几下，让接缝关闭，不留下一点针线痕迹。

剩下的全部工作，就是让公司实验室测试靠垫。然后把皮椅发送到皮卡迪利最好的俱乐部休息室。

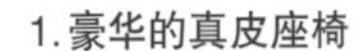

1. 豪华的真皮座椅
2. 两块木板钉在一起按模板切割
3. 座椅框架安装到位
4. 真皮座椅选用印度水牛皮革
5. 每个座椅最好用同一水牛皮革
6. 裁剪如不小心将毁掉整张皮革

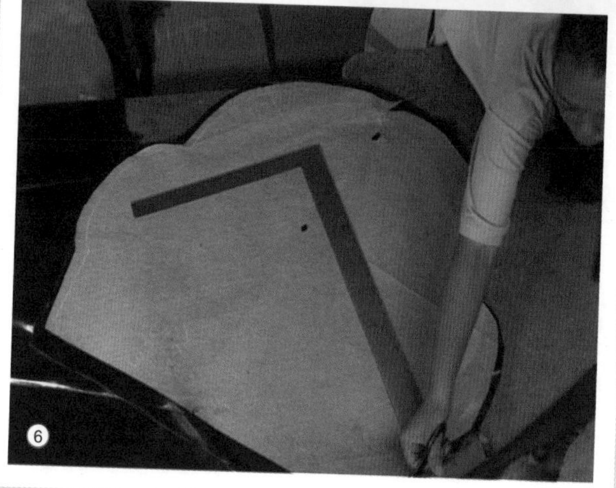

7. 绳结体系使弹簧更牢固
8. 每个座垫内装 5 千克填充物
9. 缝线能补充皮革的颜色
10. 椅子的皮革下都有一层垫子
11. 定制的外皮套在垫子上面
12. 多次跳跃测试坚牢和舒适度

Luxury Leather Chairs

You know you've arrived, when your night out involves a hot date in an exclusive club. The service and the ambience give off a feeling of luxury and success, but there's something else. Ah yes, a luxury leather chair to sit in. They may be one of the traditional parts of an exclusive members club in England, but these high quality leather chairs start life here in South Africa.

To make one of these comfortable seats, you'll need a design. The templates is where the workers begin. All chairs have one element in common. A structure. The worker will mark up his raw material and this can then be cut into shape. Here he nails two pieces of wood together so that when he cuts the pieces they are identical. This will give the chair perfect balance.

Wood working can be a very dangerous business so he has to take great care when using these saws. This worker has been trained to use this equipment because it could easily remove a finger if he didn't know what he was doing.

With all the right pieces, he now has to assemble the frame. This calls for staples, screws and even wood glue to hold everything in place. He will now measure his work to make sure he has the dimensions right, otherwise the cushions won't fit. Traditional club chairs are deep with high arms and backs. This makes them much more comfortable for enjoying your brandy and cigars in.

No, this isn't the well-lunched businessman. This is an Indian Water buffalo and its his skin we're interested in. A quality club chair feels old and well-used. If you look at an inferior quality leather that has been dyed you can see it's plain, boring and not very inspiring.

For this luxury chair the superior, raw natural leather is preferred. It emphasises the antique look adding to the exclusive ambience that comes with a member's only club chair. Ideally each chair is made using leather from the same animal. This helps keep the natural colour consistent. It takes 17 pieces to make one of these chairs. However, the seamstress has to be very careful because if she makes a simple mistake and damages one piece, the whole hide can become useless.

To be able to seat the well-lunched businessman the club chair needs good solid support. The upholsterer who works on the chair next uses tough fibre bands which he weaves back and forth. This creates a solid foundation for the next stage which is the springs. Made from wire, they appear very wobbly, but they will help improve the support for the cushions on top. The springs are made even stronger when the worker ties them together with an elaborate series of knots.

Whilst he's busy tying knots in his springs, the cushion man is also hard at work. A luxury seat needs a good cushion and each one contains up to 5 kilograms of stuffing to ensure it's as comfy as possible.

Here we see the covers being made out of the leather pieces. Thread is used that complements the leather's colour and great care is taken to not leave any marks. Now although everyone's taking care not to damage the material, the workers have to actually beat the leather to get it into place.

The underside of the leather sticks to the foam padding. This is good because it means the covers don't slip. But it does mean the workers have to be quite forceful in the first place. The whole chair receives a layer of padding underneath its leather skin. This protects it from damage on the wooden frame, but also makes the seat much more comfortable. The custom made covering is placed over the top of the padding and then beaten into place as well. Everything should fit perfectly assuming everyone's measurements have been correct, but there are always extra bits to trim and tuck away before the chair is finished.

The final trick for the perfect handmade club chair is the seamless seal. These metal strips act like a zip and mean no one has to sew the leather into place. The workers tuck the edges into the seals which are hidden underneath. A few whacks from his hammer closes the seals without a single stitch.

All that remains is for the company's test lab to try out the cushions, before it can be sent on to the lounges of the finest clubs in Piccadilly.

Did you know?

In 2007, Marc Newson sold his fibreglass and aluminium Lockheed Lounge Chaise for £ 748,500, a record for a living designer.

膨胀螺丝

扫描二维码，观看中文视频。

在家居装饰中，挂东西会成为棘手的事。比如挂一张画。一般的水泥墙对普通铁钉并不买账，但有了合适的工具，事情就简单了。

膨胀螺丝是墙上挂东西的一场革命。1919年，英国发明家约翰·罗林斯产生了一个革命性的想法，能把东西牢牢固定在墙上，今天全世界都在使用这种产品。最早的膨胀螺丝是用黄铜、纤维和线制作的。现在的膨胀螺丝由塑料制成。

大罐车把塑料运到重型的生产工厂。你可能会吃惊，制造这些小塑料珠子的材料和丝袜的材料相同。不错，这是尼龙。现在有种类繁多的膨胀螺丝，适应不同的功能、材料和负荷。

生产过程的核心，是一座这样的工厂。每个管子把尼龙原料颗粒送到机器里，将它们熔化并变成膨胀螺丝。这个系统使用注模，将熔化的尼龙挤入一定形状的空腔，冷却成所需样子后被取走。每种类型的膨胀螺丝都需要专门的模具。

当关闭机器，注入液态尼龙时，插进这些金属棒。尼龙被挤到周围空间里，金属棒造成最后用来拧螺丝的孔。

为这些机器制造模具是个耗时的过程，需要长达4个月完成。这位工程师在雕刻铜型板。它们用来制造新模具，供生产膨胀螺丝的机器使用。和钢制模板一起，铜件被小心浸到油中。启动开关，380伏电压下，电流通过铜件，在钢板上刻出正确形状的洞。

模具现在可以安装到50台制作膨胀螺丝机器中的任何一台上。液体塑料被压入，当塑料冷却后，新的膨胀螺丝就做出来。每台机器每天能生产约30万个膨胀螺丝。新的膨胀螺丝需要从制作框架上分离。膨胀螺丝落到下方的输送带，被运往包装部门。

设计新的膨胀螺丝是个复杂过程。工程师使用高科技图形软件和电脑动画，有助于分析膨胀螺丝支撑重物时经受怎样的应力。最初约翰·罗林斯开发的膨胀螺丝是圆柱体纤维，让螺钉拧紧。现代的膨胀螺丝设计具有各种形状的凸缘和形状特征，以利于螺钉抓紧周围墙壁。

实验室测试表明，一个小的变化可以造成多么大的不同。这是新的膨胀螺丝，有4个扩展部分。白色的是老式膨胀螺丝，只有2个扩展部分。看看它们是如何表现的。每个膨胀螺丝必须吊起超过280千克重的洗衣机。旧式的白色膨胀螺丝先来，混凝土块把缆绳逐渐绷紧，但洗衣机并没有升起，膨胀螺丝直接滑脱出来。

下面是新的设计，做起来似乎毫不费力。洗衣机和框架，连同所有的重量都抬离了地面，并保持悬停。诀窍很简单。右边是新的4片设计，使得4个表面与墙壁接触，是老式设计接触面积的两倍，因此抓力也是左边老式双片设计的两倍。额外抓力背后的秘密，就是这些微

你知道吗？

一个10毫米的尼龙膨胀螺丝，能安全地悬挂247千克重物。差不多四分之一吨。

小的塑料零件。

每天成千上万的膨胀螺丝从生产线倾泻而出，剩下的工作，就是把它们装箱和发送到需要的顾客那里。

这台机器小心称出一定的重量，然后把它们送到下面预先做好的盒子里。但这个最初的称重过程不完全准确，有些盒子装的不足。下一个设备会解决这个问题。再次称重后，机器能确定哪个盒子装了正确数量的膨胀螺丝。半满的盒子会被拒绝，而余下的被送到客户手中。

这个“抓人”的故事，讲述了简单而有效的膨胀螺丝。

1. 膨胀螺丝是在墙壁上挂东西的革命产品
2. 膨胀螺丝现在多由尼龙制成
3. 插入金属棒挤入熔化的尼龙
4. 金属棒造出拧螺丝的孔
5. 制造新模具花费大量时间
6. 制造出新的铜模件

7.铜模件和钢制模板一起浸到油中

8.将模具安装到机器上开始生产

9.液体尼龙冷却变成新膨胀螺丝

10.用计算机分析新设计的产品

11.新型膨胀螺丝足以辅助吊起洗衣机

12.膨胀螺丝从生产线上倾泻而出

Wall Plugs

Getting to grips with decorating your home can be a tricky business. Take mounting a picture for example. The average concrete wall doesn't take too kindly to ordinary nails, but with the right bit of equipment the job is simple. The wall plug revolutionised wall hanging. Invented in 1919, British inventor John Rawlings hit on a revolutionary idea that would help attach things to walls and today they're used all over the world. The original wall plugs were made out of brass, fibre and string, but today's wall plugs are made out of plastic.

It's transported to massive production plants in big tankers like this one. It may surprise you to learn that these tiny plastic beads are made out of the same material as stockings. That's right, it's nylon. There are a huge variety of wall plugs on offer today to suit different jobs, materials and loads.

And its factories like this one that are the heart of the production process. Each of these tubes brings the raw nylon granules to the machines so they can be melted down and turned into the plugs. The system uses injection moulding to squeeze the molten nylon into the right shaped cavity. Once it's cooled into the desired shape it can then be removed. Each type of plug needs its own unique mould.

When the machine is closed and the liquid nylon is poured in, these bars will be inserted. This forces the nylon into the spaces around the edges. The bar itself creates the hole the screw will eventually be screwed into.

Making moulds for these machines is a time consuming process and can take up to 4 months to complete. This engineer is carving the copper templates. They will be used to make a new mould for the wall plug production machines. The copper model is carefully lowered into a petroleum bath along with the steel mould plate. A switch is flicked and a massive 380 volts of electric current pass into the copper plug. This cuts the right shaped hole into the steel plate.

This mould plate can now be installed in any one of the 50 machines making wall plugs. Liquid plastic will be forced in and when it's cooled new wall plugs can be removed. Each one of these machines can produce around 300,000 every day. The new plugs need to be separated from the frame work they are made on. The plugs themselves fall onto a conveyor below and are carried off to the packing facility.

Designing new wall plugs is a complex process. The engineers will use high tech graphic software and computer animation. This helps them analyse how the plug will experience the stresses of supporting heavy objects. The original plugs developed by John Rawlings were a cylinder of fibre for the screw to grip. Modern designs have a variety of flanges and elements that help grip the surrounding wall.

Laboratory tests show the difference that one small change can make. This is the new wall plug with four expanding elements. The white one is the old style with just two. Let's see how they cope. Each plug must lift a washing machine weighing over 280 kilograms. The old style white plug is first. Tensions rise as the block takes up the slack, but nothing's rising and the plug slips straight out.

Next, comes the new design, Which seems to work effortlessly. The washing machine and frame with all its weight lifts off the ground and remains suspended. The trick is simple. On the right we have the new 4 bladed design. This puts 4 surfaces in contact with the wall. That's double the contact, and therefore gripping power, of the old 2 bladed model on the left. That extra grip is the secret behind these tiny plastic helpers.

With hundreds of thousands of plugs pouring off the production line every day, all that remains is for them to be boxed up and sent out to the waiting customers.

This machine carefully weighs up the right amount which then sends them on to the pre-formed boxes below. But this initial weighing process isn't perfectly precise, and some boxes don't get enough plugs in them. But the next device takes care of that. By weighing them again, the machine can determine which boxes have the right amount of plugs inside. Any half-filled boxes are rejected whilst the rest are sent to out the customers.

It's the gripping tale of the simple but effective wall plug.

Did you know?

A 10 millimetre nylon wall plug can safely suspend a load of up to 247 kilograms. That's nearly a quarter of a ton.

扫描二维码，观看中文视频。

斧头

无论你想到呼啸的篝火，野外的生活，或者最喜欢的恐怖片场景，斧头都是人类最古老的工具之一。这位瑞典斧匠能使用钢材、洪炉，在工业气锤的几次重击下做出新斧头。要让钢材变软适于造型，他把实心钢条放进锻铁炉。必须用钳子操作，因为炉子在骇人的1200摄氏度燃烧。一旦变得炽热，就可以把它像太妃糖一样切开。拳头大小的钢块足够制作一个斧头，但他知道不能耽搁。在第10次锤打下，斧头已开始成形。连续锻打使它完善，并压缩金属增加强度。

即使最彪悍的维京人挥舞斧头也要一个手柄，所以铁匠必须做出一个洞。他争分夺秒地工作，因为金属会很快冷却下来。他把一根钢条插进去，锤打到位。做出足够宽的孔洞容纳斧柄。另一边也同样处理，确保孔洞整个贯通。

当斧头冷却下来，红光消退了，变得越来越难锻造，但最后一锤并不需要多大的力量。只是把他个人的标记印在斧头上，工作到此完成。每年有超过50000把斧头从这里生产出来，因此公司必须维持声誉。所以除了制造坚固的斧头，还有件轻活是把斧刃磨利。用全部体重，磨刀匠把每一面斧刃靠到砂轮上，磨出锋利的边缘。顺便也打磨了斧刃，让它在阳光下发亮。一旦打磨完成，斧刃必须硬化。钢感觉很硬，但根据含碳量不同，它可以相当柔软。

斧头通过一台机器加热，然后投进冷水浴。这使钢的强度倍增，并更加经久耐用。所有这些处理激活了金属。虽然没有上下跳跃和大声呼喊，但金属分子都变得十分活跃。在这种情况下斧刃可能会变得脆弱，所以它们被放入一个所谓的退火炉。斧头被加热到195摄氏度，停留一小时。当斧刃冷却下来，被集中起来准备进行双击安全测试。它们在伐木之前还需要做些修整。

快速磨光使斧头更加有力，看上去八面威风。在这里，可以看到打磨带来的区别。第一个斧刃是原材料金属，第二个已经被打磨锋利，最后一个经过了抛光处理，并准备好伐树砍柴了。最后斧刃上油，可以保护它们免于生锈和腐蚀。

唯一还缺少抓手部分。反讽的是砍伐树木的斧头最终却安装一个木柄。木头结实并足够柔韧，能够经受住冲击。安装手柄的工匠把钢套插入斧头。木柄为了配合斧头已经修整过，但仍需要使用空气压缩机把两部分安装到一起。用快速测试确保斧柄安装正确，然后把木头楔子粘进末端。虽然看起来很简单，但它使手柄膨胀并能防止滑脱。如果斧头飞出去打到别人，强悍的樵夫就无法这样强悍了。最终把长出的木头修整掉，再用砂轮机快速打磨，一个崭新的斧头准备问世了。和恐怖片中挥舞斧头的演员一样，伐木工的斧头也备受呵护，随时能在特写镜头中露脸。

你知道吗？

用斧头处决的中世纪罪犯，斩首后从技术角度上看仍然活着。专家认为他们的心跳还能持续10秒。

1. 斧头是人类最古老的工具之一
2. 在气锤的锤打下斧头成形
3. 一根钢条插进斧头锤打到孔洞贯通
4. 在砂轮上磨出锋利的斧刃
5. 斧头加热后投进冷水增加强度
6. 斧头在退火炉中停留 1 小时

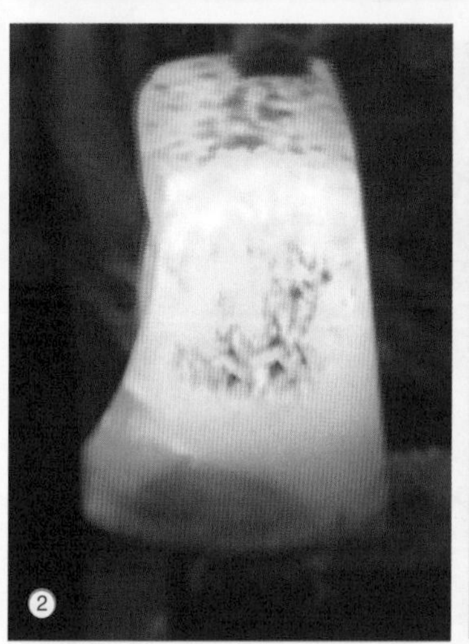

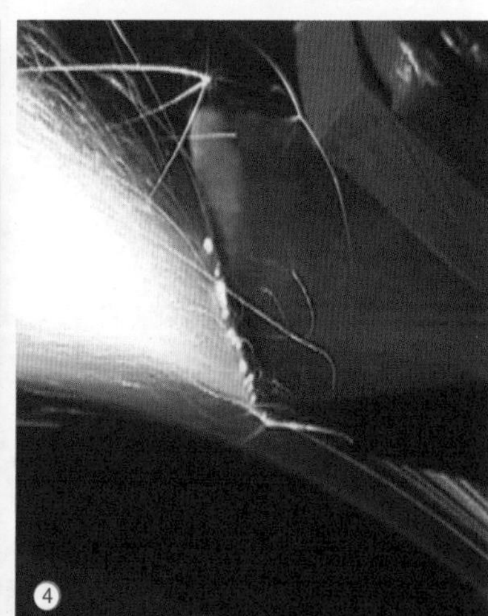

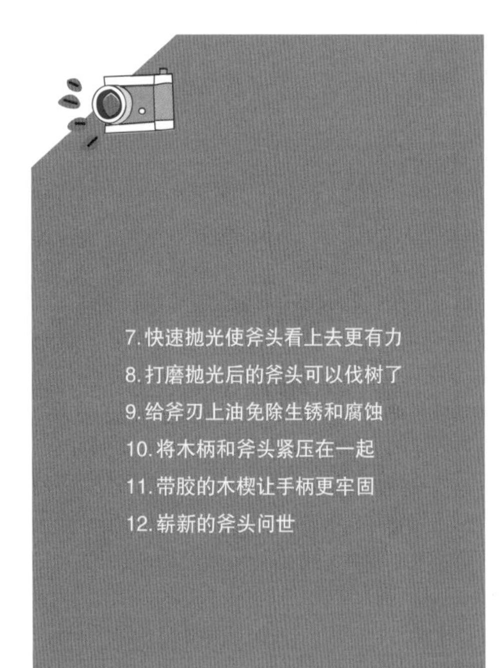

7. 快速抛光使斧头看上去更有力
8. 打磨抛光后的斧头可以伐树了
9. 给斧刃上油免除生锈和腐蚀
10. 将木柄和斧头紧压在一起
11. 带胶的木楔让手柄更牢固
12. 崭新的斧头问世

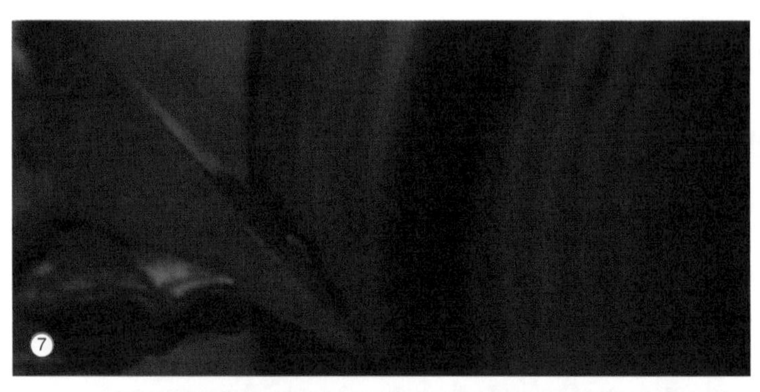

⑦

⑧

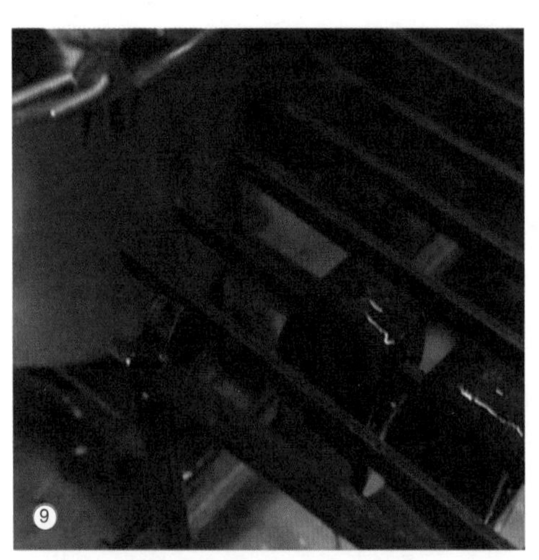

⑨

⑩

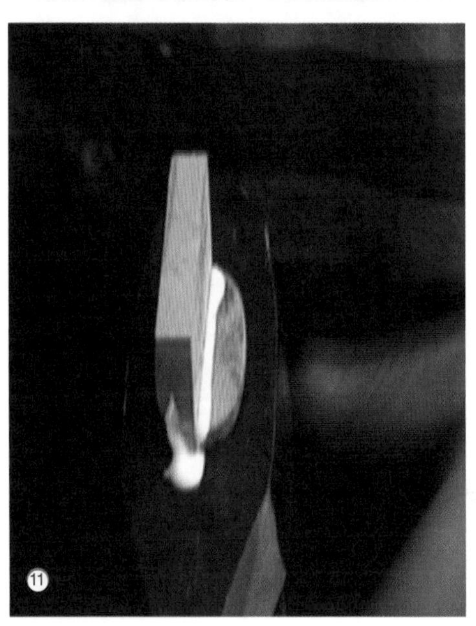

⑪

⑫

Axes

Whether you think of roaring log fires and the outdoors way of life, or a scene from your favourite horror film, the wood axe is one of the oldest tools used by man. This Swedish axe master can make a new axe head using solid steel, a blast furnace and a few whacks from his industrial hammer. To get the steel soft enough to shape it, he feeds solid bars of it into the heart of his forge. He has to use pincers because it burns at an incredible 1,200 degrees Celsius. Once it's red hot, he can slice through it like soft toffee. A lump the size of a fist will be enough to make an axe head, but he knows he can't hang around. By the 10th whack his axe is already starting to take shape. Successive blows help to finish it off, and compress the metal for strength.

Now even the toughest of Vikings needed a handle to wield an axe, so the blacksmith has got to include a hole for one. But he's working against the clock here because the metal is cooling fast. He inserts a steel peg which is hammered into place. It'll press the hole wide enough for the handle. The same is done from the other side to ensure the hole goes all the way through.

As the blade head cools, the red glow dies and it's becoming harder to shape it, but the last blow doesn't need much power. The final whack puts his personal stamp on the axe and his work is done. Over 50,000 axes are made here every year, and that means the company has a reputation to uphold. So, as well as making strong axes, there's the small matter of making them sharp too. Using his full bodyweight the sharpener leans each blade into the grinding wheel giving it a razor sharp edge. Conveniently it also shines up the blade so it'll glitter nicely in the sun. Once the sharpening is complete, the blade must be hardened. Steel may feel tough, but depending on its carbon content, it can be quite soft.

The axe head is passed through a machine which heats it up and then it's dropped into a bath of cold water. This process doubles the hardness of the steel, which will help it to last longer. All this activity has excited the metal. It's not jumping up and down and shouting a lot, but the molecules in the metal are now very active. When this happens the blade can become fragile so they're placed in what's called a relaxing furnace. Here they are heated for an hour at 195 degrees Celsius. Once the blades are cool, calm and collected and ready to face the double tap safety test. Before they are ready to chop some trunks though they need a quick make over.

A quick shine helps to enhance that powerful, menacing look. Here you can see the difference it can make. The first blade is the raw metal, the second one has been ground down and sharpened, and the final one here, has been polished and is ready to chop some firewood. Finally the blades are oiled. This protects them from rust and corrosion.

All that's missing is something to hold it with. Funnily enough, the axe that will cut down trees ends up with a wooden handle. It's strong but also flexible enough to withstand the impacts it experiences. The handle guy inserts a steel collar into the blade. The wood has been cut to fit this but he still needs to use an air compressor to get the two together. A quick test makes sure it's seated properly and then a wedge of wood will be glued into the end. This looks quite simple, but it spreads the handle out and stops it from slipping free. The tough woodsman wouldn't look so tough if his blade was to fly off and hit someone. There's just time for a final trim to remove the excess wood and a quick visit to the sander, before a brand new chopper is ready for action. So like the actor wielding it in the howor film, the woods man's axe has been very well pampered to get it readly for its close-up.

Did you know?

Medieval criminals executed by axe were technically still alive after being beheaded. Experts believe their hearts continued to beat for up to 10 seconds.

割草机

扫描二维码，观看中文视频。

修整草坪是一回事，但把所有剪下来的东西收拢和处理，则是另一回事。不过在大多数割草机里隐含着一些出色的设计，能帮助解决这个问题。

割草机的生命旅程从这家工厂开始。实际上割草并不那么困难，所以割草机用金属和塑料制成。但草地经常是湿的，水会腐蚀金属零件。所有零件一旦做好，便来到这台机器，清除污垢并涂上防腐和耐磨的油漆层。

割草机真正好用的秘密其实是刀片。它们从含碳约6%的钢带上切割下来，这种材料非常坚实，很适合这项工作。巨大的工业压床把刀片从钢铁上切下来。抛掉剩余部分，但不会浪费，所有废料都将回收。从压床上切出的刀片将进入下一个程序。虽然有了刀片的形状，但此刻实际上还很钝，因此需要磨利。

剃须刀大约有11度的切削角。割草机刀片只需要30度角就可以完成工作。它们不会像剃刀一样锋利，但面对坚韧的杂草甚至石头，这种刀片需要更结实。

弯曲每个刀片形成36度的翼角。它确实能让草飞起来，听上去不可思议，但却是真的。每台割草机有两个刀片拧到中间主轴上。刀片每分钟飞旋2800转，这就是翼尖发挥功效之处。当高速旋转时，刀片被压出的角度会在下方产生一个低压区，把切下来的碎草吸起来，直到它们穿过唯一的出口，跑进收集盘里。

于是工作完成。刀片差不多可以安装到割草机上了，但还需要先经过一个安全环节。人们知道应该在割草之前清理草坪，但有时会忘记。如果碰到什么固体的东西刀片可能会打碎，所以需要做硬化处理。首先刀片被加热到840摄氏度并保持将近10分钟。随后从高温中移出并迅速降温500摄氏度。这种急剧冷却使金属重新排列分子并变得坚硬。刀片不再那么脆弱了，即使碰到岩层石块或遗忘的园林工具也不会破碎。为完成最后部分，工人把一个中心枢轴和两个硬化好的刀片安装进来。机器按照规范自动上紧螺栓。接着在工厂生产线的另一端，工人把枢轴安装到初具模样的割草机上。

质量控制当然非常重要，成品割草机需要经过几项测试。想让旋转的刀片为你代劳，发动机必须运转，所以要进行开关测试。为了安全考虑，刀片的减速功能也非常重要。它必须在按下开关3秒钟内停止转动，一个伙计干的就是这种令人兴奋的工作，他来回按动开关并对刀片计时。

最后一关，花园里充满了棍棒和碎石，还有随处丢弃的铲子、耙子，割草机应能应付它们。每个新的型号要花费超过100小时在这个木头轨道上行驶，看它的设计能否经得住磨损。一旦通过考验，割草机就能在花园的棚屋里待命了，直到草坪需要修剪。所以请记住，棍棒和石子可能会打断你的骨头，但不会给割草机带来麻烦。

你知道吗？

英国每年都会举办12小时割草机挑战赛，1998年的获胜者在机器上行驶了313.6英里。

1. 割草机把剪下的东西顺带着收集处理了
2. 割草机用金属和塑料制成
3. 零件都要涂防腐耐磨的油漆层
4. 刀片从钢带切下来需要磨利
5. 割草机刀片需要做出 30 度的切割角
6. 弯曲每个刀片形成 36 度翼角

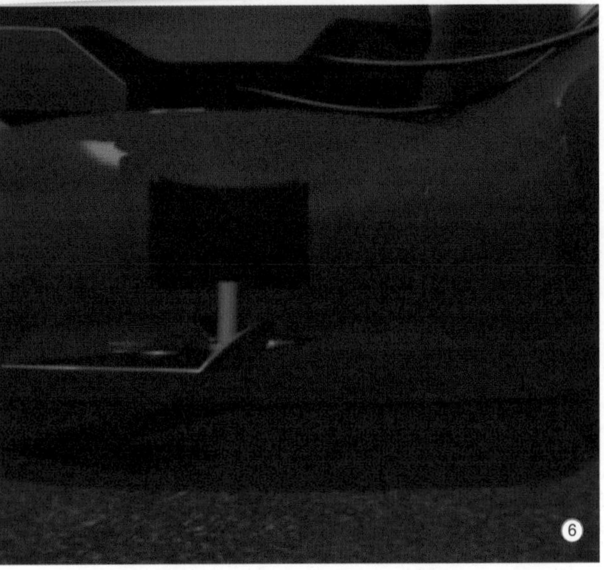

7. 割下的草被吸起进入收集盘
8. 刀片加热后迅速降温做硬化处理
9. 中心轴及刀片被安装到割草机上
10. 对发动机的运转进行测试
11. 刀片的减速功能是必检项目
12. 经得起颠簸行驶方可出厂

Lawn Mowers

Mowing the lawn is one thing, but having to rake up all the cuttings and scoop them up yourself is quite another. However hidden inside most lawn mowers there's a brilliant little bit of engineering which is designed to help.

The life of your lawn mower begins here in the factory. Cutting grass isn't actually all that tough, so they're made from metal and plastic. However grass is often wet and all that water can corrode the metal bits. Once all the bits have been made, their next stop is this machine where they are degreased and coated in a hard-wearing layer of anti-corrosive paint.

But the real secret to the labour-saving lawn mower actually begins here, with the blade. Cut from a band of steel that contains about 6% carbon, this really is tough and ideally suited to this job. A massive industrial press cuts fresh blades from the steel while discarding all the rest. It's not wasted, everything gets recycled. The freshly cut blades emerge from the press here ready for the next step. So you've got a blade shape but at the moment it's actually quite blunt so it needs to be sharpened. The razor you shave with has a cutting angle of about 11 degrees. Lawn mower blades only need an angle of 30 degrees to do their job. They won't end up as sharp as razors, but they will be far stronger when facing tough weeds or even stones.

Each of the blades is bent to include a 36 degree angle of wing. It actually makes the grass take off, which sounds amazing, but its true. Each mower has two blades screwed onto a central pivot. This spins at 2,800 revolutions per minute and this is where the wing comes in. The angle pressed into it creates a low pressure area beneath it as it spins. This sucks up the cuttings which have spun around until they pass through the only exit route available, to the collection tray.

Then the job's done. The blade has nearly ready to be attached to a mower, but there's one safety measure that it has to go through first. You know you're supposed to clear your lawn before you mow it, but sometimes you forget. If the blade was to hit something solid it could shatter so they're all sent off to be hardened. First they are heated to over 840 degrees Celsius for almost 10 minutes. They are then removed from the heat and transferred to be rapidly cooled by almost 500 degrees. This quick cooling realigns the metal's molecules hardening it. It's now far less fragile and more capable of hitting rocks, stones or forgotten garden tools without shattering. To build the final unit, a worker will place a central pivot, and two freshly hardened blades into this machine. It tightens the bolt to the right specification automatically. The pivots are then added to the lawn mower which has been taking shape on the production line in another part of the factory.

Quality control is, of course, important, so the finished mowers are run through several tests. To get that blade spinning and saving your back, the engine has to work, so that's switched on and off. And for safety's sake the blades ability to slow down is also pretty important. It must stop within 3 seconds of the operator hitting the kill switch so this guy has the exciting job of turning it on and off and timing the blade.

And finally gardens are tough places full of sticks and crazy paving and littered with un-used rakes so that mower better be able to handle it. Each new model spends over 100 hours driving on this wooden track to see whether the design will survive some hard wear and tear. Once it's been put through its paces, it's ready to be left in a garden shed until your lawn needs a quick trim. So just remember, sticks and stones may break your bones, but they won't be bothering this lawn mower.

Did you know?

In 1998 the winners of the UK's annual 12–hour lawn mower challenge travelled 313.6 miles on their machine.

园林工具

扫描二维码，观看中文视频。

英国人以喜爱乡村花园而闻名，所以 2003 年在英国销售了 400 多万把铲子和铁锹就不足为奇了。它们都是简单的工具，但你是否曾停下来想过，耙子或铲子是怎样做成的？

种地是件辛苦活，因此，园林工具要相应地结实。工厂里的生产从大张钢板开始。钢板用一种不寻常的刀具切割成型，它压根没有金属刀片，而代之以高压喷射水流切割钢板。

接下来钢坯在炉子上煅烧。燃烧的煤在钢料中留下碳的成分，使钢变得更坚硬和结实，从而生产出高质量的手工锻造铁锹。强化之后就该锻打了。80 千克的锤子击打大约 30 次，工匠把铁锹的刃做出来。大铁锤完成重头任务后再用小锤敲平凹痕。然后将铁锹送回火炉进一步硬化。这个过程反复多次，直到工匠满意为止。

下一步是修齐铁锹刃。锤击过程使金属延展变形，所以在铁锹上画出刃的轮廓并切成标准尺寸。最后做成的铁锹再次回火。工匠把锹头加热、刮净，放到压床上，用 120 吨压力使它成为特有的形状。下一步焊工将添加插孔。这块金属是软性钢材，充当减震器以缓和冲击。它有一个孔，稍后用来安装手柄。

这些生产过程会给铁锹留下污损，因此要用肥皂清洗并进行超声波沐浴。声波会抖掉尘垢。新的锹刃会进行一些修饰工作，准备摆上商店货架，同时提高它的功能。在这里锹刃被磨利，让园丁在旧玫瑰花床的硬土上劳作时能省些力。

回到火炉室，工匠正在对更多钢材做硬化处理，用于制造耙子而不是铲子。

这次工匠在硬化预制钢条。这将做成耙子的齿和叉。用不同的压床把钢条前端压成尖形，但这一过程会在尖端留下相连的多余金属。用铡刀修剪掉多余的金属材料，锋利的齿尖正适合做耙子。再进一步磨尖，让勤恳的园丁干起活来更轻松。

最后将耙齿切成适当长度。专门的压床从钢条末端把齿尖切掉。做一个新的耙头，需要精心准备 8 个齿尖。把它们装进一个特殊的虎钳里排列成行，这里不用机器人焊接。手工制作的工具需要更多时间、精力和专业知识。接下来加上手柄插孔，焊接停当。把耙尖折成弯形，这样耙拢叶子就容易多了。

无论铲子、耙子乃至干草叉，头部完成后都被送到手柄安装部门。用岑树或樱桃之类高品质木材制成的手柄，不会在压力下裂开。木柄插到孔里并接受检查。如果不直就难以使用。把这两部分夹在虎钳上，敲击到一起并用螺丝拧紧。铁锹需要更牢固的手柄应付沉重的工作。压力虎钳把手柄和锹头挤压到一起，再将两部分用铆钉连起来，确保更高强度。无论你在秋天耙落叶，还是在春季铲肥料，一把制作精良的园林工具都是必不可少的。

你知道吗？

世界上最大的花园铲高过 3 米，重达 45 千克……在英国制造。

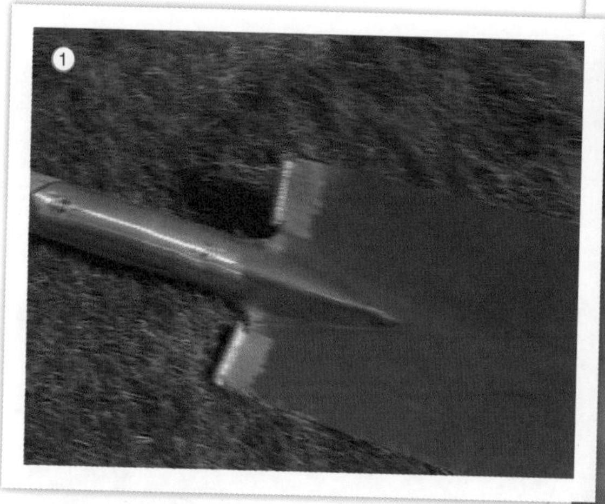

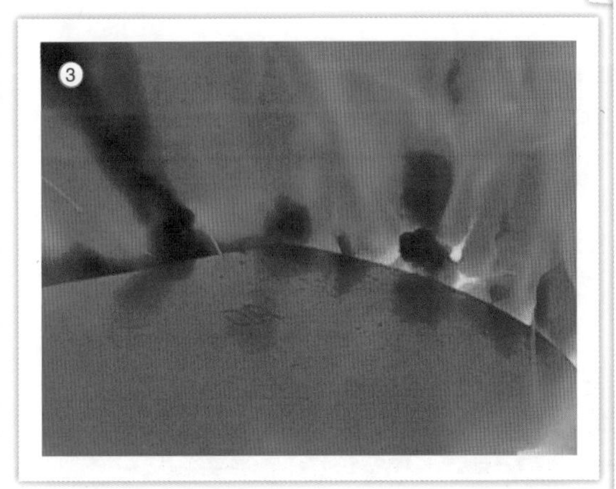

1. 铁锹和耙子是简单重要的工具
2. 大张钢板用高压喷射水流切割
3. 煅烧钢使钢变得更结实
4. 锻打出锹刃后用小锤敲平凹痕
5. 铁锹上画出轮廓切成标准尺寸
6. 铁锹在压床上成为特有的形状

7. 焊工添加插孔安装以木质手柄
8. 整饰铁锹并将其磨得更加锋利
9. 钢条压成尖形后切掉多余金属
10. 焊接 8 齿钉耙
11. 检查手柄是否装直
12. 手柄和手柄插孔用螺丝拧紧后完工

Garden Tools

The English are famous for the love of their country garden, so it's little surprise that in 2003 we bought and sold over 4 million spades and shovels. They're simple tools, but have you ever stopped to wonder how a rake or even a shovel is put together?

Working the land is a tough job, so garden tools need to be equally tough to cope. That's why here at the factory, production starts with large sheets of steel. These are cut to shape using a rather unusual knife that doesn't have a metal blade at all. Instead it cuts the steel using a high pressure jet of water.

Next the steel blanks will be fired in this furnace. The burning coal leaves carbon deposits on the steel which makes it harder, and more robust. This will produce a good quality hand-forged shovel. Once it's been toughened its time for a beating. With around 30 blows from his 80 kilogram hammer, the engineer shapes the blade of the shovel. Once the large hammer has done the heavy-duty work, a lighter hammer is used to smooth out any dents. The blade is then returned to the fire for more hardening. This process is repeated a couple of times until the engineer is satisfied.

The next step trims the blade down. The hammering process has spread the metal out of shape so the blade is marked and cut down to a standard size. And finally the finished blade is returned to the fire. The engineer will heat the shovel head, scrape it clean and place it in the press. Here it will be given its characteristic shape using 120 tons of pressure. Next the welder will add the socket. This is a piece of metal made of softer steel which acts as a shock absorber, damping the impact of heavy blows. It's also got a useful hole where the handle will be attached later.

All of the work has left the blades a little dirty, so they're rinsed in soap and sent to an ultrasound bath. Sound waves will shake loose any unwanted dirt. The new shovel blade now gets some cosmetic work to make it ready for shop shelves and to improve performance. Here it's being sharpened. This will save the gardener some effort when working an old bed of roses in tough soil.

Back in the furnace room, the engineer is hardening more steel, but this will be used to make rakes instead of shovels.

This time, the engineer is hardening pre-formed bars. These will be used to form the rake's tines or prongs. Using a different press he will crush each bar so the end resembles a point, but this process leaves extra metal attached to the tips. Using a guillotine he can now trim away the excess, giving the tines sharp points perfect for a rake. Then they are sharpened even further which will help the dedicated gardener to do his job more easily.

Finally the points are cut to length. A dedicated press will cut the spike from the end of the bar. It takes 8 of these carefully prepared tines to make a new rake head. They're mounted in a special vice that will hold them all in line, but there's no robotic welder here. Handmade tools require more time, effort and expertise. Next a socket is attached and welded into place for the handle. And now, the rake is bent into shape which will make raking up all those leaves so much easier.

When all the heads have been completed, whether for a shovel, a rake, or even a pitchfork, they're sent to the handle department. High quality woods like ash or cherry are used, to make a strong handle that won't crack under pressure. The wooden shaft is fitted to the socket and checked. If it isn't straight, it won't be up to the job. The two parts are then fitted into a vice, hammered together and screwed up tightly. For the shovel a stronger handle is needed to cope with heavier work. A pressurised vice squeezes the blade and handle together, and the two parts are riveted together for extra strength. So whether you're raking leaves in the Autumn, or shovelling fresh compost in the spring a well made garden tool, is a must.

Did you know?

The world's largest garden spade is over 3 metres high and weighs almost 45 kilograms...and it's made in Britain.

购物车

扫描二维码，观看中文视频。

不论是每周大批采买，或是到商店快速购物，消费者在超市四处寻找商品时离不开购物车。有的购物车甚至带一个额外的乘客座位。普通购物车的生产从这种大卷轴上的钢丝开始。每个卷轴上的钢丝有1.5千米长。首先将钢丝切成一定尺寸。它被送入机器拉直并切割成特定长度。尺寸取决于用来制作推车的哪一部分。

小推车制作的第一个部分是货篮。钢丝放进槽中形成格栅，互相跨越交叉。然后将格栅送入多点焊机把钢丝焊在一起。购物推车会经历很多的磨损。从装载大量的杂货到运送醉酒的学生，因此需要足够结实才能胜任工作。

焊接钢丝格栅时，更多的钢丝切了出来，但这次钢丝更长，用来做支撑结构。压床把钢丝弯成长方形。它们用来支撑篮筐，提供额外的强度。成品钢丝格栅被送入压床，把它们折叠成熟悉的小车篮筐形状。加上两或三个长方形支撑，然后篮筐各部分放在夹具里固定，准备进一步焊接。永久固定这个结构需要204个焊接点。机器人只用2分钟就能完成，比人工时间少得多。机器人完成任务后，人工再度接管。把成品篮筐上突起的钢丝修剪掉，送到生产线的下一步加工。

购物车大量的时间都在户外并受到相当多的碰撞和击打。为进行保护篮子要经过19次浸泡处理。第一次是金属脱脂消除污垢，接着还有几次，包括一次酸浴。

购物车面临的最大问题之一是腐蚀或生锈。暴露在自然环境下会严重损坏金属。为了防止生锈，钢丝结构需要镀锌。来自这些球上的锌附着于钢表面，形成结实的防锈层。一般购物车预期至少使用8年，所以需要尽可能的保护手段。

最后给篮筐上清漆。现在手推车可以装配了。框架安装到篮筐上，车轮用螺栓固定到位。轮子磨损之后将会给消费者带来各种问题。但现代购物车都通过严格设计和冲击测试程序。购物者最讨厌的事情之一是轮子不往同一方向走。

这个工厂对轮子性能进行全面测试。最苛刻的测试之一是购物车装载65千克的重物，沿着这条过道来回运行6000多次。车轮必须能经受各种劣质的路面包括砂砾和鹅卵石而不被卡住。即使折叠式婴儿座椅的性能也要严格监测。

测试之后，制造商确信购物车的轮子都会朝你需要的同一方向走。篮筐现在可以装配在底座上并夹紧到位。然后安装购物车的后部包括婴儿座椅。它可以折叠起来更便于同时存放和移动几个购物车。

接下来安装带有投币锁的把手。这个设备鼓励消费者用完后返还购物车。重新添置是很昂贵的，最后进行全面的质量测试，以确保一切安装停当。

没人确切知道英国和世界各地到底有多少购物车，但我们大多数人都离不了它。

你知道吗？

据估计，每年英国购物者推购物车行进9.28亿千米。相当于伦敦外环公路的500万倍。

1. 消费者在超市离不开购物车
2. 把钢丝拉直并切割成特定长度
3. 多点焊机把钢丝格栅焊牢
4. 将格栅折成货篮形状
5. 在货篮外加上长方形加固框
6. 完成固定结构的 204 个焊接点

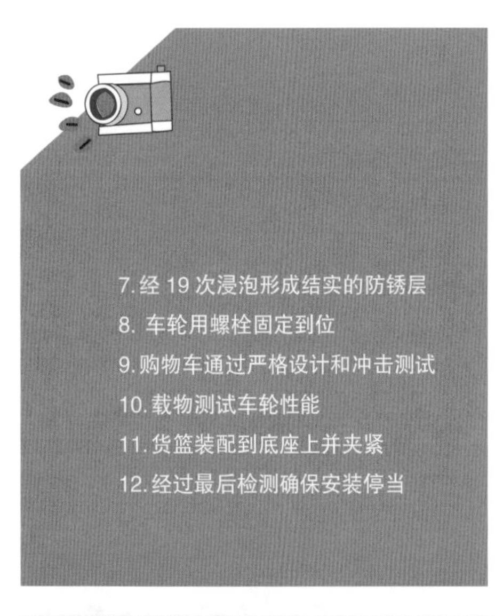

7. 经 19 次浸泡形成结实的防锈层
8. 车轮用螺栓固定到位
9. 购物车通过严格设计和冲击测试
10. 载物测试车轮性能
11. 货篮装配到底座上并夹紧
12. 经过最后检测确保安装停当

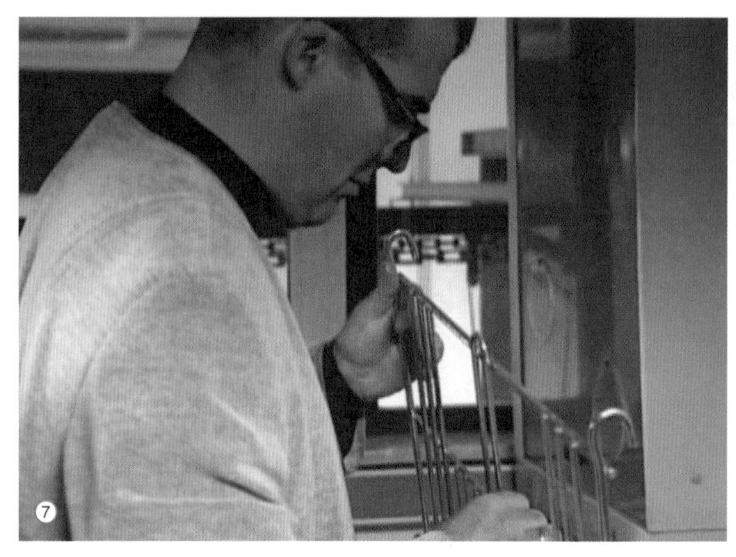

Shopping Trolleys

But first, whether it's a big weekly shop, or a quick trip for the groceries, shoppers just can't do without a trolley to get them and their supplies around the supermarket. Some trolleys even come with handy seats for an extra passenger. Life for the average shopping trolley starts out on a huge spool of steel wire like this one. There's 1.5 kilometres of wire on each spool. First the wire is cut to size. It's fed into a machine which straightens and cuts it to a specific length. The size depends on which part of the trolley it will form.

The first part to building a trolley is to make the basket for your shopping. The wire pieces are slotted into a grid so they lie across one another. The grid is then fed into a multiple welder that joins the wire pieces together. Shopping trolleys take a lot of wear and tear, from carrying huge amounts of groceries to being used as transport for drunken students. They need to be tough enough for the job.

While the wire grid is being welded, more wire is being trimmed, but this time longer pieces are cut and they will form the support structures. This press bends the wire into rectangles. They will be used to brace the basket, giving it extra strength. The finished, wire grids are fed into press which folds them into the familiar trolley basket shape. Two or three of the rectangular supports are added and the wire basket is then enclosed in a clamp which holds the parts in place ready for more welding. There are 204 separate welds to be done to fix this structure permanently. This robot takes just 2 minutes to do the job. Far less time than it would take a human. When the robot has done its work, the human workforce takes over again. The finished basket is trimmed of any protruding pieces of wire, and then sent on down the production line to be treated.

Trolleys spend a lot of time outdoors and get bumped and bashed around quite a bit too. To protect them, the baskets now get 19 baths. The first one degreases the metal, removing any unwanted dirt. This is followed by several more including one full of acid.

One of the biggest problems the trolleys face is corrosion or rust. Exposure to the elements can severely damage the metal. To protect the trolley from rust the wire structure is galvanised. The zinc from these balls attaches itself to the steel giving it a tough rust proof layer. The average shopping trolley is expected to last for at least 8 years, so they need all the protection they can get.

Finally the basket is varnished. The trolley can now be assembled. A frame is added to the basket and wheels are bolted into place. These wheels used to cause shoppers all sorts of problems as they wore out. But modern trolleys are put through rigorous design and shock testing procedures. One of the most annoying things for a shopper is a trolley with wheels that won't go in the same direction.

But at this production plant the wheel's performance is tested thoroughly. In one of the toughest tests, trolleys are loaded with 65 kilograms of weight and run up and down this track over 6000 times. The wheels must endure all of kinds of rough surfaces including grit and cobblestones without seizing up. Even the performance of the fold-up baby chair is closely monitored.

After all this testing, the manufacturers are convinced the wheels of their trolleys will all drive in the same direction when you need them to. The baskets can now be fitted to the chassis and clamped into place. The backs are then attached. This includes the baby seat and it folds up to make storing and moving several trolleys at a time, much easier.

Next a handle is added with the all-important coin slot lock. This device encourages shoppers to return trolleys which are expensive to replace. And finally a comprehensive quality test is run to make sure everything's fitted securely.

No one knows exactly how many shopping trolleys there are throughout the United Kingdom and the rest of the world, but most of us would be caught short without them.

Did you know?

Every year, British shoppers clock up an estimated 928 million trolley kilometres. That's the equivalent of 5 million times round the M25.

暖气片

扫描二维码，观看中文视频。

家用暖气片是一种普通的日常设备。看上去挺乏味？其实并非如此！大多暖气片是钢制的，所以首先要生产钢材。途径之一是通过铁矿石和高炉。铁矿石熔化的温度大约2000摄氏度，这足以将它变成液态。这位工人正在检测熔化的金属。它必须达到一定的流体均匀度，否则无法顺利加工成钢。这是炼铁的高炉。初始形态的铁含有碳，使它坚硬而易碎。钢必须碳含量较低，具有更好延展性而易于成型。把熔化的铁矿石变成钢，要泵入空气以烧掉多余的碳。

完成加工后的钢仍然炽热。在超过1200摄氏度的温度下，能用工业压力机压延成形。巨大的钢辊把工业钢锭压到3毫米厚。钢板将用来制造暖气片。

金属来回移动时，不断往上喷水。制造出来的钢板有1800米长。如果平放是无法进行运输的，所以要卷起来，运到生产工厂。暖气片的需求是巨大的，数字表明在英格兰和威尔士的家庭里，有差不多1.06亿套暖气片。每卷钢材大约能制造90套。

传统暖气片具有波纹状的表面，其中的科学道理你将会看到。暖气由3个主要部分构成。在前面有一个散热器，接下来是瓦楞形金属片，最后是第二个散热器。

生产过程从钢板送入自动压床开始，冲压出暖气片的两面。商标性的楞脊压好后，两面钢板在这里合而为一。它们先被点焊到一起，直到牢固地连接起来。

机器仍在对连续不断的钢板加工。下面要把每一片切开。它们一旦分离后就可以密封了。每两片对应的脊部被12个点焊连在一起，接着滚焊机把暖气片的顶部，底部和侧面密封。气密密封是至关重要的，因为散热器中将充满连续流动的沸水，从顶部进入，底部流出。

这时波纹就显得重要了。热水流过使金属变暖。波形起伏扩大了表面积，从而增加了向房间热辐射的面积。逻辑上说，如果一个暖气片很好，那么两个就更好。而把两个暖气片放在一起，还有另一个重要的额外好处，便是两个暖气片间的空隙。在间隙中填充瓦楞状金属片。其间的冷空气受热上升。便会形成向上的气流，冷空气将从下方进入空隙，填补暖空气的位置。

这个隔离板和暖气片用同一种钢板制作。经过锻压，钢板焊接到一个暖气片上。翻转过来并配上另一个暖气片，但并不马上焊接。首先要加进管道设备。水将通过管道流入，并在系统内循环。管道挤在两个暖气片中间，最后的焊接完成了整个暖气片的组装。

新做好的暖气片可以上涂料了。这道工序第一步是清洗。清洁的表面能更好附着涂料。然后将它们浸入装满涂料的巨大容器里。生产出的暖气片具有清新洁净的白色，但这并不是它唯一的涂层。为最初的底层再喷上第二层涂

你知道吗？

据估计，平均每间房屋每年排放的碳比平均每辆汽车排放的多。

料。这种双重保护使暖气片有了坚固的外衣，能经受住可能遇到的严重损害。

最后把顶部和底部安装上去，这样就不会把东西掉进中间，比如你的车钥匙。

暖气片现在可以安装到位，并和中央供暖系统连接起来，准备给房间很好地供暖了。

如果你曾经好奇为什么暖气片装在窗户下面，秘密就在于此。上升的温暖空气和窗口的冷空气相遇，在房间内流通。这样暖气设备产生不断的热气流，使房间和你暖和得更快。

1. 暖气片大多是钢制的
2. 转炉过程泵入空气烧掉多余的碳
3. 钢锭用工业压力机压延成形
4. 钢辊把钢锭压成 3 毫米厚的成卷钢板
5. 波纹状表面增加了暖气片散热面
6. 瓦楞形金属片和两个散热器

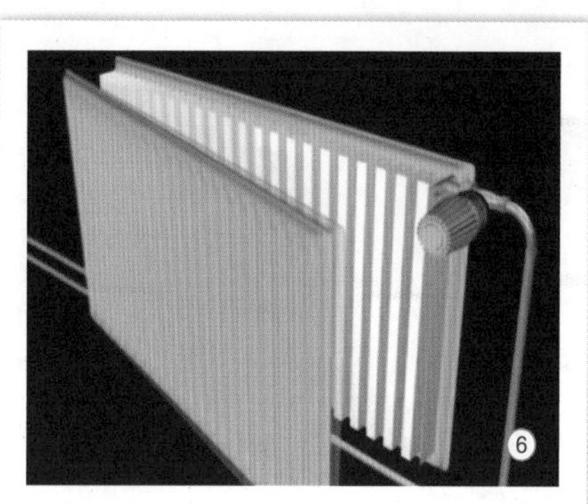

7. 楞脊压好后两面钢板合二为一
8. 切开后脊部点焊并把暖气片密封
9. 翻转过来配上另一个散热器
10. 两个散热器中间加进管道
11. 做好涂层后安装顶部和底部
12. 暖气片安装在窗户下面

Radiators

But first, the household radiator, an ordinary everyday device. A bit boring maybe? Not at all! Most radiators are made of steel so the first thing to do is to produce some. One way of doing this is with iron ore and a blast furnace. First the iron ore is melted at about 2,000 degrees Celsius which is enough to reduce it to a liquid. This worker is testing the molten metal. It must reach a specific fluid consistency otherwise it can't be processed into steel properly. And this is the blast furnace. In its original form the iron contains carbon which makes the metal hard but brittle. Steel has to have a lower carbon content, making it more malleable and easier to shape. To convert the molten iron ore into steel, air is pumped in which helps to burn off the excess carbon.

After processing the finished steel is still searingly hot. At a temperature of over 1,200 degrees Celsius the metal can be shaped using this industrial press. Giant steel rollers will flatten this ingot of industrial steel to just 3 millimetres thick. Steel sheeting which will be used to make radiators.

As the metal is passed back and forth it's sprayed with water continually. The plate of steel that emerges is over 1,800 meters long. This would be impossible to transport if it was laid out flat, so it's rolled up, ready to be sent to the production plant. Demand is huge and average figures suggest there almost 106 million radiators in homes in throughout England and Wales. Each one of these rolls contains enough steel to make about 90 units.

Traditionally radiators have a corrugated surface and there's a good science behind that as you're about to find out. They're made with three key parts. At the front there's a radiator, next there's a sheet of corrugated metal, and finally a second radiator.

The construction process starts with the sheet steel which is fed into an automated press. This will stamp out the two sides of each radiator block. When they have received their trademark ridges, the two sides will be brought together here. They will then be spot welded to hold them together until they can be properly joined.

The machine is still working on the continuous sheet of steel, so next each block must be separated. Once they've been isolated they can now be sealed. Each pair of opposing ridges is held together with 12 spot weld points before the roller welder takes over and seals the radiator's top, bottom and sides. An airtight seal is vital, because the radiator will be filled with boiling water which flows continually in at the top and out at the bottom.

And here's where the corrugation becomes so important. As the hot water passes through, it warms the metal. The corrugation increases the surface area which increases the area from which heat radiates around the room. Logic says that if one radiator is good then two are better. However, putting two radiator blocks together has another important bonus, the space between them. This gap is filled with a sheet of corrugated metal. As the cold air between the sheet warms, it rises. This creates air-flow which travels upwards as cool air is drawn into the space to replace the rising warm air.

This separating sheet is made from the same sheet steel as the radiator blocks themselves. Once pressed, the sheet is welded onto one of the blocks. It's then flipped over and the other side is introduced. But that isn't welded on just yet. First the pipe work needs to be added. This is how the water will be introduced and circulated around system. Once, the pipes are squeezed between the two blocks the final welds are made to fix the radiators into place.

The new radiators can now be painted. The first step of this process is to clean them. The paint that's used adheres better to clean surfaces. They're then immersed in a huge bath full of paint. The emerging radiators are a fresh clean white, but this isn't the only paint layer they will receive. After the initial undercoat, the second layer is sprayed on. This double protection gives the radiator a tough coating which will resist most damage that may occur.

Finally the units are fitted with tops and bottoms so you don't lose anything down the middle like your car keys.

They can now be fitted into place and plumbed into the central heating system, ready to warm a room nicely.

And if you've ever wondered why radiators are fitted under windows, here's the secret. The warm, rising air meets the cold air of the window and passes away around the room. So the radiator creates a continual flow of heat to warm the room and you much faster.

Did you know?

Estimates show the average house produces more carbon each year than the average car.

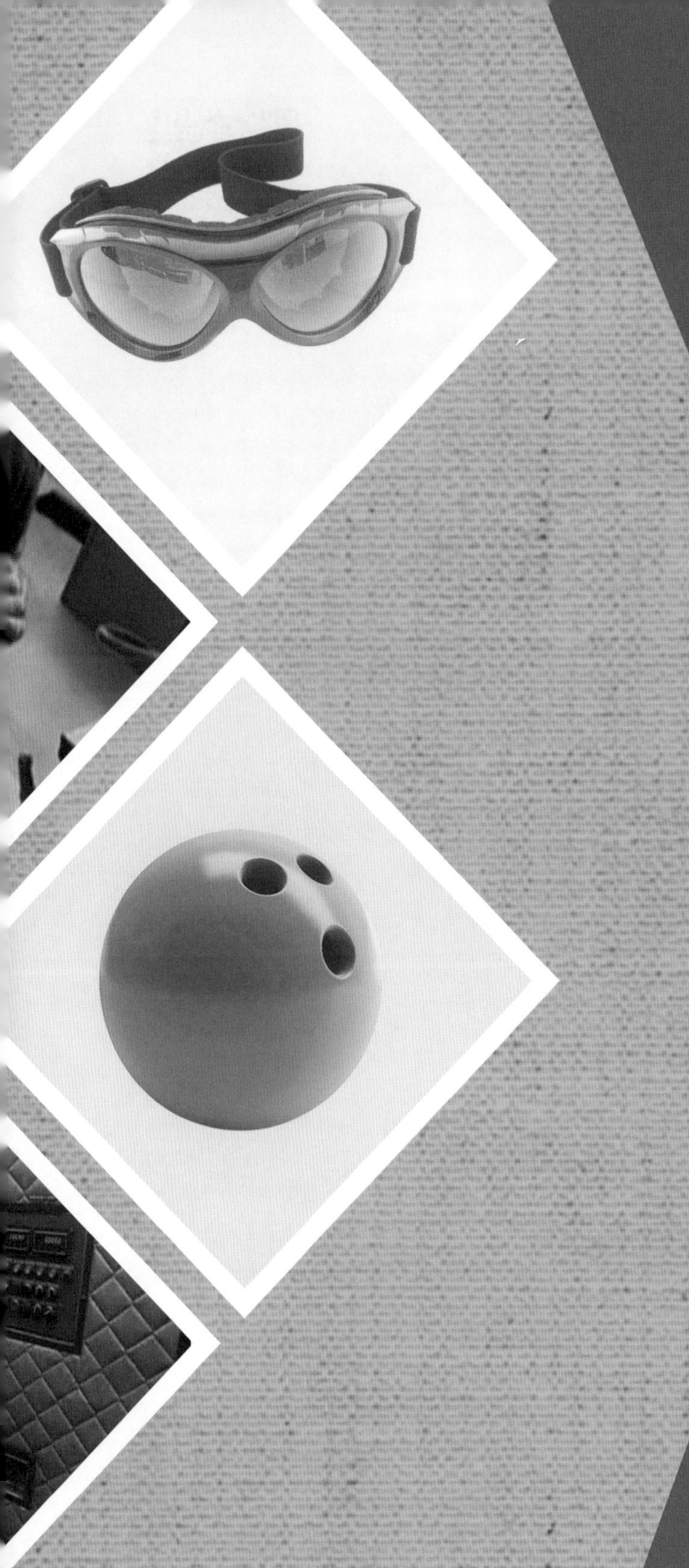

足球

足球门

保龄球

台球桌

登山鞋

山地自行车

体育篇

自行车头盔

高档自行车座

自行车打气筒

自行车轮胎

哈雷戴维森摩托车

冰球靴

有舵雪橇

滑雪镜

足球

扫描二维码，观看中文视频。

说起体育用品公司，都为设计出更圆更快的足球进行着永不停息的竞争。大卫·贝克汉姆也许能告诉你如何踢进一个漂亮的任意球，但如果你想知道一个好足球是如何做出来的，就必须得去制造商那里了。

球的表面由泡沫片材制成。它外部附着一层聚氨酯来防止磨损。

下一步是做好准备和内层连接，所以泡沫材料表面被涂上胶水。他们使用具有弹性的乳胶，能让球在以 80 英里的时速踢出去时，各层之间不会散开。

此时泡沫片材被放到架子上，直到胶水变得黏稠。下一步，内外两层被粘到一起。泡沫材质的外层被放到内层上面。内层由一种特殊纤维制成，使球保持形状。一个工人用滚轮把它们碾到一起，接着用机器压实，形成永久性黏连。

泡沫的材质让球踢出去时获得爆发力。形象不仅对足球队员重要，对足球也很重要。球上的设计图案由丝网印刷机完成。这台机器把材料冲压成制造足球所需要的形状。每个球由 32 个片板构成，包括 20 个六边形和 12 个五边形。一个工人把片板码放整齐并捆扎起来，送去用针线缝合拼接。现在看起来全部还都是平面的。但这个盒子里的材料很快就会变成 20 个足球。

它们用手工缝制。针线工在每个球上要花费 15 米的线和 3.5 小时的劳动。工人刻意留出几个片板不完成拼接，这样球就可以翻过来。接着，球被加入内胆，然后就可以完全缝上。

这时球被打满气，看去已能够送到足球场了，但却还有几个步骤需要完成。机器将球置于高压下，并加热到 70 摄氏度。这会让球的质地均匀和形体浑圆。

此后，还必须对足球做几项检查，以保证产品符合国际足联的要求。球的重量必须在 420 克到 445 克之间。球的周长必须是 69 厘米，国际足联允许的误差仅 5 毫米。

这家工厂每天只能生产 750 个足球。在每个足球上要花费数小时，确保达到严格的标准。如果英国队能花相同的时间练习点球，他们可能会更多地射进球门的上角。

你知道吗？

世界上尚存的最古老的足球已有 450 多年的历史。它是由皮革包裹的猪膀胱制成的。

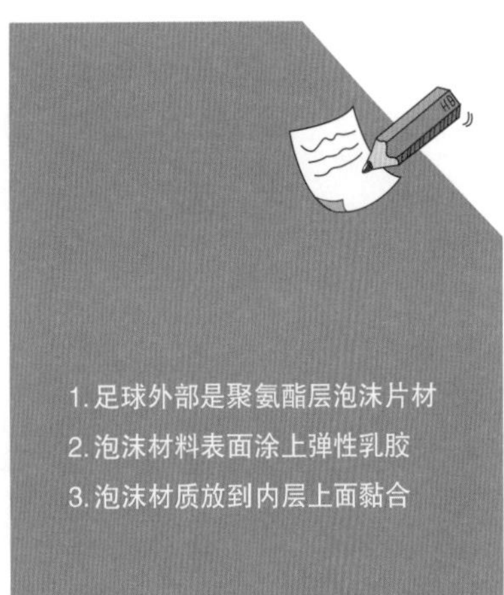

1. 足球外部是聚氨酯层泡沫片材
2. 泡沫材料表面涂上弹性乳胶
3. 泡沫材质放到内层上面黏合

4. 足球上的图案由丝网印刷机完成
5. 足球由 20 个六边形和 12 个五边形构成
6. 足球用手工缝制

7. 刻意留出几个片板不拼接
8. 足球从开口处翻过来并加入内胆
9. 缝合好所有片板并给足球充气

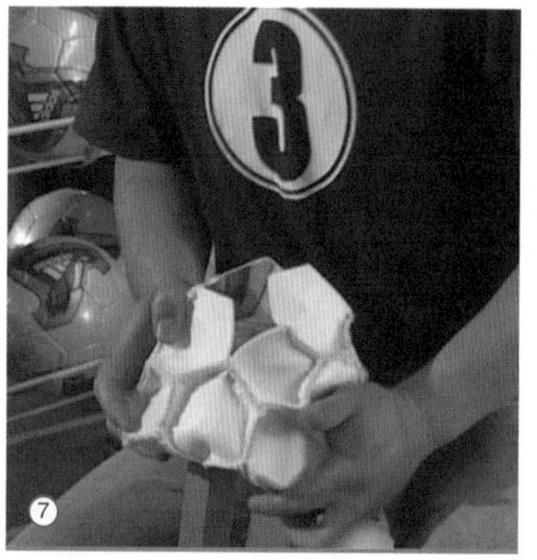

⑦

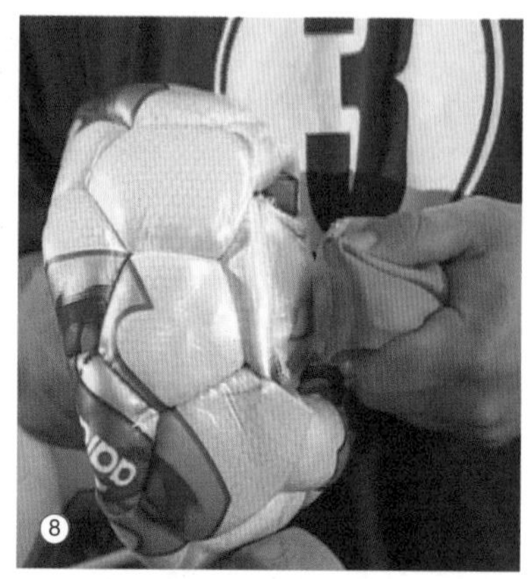

⑧

⑨

10. 将球置于高压下加热到 70 摄氏度
11. 足球的重量必须符合要求
12. 按国际标准足球周长 69 厘米

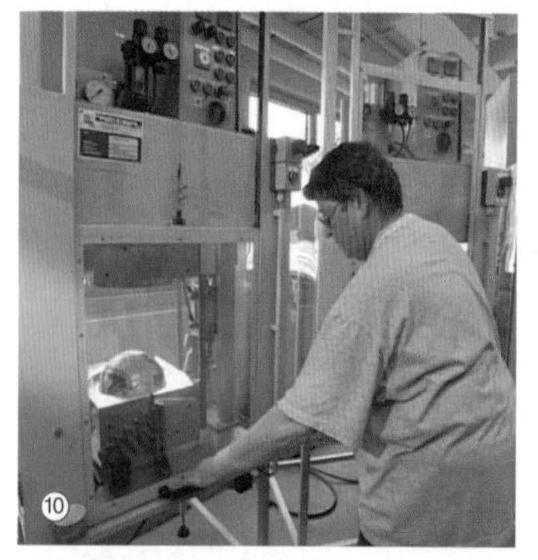

⑩

⑪

⑫

Football

But first, sports companies are constantly competing with each other to design footballs which are rounder and faster than ever before. And although David Beckham might be able to tell you how to bang in a great free-kick, if you want to know how a great ball is made then you've got to go to a manufacturer.

The outside of the ball is made from a sheet of foam. It's been coated in polyurethane which will help to stop it from getting scuffed.

It'll be attached to an inner layer next so they prepare the foam by coating it in glue. They use a latex glue which is elastic, this'll help the layers to stay together despite the strains of being kicked at 80 mph.

The sheets are then left on a rack until the glue becomes tacky. Next they're going to join the two layers together. The foam is placed on top of the inner layer which is made from a special fibre that will help the ball keep its shape. A worker uses a roller to press them together, then the layers pass through a machine which applies pressure to the sheets, and they form a permanent bond.

The foam gives the ball the explosive energy to make it fly off the boot. Image isn't just as important for footballers, it's important for footballs too. The ball's design is put on using screen printer. This machine punches out the shaped panels that will make the football. Each ball is made from 32 panels, 20 hexagons and 12 pentagons. A worker bundles them together into stacks to be sent off for stitching. It's all looking a bit flat at the moment. But the contents of this box will soon become 20 footballs.

They are sewn together by hand. The stitcher will use 15 metres of thread per ball and it will take him 3.5 hours to finish each one. He leaves a few panels unstitched so the ball can be turned the right way out. Then the inner tube gets put in. and the ball can be completely sewn up. It's inflated and looks ready for the pitch, but there are a few more steps to go. This machine puts the ball under high pressure and heats it up to 70 degrees Celsius. It makes the ball even and perfectly round.

扫描二维码，观看英文视频。

Then there are a couple of checks that have to be done to make sure that it meets FIFA regulations. It needs to weigh between four hundred and twenty and four hundred and forty five grams. And the circumference has to be sixty nine centimeters. FIFA only allow five millimeters leeway.

The factory only produces seven hundred and fifty balls a day. Hours are spent making each one to exacting standards. If only the England team spent as long practicing penalties then maybe they'll hit the top corner a bit more often.

Did you know?

The oldest surviving football in the world is over 450 years old. It was made from a pig bladder surrounded by leather.

足球门

扫描二维码，观看中文视频。

点球——一切都取决于压力下的表现。就像成功的一脚，生产一个足球门也需要安排时序和流程，这里是由塑料生产球门网的流程。

在这些巨大的袋子里，装的是颗粒状塑料聚丙烯，用来制造球门网的原料。真空泵把颗粒吸入机器，在230摄氏度将它们熔化成液体。在熔化之前，将抗紫外线材料加入混合物。当球网在室外使用时，它能提供防晒保护。再添加进颜色。

一旦塑料熔化，被压出这个筛子，形成长长的绿色的意大利面条般的线材。这种塑料线材以每分钟22.5千米左右的速度绕到小卷轴上。这种线的强度，远远不能阻挡大力射来的足球。要做出像样的足球网，单股线必须编织到一起。

将塑料线缠到大线轴上，然后装入制网机。在编织开始前需要装载18个大线轴。

整个英国约有45000个足球场经过足球总会认证，它们都需要足球网。但还有很多未经认证的球场，所以需求量总是很大的。

编织的第一阶段是将单股线组合起来，形成一个网状结构。这种紧密结合的材料看起来像张网了，但它仍然没有结实到兜住贝克汉姆射门的强度。正是编织过程，让材料具有真正的高强度。

这台机器让材料围绕自己来回编织，每分钟做出500个环。它不给材料打结，因为绳结会在网状结构中形成薄弱环节。

这个生产过程做出锁子甲一样的结构层，可以承受有力的凌空抽射。这18个卷筒的塑料线，足够制作5000个球门的网绳。

生产出来的巨大球网被切成正确尺寸，然后取样测试。每个网环必须能抵挡150千克的力，才可以通过测试。每个球网设定的球门区为7.5米宽，2.5米高，2米深。要在这样的空间范围架设球网，工厂现在要生产门柱。

金属基本成分是铝，其中掺有镁和硅。将巨大的金属锭送进加热器加热到520摄氏度。使金属具有延展性，为下一工序做准备。

高温柔软的金属通过定形的筛子，分成4个分离的长条。这些长条接着被送入另一个定形的管子。它们再次被压在一起，形成的薄圆管将成为球门柱结实而轻巧的外壳。

长而薄的管子生产出来后，连接到一台机器并被拉直。大功率的电锯把圆管末端切掉，圆形的管子被压扁，变成熟悉的椭圆形门柱，任何足球运动员都认得出来。

一个连接件被焊接在底部，这样门柱才能装到球场的洞里。用于连接网的钩子也焊接到位。这个德国工厂仍然使用金属连接件，但英国的安全规范要求生产厂家必须用塑料或是橡胶，以保护球员，使其免受潜在的伤害。

球门准备好可以使用了。只需要几个步骤，就可以竖起这个简单而有效的结构。球场管理

你知道吗？

最快的进球纪录属于考斯足球俱乐部的麦克·布洛斯。他从开球到得分只用了2秒钟。

员工作完毕后，球门就能为千万球迷提供娱乐了。如果英格兰与德国对阵，也许会在另一场可怕的点球罚射中伤透他们的心。

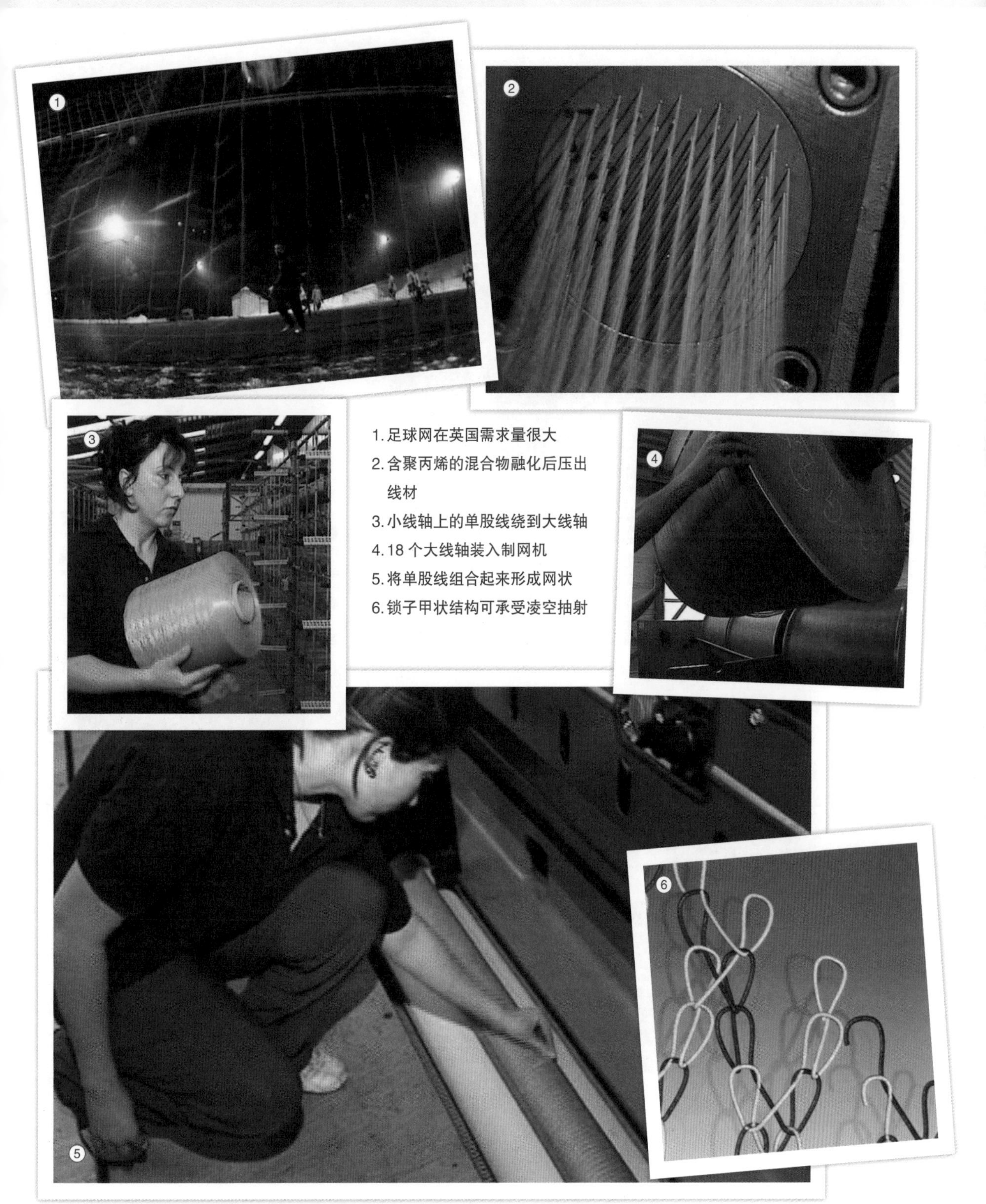

1. 足球网在英国需求量很大
2. 含聚丙烯的混合物融化后压出线材
3. 小线轴上的单股线绕到大线轴
4. 18 个大线轴装入制网机
5. 将单股线组合起来形成网状
6. 锁子甲状结构可承受凌空抽射

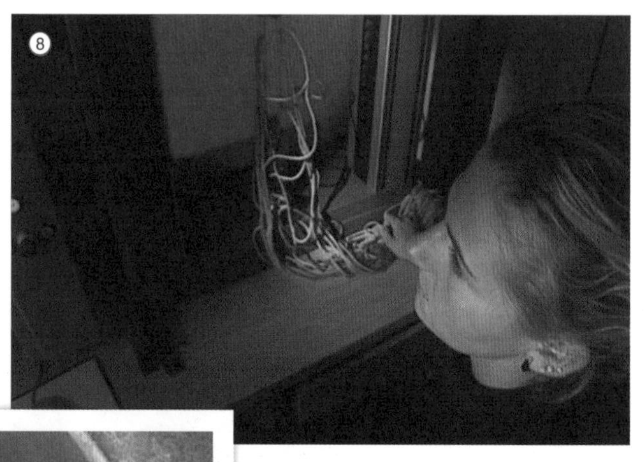

7. 把大网分割成合适大小
8. 每个网环须抵挡 150 千克的力
9. 高温柔软的金属进入定型管
10. 长而薄的管子被拉直
11. 焊接门柱装入球场的连接件
12. 挂好球网门框插入场地

Football Goals

The penalty kick, it's all about performance under pressure, and just like a successful strike, making a set of football goals requires timing and flow … in this case, it's a flow of plastic to make the netting.

Inside these huge sacks is granulated polypropylene plastic - the raw material used to make nets. A vacuum pump sucks the granules into machinery which will melt them into a liquid at around 230 degrees Celsius. Before it's melted down UV resistant material is added to the mix. This will act as a sun-block to protect the netting when it's used outdoors . Colour is also added.

Once the plastic is melted it's forced out of this sieve to form long, green spaghetti-like strings. This plastic string is rolled onto small spools at a rate of around 22 and a half kilometres every minute. In this form it's far too weak to stop a well struck football. To make a proper net, the individual strands must be woven together.

The plastic is fed onto larger spools which are then fitted into a net making machine. 18 of the large spools need to be loaded before weaving begins.

Throughout the UK there are around 45,000 FA approved pitches which all need nets. But there are plenty of unapproved pitches as well , so demand is always high.

The first stage of the weaving process is to combine the individual strands to form a net-like structure. This tightly bound material looks like netting, but it still wouldn't be tough enough to stop a Beckham special. It's the weaving process that gives the material its real strength. This machine creates 500 loops a minute by twisting the material back and forth around itself. It doesn't use knots, as they would cause weak points in the net structure.

The process creates chain mail like layers which could withstand the power of a well-struck volley. Altogether the 18 spools of plastic strand can turn out enough woven fabric for around 5,000 goals.

The enormous net that has emerged is cut down to the right size and then a sample is taken for testing. Each loop must resist 150 kilogrammes of force to pass. Each net will define the goal area which is 7 and a half metres wide, 2 and a half metres high, and about 2 meters deep. And to frame the net around that space, the factory must now produce the goal posts. The base metal is aluminium, combined with magnesium and silicon. Huge bars of it are fed into a heater which warms them up to around 520 degrees Celsius. This makes the metal malleable ready for the next stage.

The warm pliable metal is passed through a shaped sieve which divides it into 4 separate strips. These strips are then fed through another shaped tube. They're forced together again to form a thin tube that will become the tough but lightweight shell of the goal posts.

As it emerges the long thin tube is connected to an engine which pulls it straight. A high powered saw slices off the end and the round tube is then flattened out to make the familiar oval shaped post any footballer would recognise.

A connection is welded to the bottom so the post can be fitted into holes on the pitch. The hooks for attaching the netting are also welded into place. This German factory still uses metal connections, but safety requirements in the UK mean producers would have to use plastic or rubber to protect players from potential injury.

And the goal is ready to be put into action. It only takes a few steps to erect this simple but effective structure. And when the groundsmen are done, the goal is ready to help entertain thousands of adoring fans. Or if England is playing against Germany, perhaps break their hearts in another dreaded penalty shoot out.

Did you know?

The record for the fastest goal ever belongs to Marc Burrows of the Cowes Sports Reserves. He took just 2 seconds to score from kick–off.

保龄球

扫描二维码，观看中文视频。

来吧！保龄球是一种几千年来就存在于不同文化中的运动。人类学家甚至在5000多年前的埃及男孩墓葬中，发现了球和球瓶的早期版本。当代最流行的保龄球演变之一，是10瓶制保龄球，起源于美国。

生产过程从树脂混合物制作的球芯开始。混合物被倒进模具静置。取出时，每个球芯必须恰好4.85千克或10.7磅重。不多也不少。

虽然球芯主要是树脂，内部还有两个重要的附加部分。一个金属球，如这里所示，赋予保龄球质量。另一块金属偏离中心，如金色部分，帮助专业保龄球选手打出弧线球。

最理想的保龄球投掷不是笔直的。曲线投球一次打倒所有10个球瓶的机会更高。这是被渴望的全中，保龄球的最佳投球。球芯用特殊的石头洗磨。能对表面产生挤压，让树脂外层更紧地黏合。

下面要添加的是外层。把球芯小心放到新模具中。每个模具把球芯放置在正中心，让表面材料能均匀分布在球芯周围。随后球芯和模具将被送到生产线上造出外层。

外层是一种聚氨酯混合物，传统上有丰富的色彩。但有种古怪的趋势，一些保龄球手还希望自己的球发出特殊气味。现有品种包括肉桂苹果、柠檬，甚至杏仁味。色彩缤纷的聚氨酯泵入模具，但不同的保龄球有不同的重量，这是如何实现的？

为了使球更重，工人只需增加聚氨酯混合物的浓度。球芯的重量都完全一样。通过改变外层材料的浓度，可以产生出不同尺寸和重量的保龄球。多余的外层材料用空气喷枪吹掉，整个装置送到下一步，模具可以用机械方法拿掉。

保龄球排起队，等待修整到合适的大小。这家工厂每年用这种方式生产五十万个保龄球。

注塑和滚动使新球的边缘相当粗糙。下个机器把球研磨到位，而不会耗去过多材料。外形恰到好处时，就该修饰收尾了。将商标刻上去。最后加一抹额外的色彩，在磨掉的地方填上颜料。

据估计，全世界超过4000万人定时打保龄球，高需求加上公司希望维护形象，因此外观和质量一样重要。

下一阶段是抛光。球放置在碗中，抛光辊刷开始工作。

现在可以打保龄球了。不，还差一点。没有关键的指孔，是不可能抓住这些球的。

使用特定的测量装置，精确标出需要钻指孔的地方。钻孔通过自动化过程完成。吸走碎片，测量孔的深度并确保恰到好处。

最终保龄球整装待发。用塑料封装包裹，为用户保护好光滑的表面，然后放进盒子。

对一个深谙门道的专业选手，打好所有重要的球，成为一个“全中王”，优质的保龄球是必不可少的。

你知道吗？

“挪威明珠”是世界上第一个为旅客提供全尺寸保龄球馆的邮轮。在波涛汹涌的大海中，每个人投球的轨迹都是弯曲的！

1. 保龄球生产从制作树脂球芯开始
2. 球芯从模具中取出时恰好 4.85 千克
3. 球芯内有两个重要附加部分
4. 球芯用特殊石头磨洗
5. 模具中保证球芯放置在正中心
6. 聚氨酯混合物注入模具中

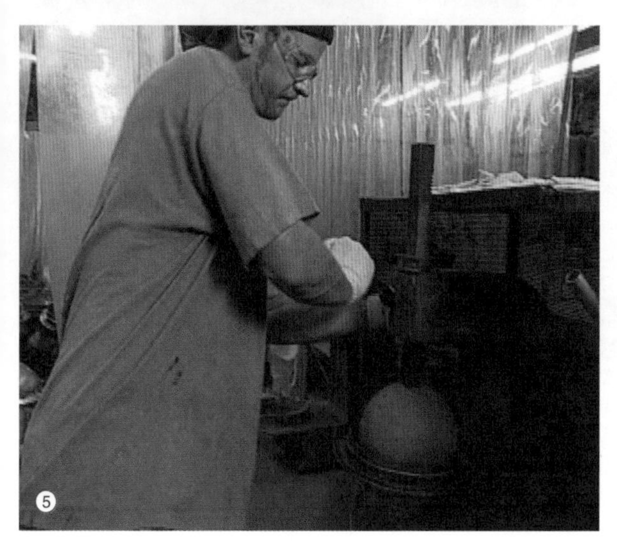

7. 聚氨酯浓度决定球的重量和尺寸
8. 机器把保龄球研磨到位
9. 刻好商标填上颜料
10. 保龄球抛光
11. 精确标记出指孔位置
12. 为选手提供优质的保龄球

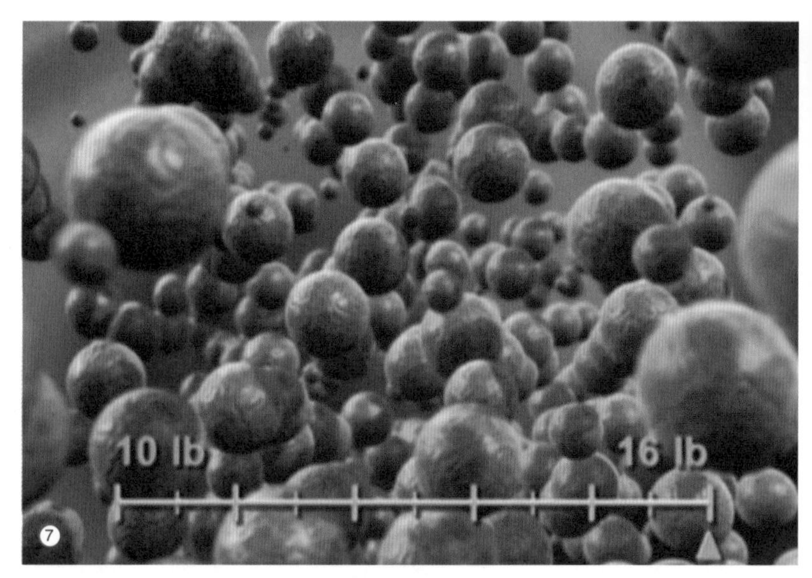

Bowling Balls

Let's roll. Bowling is a sport that has been played in different cultures for thousands of years. Anthropologists have even found an early version of a ball and pins in the grave of an Egyptian boy buried more than 5,000 years ago. One of the most popular modern variations is 10-pin bowling and its traditonal home is here in the USA.

The process begins with the production of the central core made with this resin mixture. The mix is poured into the moulds and left to set. When they're removed, each core one must weigh exactly 4.85 kilos or 10.7 US lbs. No more. No less.

Now although the core is mostly resin, there are two additional important pieces inside. A sphere of metal, seen here, gives the bowling ball mass. Another piece placed off-centre, seen here in gold, helps professional bowlers to curve their shots.

The ideal shot in bowling is not a dead straight one. A curved shot has a higher chance of knocking all 10 pins over in one go. This is the coveted strike, the best shot in bowling. The cores are now washed with special stones. This helps to key or dent the surface so that it will bond tightly with the resin exterior.

This exterior is the next part to be added. Each core is carefully placed in a new mould. Each mould holds the core dead centre so that there is an even distribution of the surface material all the way around it. The core and mould will then be passed along the line to be coated.

The coating is a polyurethane mix that traditionally is quite colourful. But in a slightly bizarre trend some bowlers also want their balls to have an individual smell. Variations available include cinnamon apple, lemon and even amaretto. The colourful polyurethane is pumped into the moulds. But different bowling balls have different weights, so how is this achieved?

To make a new ball heavier, workers simply add a higher concentration of the polyurethane mix. The cores are all exactly the same weight. By varying concentrations of the covering mixture, a range of balls can be produced to different sizes and weights. Any excess on the coating is blasted away with air guns, and the whole thing is passed along so the moulds can be mechanically removed.

The ball will now join the queue to be cut down to size. Each year this factory produces half a million bowling balls this way.

Now, all the moulding and rolling has left our new ball rather rough around the edges. This next machine grinds the ball into shape, without removing too much material. With the shape just right, it's time for the cosmetic, finishing touches. Here the logo is being engraved. And to add a final splash of extra colour, the ground out areas are filled with coloured paste.

Estimates suggest that over 40 million people bowl on a regular basis worldwide and high demand plus a company image to maintain means the look is just as important as the quality.

The next phase is the polishing. Each ball is placed in one of these cups and the polishing rollers get to work.

And finally they're ready to go bowling. Well, not quite. Without the vital finger holes, gripping these balls would be impossible.

Using a specific measuring device exact spots are marked out where finger holes need to be drilled. The drilling is done in an automated process. While the debris is vacuumed up, the depth of the holes is gauged to make sure it's just right.

And finally the ball is ready to go. It's wrapped in plastic packaging to protect the shiny surface for the customer and then it's boxed up.

For the professional who knows his game, a quality bowling ball is a must, to make those all-important strikes needed to become a ten-pin king.

Did you know?

The Norwegian Pearl is the world's first cruise liner to offer passengers a full-size bowling alley. In rough seas, everyone's shots are curved!

台球桌

扫描二维码，观看中文视频。

台球在全国的酒吧和俱乐部里随处可见。虽然具体规则有诸多不同，大部分人都懂得这个游戏的一般要点。但很少有人知道台球最初发明时是门球的室内版。

大部分的台球桌是由工厂批量生产出来的，但是这家公司仍然用手工制作台球桌。首先台球桌顶部和侧面的胶合板被切成合适的尺寸。然后涂上一层胶水再贴上仿实木面板。将它们放置到位后，各层紧紧压在一起，胶水均匀铺开，面板被永久黏合。

现在该做台球桌的外框了。就是你等待比赛时投放零钱的地方。6 个落球孔用钻机切出。要打好一场比赛，台球在桌子上的反弹性应保持不变，因此下一步制作程序非常关键。工匠把橡胶边粘到木托上。当包上呢绒后，就成了台案内沿的软垫。可以保证球每次被均匀地弹回。

用来铺台面的粗呢布来自比利时。台呢的成分甚至颜色，都有严格的国际标准规范。这就意味着，不论在巴恩斯利还是在曼谷，台球在台球桌上滚动的速度应该大致相同。他们用意大利北部采石场的石板做球桌的台面。台呢在台面被拉紧，然后被小心地粘上去。此后，工匠们理平边缘和落球袋，以确保没有褶皱。当他们往台面上粘台呢的时候，另一个工匠把台案内沿的软垫和外框安装到一起。

下一阶段的工作是对台球桌的内部进行加工。这个机关可在台球落入袋中时，对它们的运动进行控制。你有没有想过台球桌是如何判断得分球还是犯规球的？这个可以通过磁铁做到，在这张桌子上甚至更简单。白球比其他彩球要稍微大一点，所以当它碰到一个高的屏障时滚动路径会改道。于是就滚了出来而不是留在里面。这个出球装置由一个传动机械控制，需要电池来保持运行。新电池通常可以工作将近三年才更换。当系统完成后会放一组球上去检验效果，确保运转正常。

在四处做一些调整，就可以完工了。现在要把台球桌组装起来。钉上最后一根钉子，就能第一次看到台球桌的整体运作情况了。出球装置也受到测试，确保玩家可以开始新的一场游戏。

装上石板后桌子的重量会达到半吨，并因过于沉重而无法整体搬运。所以最后的装配在现场进行。工匠们首先装上可调节的桌腿。接着把台面放置到位。此后在桌子最上面加上外框。很少有哪里的地面是完全平坦的。为了保证球不会都滚到一头去，桌子将用水平尺检查，每个桌腿会被相应地调高或者降低，以确保桌面的水平状态。所以，下一次当你连续输球时，在怪罪台球桌之前，先想想制造这张桌子所需要的全部精细工作。

你知道吗？

在 19 世纪 70 年代，象牙球被赛璐珞胶片制成的球所代替。不幸的是，有时它们会爆炸。

1. 台球最初是门球的室内版
2. 把胶合板涂胶和仿实木板黏合
3. 外框做好切出 6 个落球孔
4. 制作台案内沿软垫要贴橡胶边
5. 国际标准的台面呢铺在石板上
6. 台案内沿软垫和外框装到一起

7. 对台球桌内部进行加工
8. 球落入袋后桌下的控制机关
9. 球桌下面的内部构造
10. 精准安装桌面
11. 用水平尺检查桌面
12. 调整桌腿使桌面保持水平

Pool Table

Pool is played in pubs and clubs up and down the country. Although there are many different variations on the rules, most of us know the general gist of the game. What a lot pf people don't know is that it was actually invented as an indoor version of croquet.

Most pool tables are massed produced in factories, but this company still produces them by hand. First, panels for the table top and sides are cut to size. Then a layer of glue is added so an imitation wood panel can be stuck on top. Once they're in place, the layers are firmly pressed together and the glue spreads evenly forming a permanent bond.

Now it's time to build the frame. This is the part of the table you will be leaving your change on when you are waiting for a game. The 6 pockets are cut out using a drill. For a good game of pool the balls need to bounce consistently so the next step of the process is vital. They glue rubber bumpers on to the rails. When they're covered these will be the cushions and they'll make sure that the balls rebound evenly every time.

The baize cloth used for the surface of the table comes from Belgium. There are strict international standards governing the mix of fabrics and even the colour of the cloth. That means that the pool balls should roll roughly at the same speed whether they're in Barnsley or Bangkok. They use slate from a quarry in northern Italy to form the base of the table. The baize is pulled tight on top, and then carefully glued into place. Then they smooth out the edges and pockets to make sure there are no creases. As they secure the cloth to the base, another worker attaches cushions to the frame.

The next stage of construction is the internal workings of the table. This mechanism controls the movement of the balls when they go down a pocket. Have you ever wondered how a pool table can tell a winning shot from a foul? It can be done with magnets, but on this table it's even simpler. The white ball is slightly larger than the colored balls so when it hits a height barrier it's diverted back out again instead of remaining in the machine. The ball release system is controlled by a gearing mechanism which requires battery power to keep it running, These batteries generally last up to three years before there need replacing. Whilst the system is been completed a set of balls is passed through to make sure everything is in working order.

A few tweaks here and there and it's finished. Now its time to put the body of the table together. The final nails are driven in and for the first time they can see how it will all perform together. The release system is also tested to make sure that players can start a new game.

When the slate goes in the table will weigh half a ton and be far too heavy to move as a whole unit so the final assembly takes place on site, first they fit adjustable legs. Then they slot the tabletop into place. And they add the frame on top, the ground in any one place is rarely perfectly flat so to ensure your games aren't all played at one end of the table its checked with a spirit level, and each leg is raised or lowered to even the table out. So the next time your on a losing streak before you start blaming the table just think of all the precision work that went into building it.

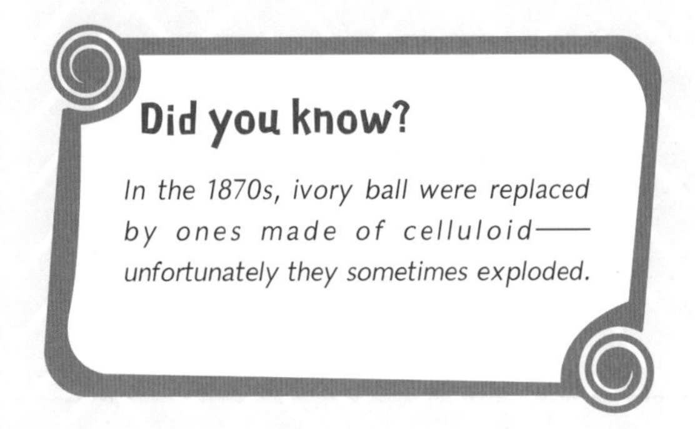

登山鞋

呼吸新鲜空气的最好办法，是到乡间走走。但是，如果你远足穿错了鞋，可能会把脚打湿，导致脚起泡和其他潜在的严重伤害。如何解决呢？

答案是一双高档登山鞋。根据最初的设计图，设计师用模板来做出一种特定形状和风格的鞋子，既好看又能支撑穿着者的双脚。一旦设计通过，工厂就可以开始生产。第一步是选取材料。为了结实，鞋子用皮革制作，但每种皮革不同，因此正确选择原料很重要。

选中合适的皮革后，裁缝切割出需要的形状。这双鞋子总共使用不同的110多块皮革、布料和金属部件。这些鞋子将承受很大的破坏力，尽管结实，它们也有薄弱的一面。鞋子里面的鞋舌是个复杂的呼吸系统，让登山者的脚保持凉爽。水蒸气被吸收到鞋舌上，步行动作把水汽从它表面的小孔泵出来。功能是把汗排走，不让水进来。秘密在于戈尔特斯面料。

最小的水滴比透气孔大20000倍，因此水不会流进鞋舌里。气态水分子比透气孔小700倍，因此可以从鞋里跑出去。这种单向系统能保持脚部干爽。对戈尔特斯内衬做取样测试，以确保做工良好和有效防水。

铆钉和鞋带挂钩都用定制的压床安好，鞋的上半部分已接近完成。除了防御自然因素，鞋子还必须穿得舒服，所以内部支撑结构也是应有之义。鞋子的上部完全做好了，现在要赋予它特别的形状。

防水和高强度是这类鞋最重要的品质，因此密封至关重要。首先机器人装置夹紧上部分，使之成为脚的形状并保持不变，同时下部分放置到位。但穿着这样的鞋子走不了多远，因为会很快散架。这层胶能够防水和密封，第二层胶水提供额外的保护，鞋子将在加热炉过一遍，把各个部分密封到一起。

接下来工人在鞋底增添一个橡胶层。这附加的保护让生产厂家更有把握，这款鞋能让穿着者足部维持干燥，他们真正敢于担保。

为了保持高标准，厂家会定期检验他们的产品。鞋子在这里待上7天，其间将行走30多万步。如果渗进1滴水，就算测试失败。

这些实验条件模拟现实生活中的磨损和撕扯，接下来是不折不扣的压力测试。测试仪把水倒入鞋中，然后放进每分钟240转的离心机。这个设计模拟在缝线和连接处施加10倍于正常值的压力。评估鞋子是否真能经受考验。发生任何泄漏都意味着测试失败。

组装鞋子的最后一步是鞋底。它必须在最恶劣的条件下保持抓地。44%的英国成人说，他们散步是为了休闲。这是非常流行的消遣活动，而安全是每个认真散步者所关注的。组装完毕，工作人员现在做收尾工作。橡胶被打磨平滑和擦拭干净，用火烧掉线头，再喷上保护皮革的涂层。

你知道吗？

赫尔大学的研究人员发明了一种登山鞋，可以在你行走时为手机充电。

穿上它们之前，剩下的是系好鞋带。单单这家工厂，每年就使用超过 250 万米长的鞋带，确保鞋子留在顾客的脚上。

它们准备出厂了。只用在盒子里待很短时间，就会有热心的旅行者买走，穿着它们向最近的山地进发。最坚毅的旅行者也需要休息，一双好的鞋子却应该经得起长久消磨。

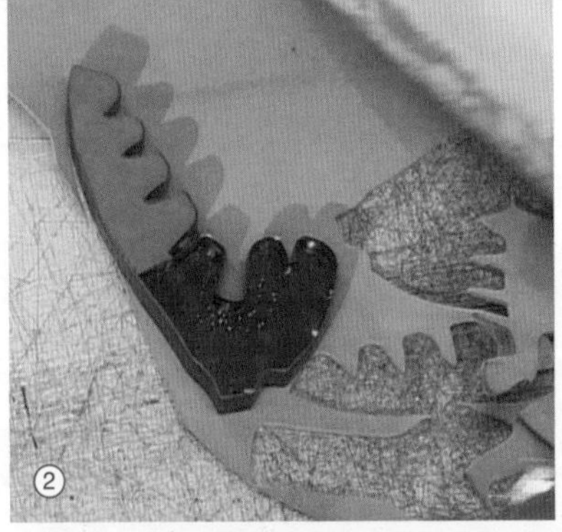

1. 设计出登山鞋图纸并做出模板
2. 选中皮革后切出需要的形状
3. 鞋舌的戈尔特斯材料排汗而防水

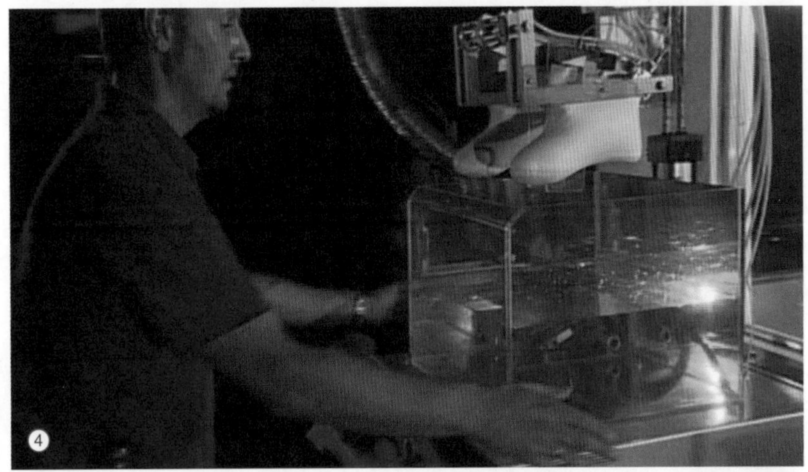

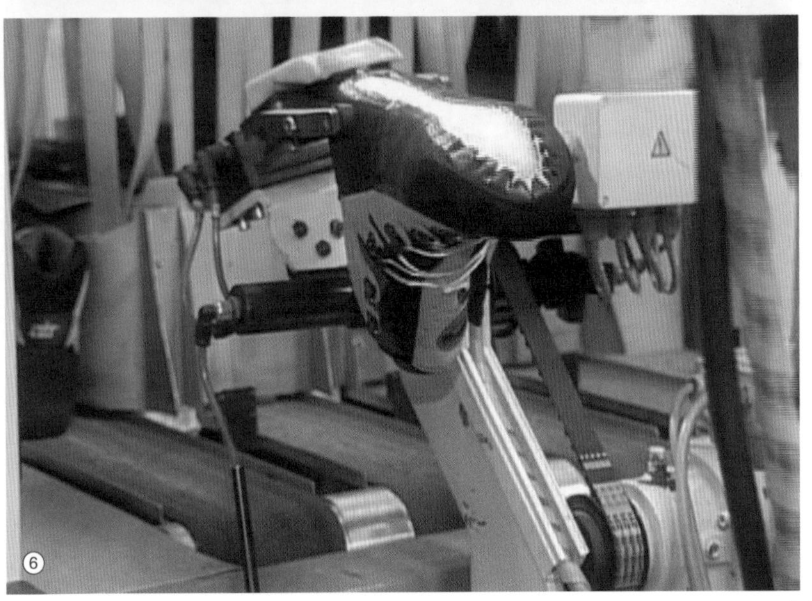

4. 取样检测戈尔特斯材料内衬
5. 压床把铆钉和鞋带挂钩安好
6. 鞋子的鞋帮完全做好

7. 鞋底第二层胶水提供额外保护
8. 鞋子经加热炉处理
9. 绕鞋帮底部增添一个橡胶层

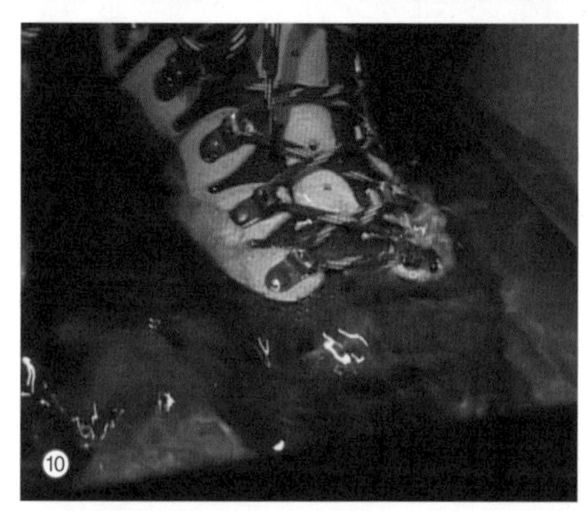

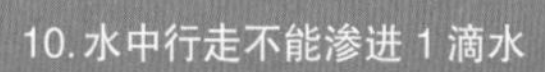

10. 水中行走不能渗进 1 滴水
11. 鞋底组装是最后工序
12. 系好鞋带准备出厂

Hiking Boots

A walk in the country is the best way to get some fresh air. But if you're hiking in the wrong footwear, wet feet, blisters and potentially serious injuries could result. So what's the solution?

One answer is a pair of top of the range hiking boots. From the initial design drawings a designer will use his templates to create a shape and style that both look good and support the wearer's foot. Once it's been approved, the production plant can get to work. The first step is material selection. The boots are made of leather for its strength, but each skin is different, so choosing the right raw material is important.

Once the right hides are selected, the tailor starts cutting out the shapes he needs. These boots use over 110 different pieces of leather, cloth and metal in all. These boots will take a lot of strain but despite their strength they also have a softer side to them. Inside the boots' tongue is a sophisticated breathing system to help the hiker's feet cool. Steam is collected in the boot's tongue and the walking action pumps it out through tiny holes on its surface. The idea is to let the sweat out, without letting water in. The secret is Gore-tex.

Smallest Water drops are 20,000 times bigger than the air holes so water won't flow into the tongue. However, steam molecules are 700 times smaller so it can escape. It's a one-way system to help keep the feet dry. Tests are run on samples of the Gore-tex inners to make sure they're well-made and water tight.

Rivets and eye hooks for the laces are fitted with a custom press and the upper part of the shoe is now nearing completion. As well as protection from the elements, the shoes need to be comfortable to wear so this internal support is included. With the upper section fully constructed the boot to can now be given its characteristic shape.

Waterproofing and strength are the most important elements for this bit of the shoe so it's vital that the seal is solid. First, this robotic device clamps the upper into the shape of the human foot and holds it, whilst the lower portion is put into place and released. Now, walking in boots like this wouldn't get you very far as they would fall apart very quickly, but this layer of glue will help waterproof the boot as well as sealing it. A second layer of glue is added for extra protection and the boots are now ready for a trip in the heating tunnel. This helps seal everything together properly.

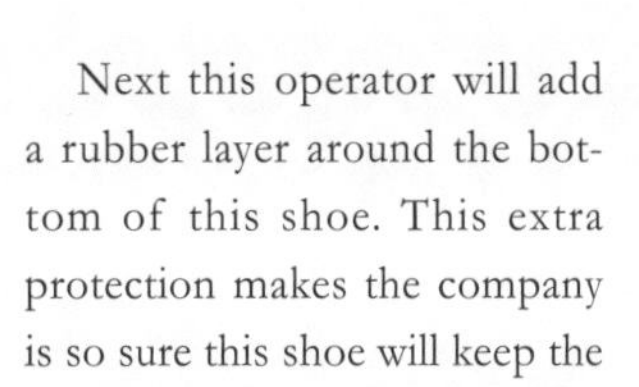
扫描二维码，观看英文视频。

Next this operator will add a rubber layer around the bottom of this shoe. This extra protection makes the company is so sure this shoe will keep the wearer's foot dry, they can actually guarantee it.

To maintain this high standard the company regularly tests their product. They're put in here and left for 7 days during which time, they will be forced to take over 300,000 steps. If they let in 1 drop of water, they fail.

These test conditions mimic the kind of wear and tear they'd experience in real life, but the next test quite literally pours on the pressure. Yes, the tester is filling the boots with water. He will then place them into a centrifugal machine which will spin them at 240 revolutions per minute. This is designed to simulate up to 10 times normal pressure on the seams and joins of the boots. This test is to assess whether really will take the strain. Any leaks and it's a fail.

The final stage of putting the boot together is the sole. This must grip in even the trickiest of conditions. 44% of the adult population in Britain say they walk for leisure. It's a very popular past time and safety is a concern for any serious rambler. With construction complete the staff can now add the finishing touches to get the boots ready. Rubber is smoothed and polished clean, hanging threads are burned off with a lighter, and a quick layer of leather protection is sprayed on.

All that remains is for them to be laced up before they can get to work. Each year this factory alone uses over 2.5 million meters of laces to keep their boots on their customer's feet.

And they're ready to go. They'll only spend a short time in this box before a dedicated rambler will buy himself a pair and then march them to the top of the nearest mountain. And while even the most hardened hiker needs a break, a good pair of boots should last for a long time.

Did you know?

Researchers at Hull University have invented a pair of hiking boots that can charge your mobile phone while you walk.

扫描二维码，观看中文视频。

山地自行车

自行车从 1816 年问世以来，有了很大的发展变化，当时一位德国男爵发明了第一辆可以转向的自行车。尽管它还没有踏板。但随着设计的改善，自行车的普及程度迅速飙升。在今天的英国，人们每年购买超过两百万辆自行车，一半以上是山地自行车。

最先进的山地车生产工艺从一堆铝管开始。这位制造商根据订单生产自行车，每一个车架可以按照未来主人的尺寸进行定制。首先用圆锯把管子切割到合适的长度。管子从两端稍稍压瘪，能增加强度和更为耐用。然后把管子在虎钳上夹紧，用圆形车刀向下切削，使之与即将连接的管子同样尺寸。精度达到 1/10 毫米，以确保车架能完美地组装起来。

所有的管子被切割和钻孔后，车架便在夹具台上进行组装了。用支架固定好形状，然后把它们焊接在一起。钢焊条被加热并在接触点熔化。这样会形成难以置信的高强度连接，让自行车使用时承受巨大应力。管子一个接一个组装起来，形成结实的车架。其坚固程度异乎寻常，尽管事实上它的重量只有 1/4 千克多点，比一包糖稍重。

回到夹具台上，工人检查它是否仍然与模板契合，因为有时候焊接可能导致车架扭曲。每一根管子都要精确到毫米。这个车架略有偏差，但它可以通过一点轻微的矫正，回到准确的形状。完美对接后，车架在 180 摄氏度的工业炉中烘烤 24 小时。这会使钢硬化，并且让管子变得不可分离。

接下来，车架的表面在这个密闭的蓝色装置里进行处理。整个过程的工作原理与喷砂大致相同。一把喷枪把微小的玻璃颗粒喷射到车架上，除去表面的任何残留物。前后对比清楚地显示了这道工序的好处。

待到清理干净后，车架被挂在钩子上并接通电流。喷枪让涂料粒子以相反的极性带电，使得车架像磁铁吸引铁屑一样吸引涂料。工人用胶带遮挡一些区域，然后喷上第二层涂料。多余的涂料用压缩空气小心吹掉。车架将返回到烤炉里再待上 20 分钟，使得清漆固化，并永久保持光泽。

出炉后工人便进行收尾工作，车架可以与自行车的其余部分组装了。

首先是后叉，它有一个带铰链的内置悬挂系统。然后加上减振器，让行驶更加平稳，保护骑手在崎岖的地形免受伤害。然后安装前叉和车把。接下来是车轮，安装有盘式车闸。和那种容易磨损、潮湿天气可能失灵的钳式车闸相比，盘式车闸更加性能可靠。

变速器安装在后面。它是把链条移动到不同齿轮上进行换挡的装置。安装工人用一根钢缆连接到变速器，并检查它是否能平稳工作。最后加上踏板，它们的设计类似滑雪板的固定器，使骑手的鞋子能毫不费力地扣紧和松脱。

你知道吗？

山地自行车跳跃最远的世界纪录是在威尔士创下的，达 41 米。

从开始到结束，这里的山地自行车像200年前最早的自行车一样，由手工装配完成。但却以轻质铝合金管材制作，配置了带铰链的车叉，并且安装了踏板。

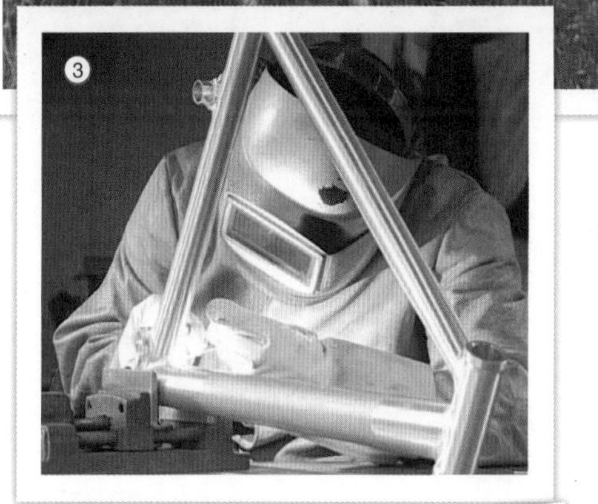

1. 英国每年售出的自行车中一大半是山地车
2. 铝管按未来车主的尺寸下料
3. 车架在夹具台上组装焊接
4. 烘烤车架使钢硬化，车架牢不可分
5. 喷完第二层涂料
6. 组装后叉及减振器

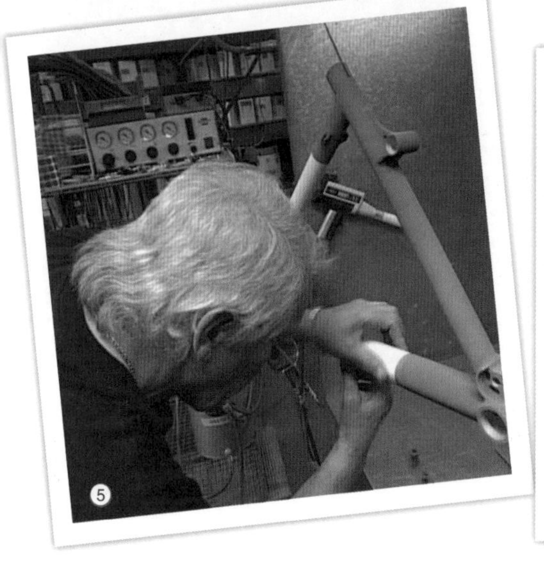

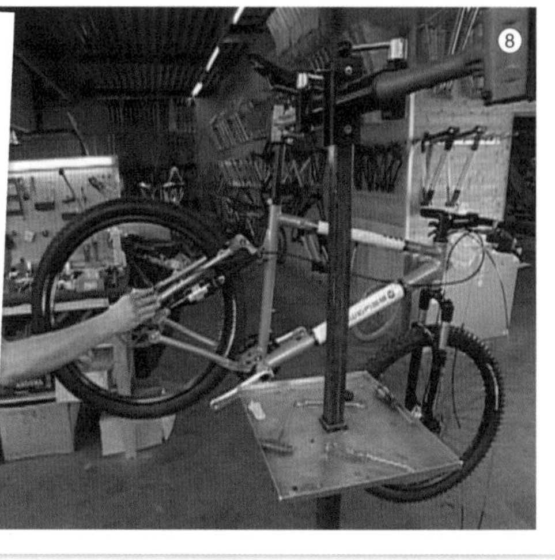

7. 安装自行车前叉和车把

8. 安装好盘式车闸

9. 安装变速器

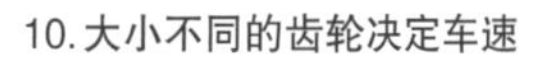

10. 大小不同的齿轮决定车速

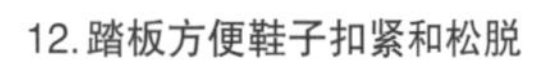

11. 一根钢缆连接到变速器上

12. 踏板方便鞋子扣紧和松脱

Mountain Bikes

Bikes have come a long way since 1816. Back then a German baron made the first one that could steer. although it still didn't have any pedals. But as the design improved its popularity soared. In the UK today we buy over two million bikes every year and over half of them are mountain bikes.

A state of the art mountain bike starts life as a pile of aluminium tubes. This manufacturer makes bikes to order so each frame can be customised to the measurements of its future owner. First the tubes are cut down to size with a circular saw. The ends are slightly flattened, this makes them stronger and more durable. Then it's clamped in a vice where it's drilled with a circular bit the same size as the tube to which it'll be connected. It's accurate to a tenth of a millimetre which ensures that the frame will fit together perfectly.

Once all of the tubes have been cut and drilled the frame is assembled on a jig bench, held into shape by brackets. Then they can weld it together. A steel welding rod is heated and becomes molten at the point of contact. It forms an incredibly strong bond that will help the bike endure the huge stresses that'll be excerpted on it. One by one the tubes are joined to form a solid frame. It's incredibly sturdy, despite the fact it only weighs a touch over one and a quarter kilos, just a bit heavier than a bag of sugar.

Back at the jig bench they check it still fits the template as sometimes the welding can cause distortion. Each tube has to fit to the millimetre. This frame's slightly out but it can be forced into shape with a bit of gentle persuasion. Once it's perfectly aligned it's baked in an industrial oven for 24 hours at 180 degrees Celsius. This causes the steel to harden and the tubes become virtually inseparable.

Next the frame's surface is treated in this airtight blue unit. The process works in much the same way as sandblasting. A gun sprays tiny particles of glass onto the frame and removes any residues which are lurking on the surface. The before and after effect clearly shows the benefit of the process.

Once it's been cleaned up the frame's hung on a hook that connects it to an electric current. The spray gun charges the paint particles to the opposite polarity and this causes the frame to attract the paint like a magnet attracts iron filings. They mask off some areas with tape and then they can spray on the second coat of paint. Any excess is carefully blasted away with compressed air. The frame will go back into the oven for a further 20 minutes and this'll fix the varnish to ensure a lasting shine. After it's been baked they add some finishing touches and it's ready to be attached to the rest of the bike.

The rear forks are first they've got an inbuilt suspension system with a hinged mechanism. A shock absorber is added which will make for a smoother ride and protect the biker from injury on rough terrains. The front forks and handlebars are fitted. And the wheels come next. They've got disc brakes which are more reliable than caliper brakes that wear easily and can fail in wet weather. A derailleur is fitted at the back. It's a mechanism that moves the chain onto different cogs to change gears. The fitter threads on a cable that connects to the derailleur and checks its all working smoothly. Finally the pedals are added, they're designed like a ski binding so that the rider can snap his shoes in and out with the minimum of fuss.

From start to finish the mountain bikes have been assembled by hand just like the first bikes nearly 200 years ago. But these ones have been crafted from lightweight aluminum tubing, fitted with hinged forks and they've got pedals.

Did you know?

The world record for the longest mountain bike jump was set in Wales at 41 mitres.

自行车头盔

扫描二维码，观看中文视频。

骑自行车是一项结合了保健与乐趣的运动。山地自行车难度更大，风险更多。不管在哪里骑，即使是专业选手也会戴上防撞头盔。

尽管头颅和西瓜有极大的不同，这个演示也能让你明白，发生事故时戴不戴头盔会有致命的不同后果。

那么，在你摔倒时能够救命的防撞头盔是怎样做成的呢？

也许你会感到惊奇，头盔的制造顺序是反的。构建从最外层开始，样式时髦又确保安全。形象设计师把图案印到塑料材料上。每层不同的颜色都需要一个不同的打印模板。设计的颜色越多，需要的模板就越多。

颜料干了以后设计师就把模板送给打印员，把每层不同颜色最后打印到塑料片上。这就是头盔的外壳。每张塑料片只有 1 毫米厚，但非常坚固，能为保护自行车骑手的脑袋提供外壳。印好的塑料片就像这样。这种形式还没有保护作用，但是看起来很漂亮。

现在该把这些塑料片做成真正的头盔外壳了。

每种形状的头盔都有一个特定的模具。将模具放到炉子里。接着塑料片也放进去，加热的模具向上推顶，就形成了特制的形状。制作出来的就是头盔外壳，但还有些工作要完成。

检验员仔细检查质量，这个外壳有点问题，但无关紧要，下一步能够补救过来。

精密切割机把无用的部分切除，有瑕疵的部位也去掉了。

现在这个外壳看起来更像人们熟悉的自行车头盔了。

没人相信这么薄的彩色塑料壳就能保护你的头颅免受损伤。它很坚固，确实也能在事故中发挥作用，但还缺少几个关键的部分。

防撞的最重要的部件是减震层，此处就是头盔的衬里。用很轻的聚苯乙烯做成，但并非这种形态。如此形状的聚苯乙烯没有用处，首先要让它松软。将热空气泵入盛满聚苯乙烯颗粒的容器中，会使它们膨胀到原来体积的 6 倍。这将改善它们的减震性能。

如果头盔衬里用这样的聚苯乙烯制作，就会破碎解体，毫无作用。让颗粒结合在一起的诀窍是超高温蒸汽。蒸汽加热使颗粒粘在一起形成固体层，可以承受很强的冲击。

现在来做头盔。首先把彩色外壳放进蒸汽炉。还可以加一些有用的附件，比如定制的纱网，防止空中昆虫钻到骑手的头发里。

每个部件放好后，蒸汽加热器开始工作。最后完成的是一个有着防撞衬里的头盔，能用来保护各类自行车骑手。

工作人员对每一批成品取样做全面的测试。尽管真实生活中不会发生这样受控制的摔落，这项沉重的冲击实验证明骑手可以承受一次很严重的事故。

蒸汽结合的衬里有稍许开裂，外壳绝对是

你知道吗？

最昂贵的定制自行车头盔花费了近 *2800* 英镑。

损坏了，但是头盔里的头颅还会是完整的。任务完成。

剩下的就是固定的带子了，用来确保头盔不会挪动。带子是防撕坏的尼龙织成的，有足够的强度保证头盔牢靠地待在头上。如果在最需要的时候断掉，那这些带子就一点用都没有了。

经过训练的行家把带子装在一起，包括可用来调整长短尺寸的调整扣。然后整个放进事先做好的头盔壳里。

万事俱备了。你和你的自行车，一段长长的上山路，然后是充满刺激的下山路。你知道就算最坏的情况发生，你的脑袋戴着头盔会安全得多。

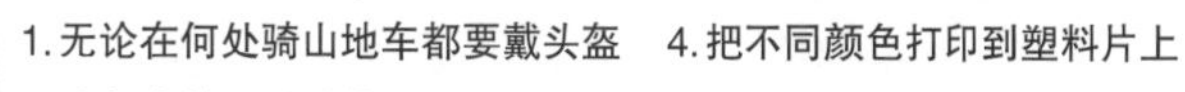

1. 无论在何处骑山地车都要戴头盔
2. 头盔中的西瓜跌落后受损轻微
3. 设计师把模板送来打印
4. 把不同颜色打印到塑料片上
5. 模具和印好的塑料片放入炉子
6. 模具顶起塑料片成特定形状

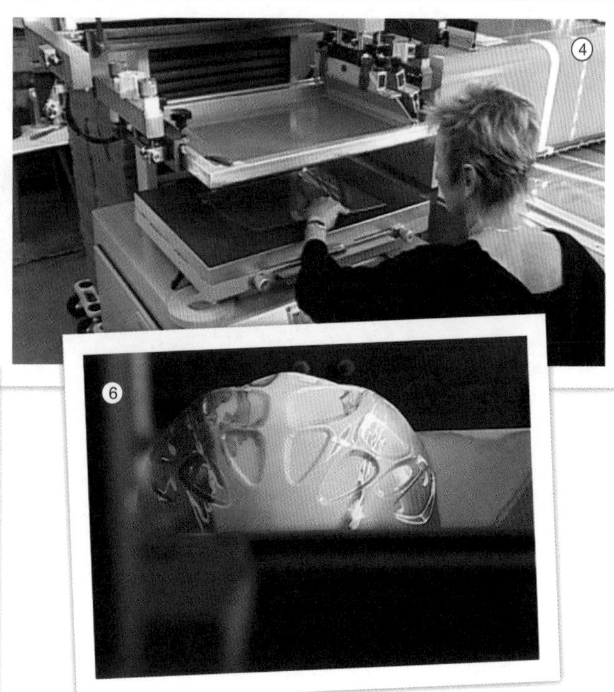

7. 精密切割机把无用的部分切除
8. 未经高温蒸汽的聚苯乙烯内衬易碎
9. 把外壳和附件放好关闭蒸汽炉
10. 内衬聚苯乙烯形成固体层
11. 把带子固定好
12. 头盔让你在骑行中增加安全系数

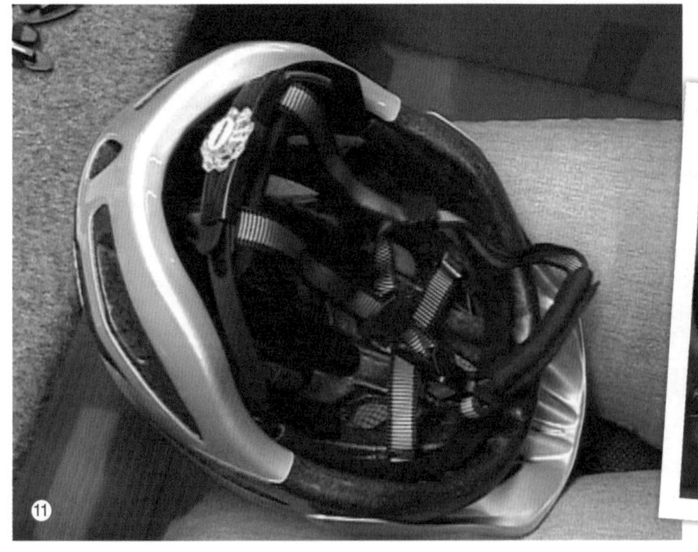

Bicycle Helmets

Cycling, it's a popular activity that combines fitness with fun. Mountain biking is even tougher, and brings more risk as well. But no matter where they cycle even professional riders wear a crash helmet.

Although a skull is very different to a melon, this demonstration gives you an idea of the critical difference a helmet could make during an accident.

So what goes into building a crash helmet that would save you if you wiped out?

Well, it may surprise you to learn that a cycle helmet is built backwards. Construction begins with the outer shell which has to be stylish as well as safe. A graphic designer is employed to add style to designs which will be applied to the raw plastic. The different layers of colour each need a different template for the printing presses. The more colours in the design, the more plates he will need.

Once they've dried, the designer will then take them to the printers who will layer each different colour onto a final plastic sheet. This will become the helmet's outer shell. Each sheet is only 1 millimeter thick but it's a remarkably strong and will provide the outer layer of protection for the cyclists' melon. The final sheet of painted plastic ends up like this. Utterly useless for protection in this form, but it does look quite good.

It's now time to mould these sheets into the protective shell that makes the outside cover.

For every shape of helmet there's a specific mould. The mould's put into the oven. And the raw sheets are loaded in and the heated mould is forced up through the plastic to give it its characteristic shape. What emerges is the outer skin of the helmet but it still needs a bit of work.

Inspectors keep a close eye on the quality and this one appears to have a problem, but it's nothing major and can be fixed during the next phase.

A precision cutter slices away the unnecessary bits of the shell, including the gap where the defect was. Now, the shell is looking more like a familiar bicycle helmet.

It's hard to believe that this thin sheet of multi-colored plastic will help save your skull from damage. It's pretty tough stuff and will certainly play its part if you had an accident, but several key elements are still missing.

The most important part of the crash protection is the shock absorber; in this case the interior lining of the helmet. It's made from this lightweight polystyrene but not in this form. Like this it'd be pretty useless … first, it has be 'fluffed'. Hot air is pumped through the bin full of granules which swells them up to 6 times their original size. This improves their shock absorbing qualities.

If a helmet was lined with the polystyrene in this form it would just crumble and disintegrate. No use whatsoever really! The trick to getting the granules to bind together is superheated steam. Steam-heating bonds them together to form a solid layer which can withstand huge impacts.

Now, to construct the helmet; first the colored shell is placed into the steam heater. Useful additions are included here like a custom made fly screen to keep airborne insects out of the rider's hair. Once everything's in place, the steam heater gets to work. What emerges is a finished helmet with a shock proof lining ready to protect cyclists of all abilities.

The staff run comprehensive tests on a sample from each batch. Although its' unlikely to be so controlled in a real life spill, this massive impact test proves these helmets would help a cyclist withstand a pretty serious accident. The steam-sealed lining's slightly cracked and the outer shell is definitely damaged, but the head inside this helmet would still be in one piece. Job done.

All that's left is the strapping. The helmet needs some added to help keep it in place. Woven from a tear resistant nylon, the straps are sewn together to provide enough strength to keep the helmet firmly in place. It wouldn't be any use if the straps snapped just when they were needed most.

Trained experts assemble the straps which included adjusters for different sized heads. And the whole unit is then fitted into the pre-built shells.

And you're ready to go. Just you, your bike, a long journey up a hill, and an adrenaline fuelled race down again. Happy in the knowledge that if the worst happens, your head is much safer inside the latest bicycle helmet.

Did you know?

One of the most expensive custom-made bicycle helmets ever produced cost nearly £2,800.

高档自行车座

扫描二维码，观看中文视频。

先来看看普通的钢制弹簧。这根弹簧很重要，作为豪华真皮手工自行车鞍座的一部分，它已经在140多年间为骑自行车的人带来臀部的舒适。

这些豪华的车座是英国一个叫布鲁克斯的公司制造的。虽然时光流逝，他们仍然使用一些百年前的初始设计。构成这种座位的第一个零件是架子。

它的设计非常高明，在1920年代，大多数环法自行车赛的选手就已经使用这些高质量的车座了。尽管时间推移和竞技水平提高，车座的制造过程仍保持不变。材料消耗计数器也是发条驱动的。在这里不会有激光制导的计算机发生故障，质量控制也沿用老办法。为了保持传统，这些手工制作的车座避免使用现代材料，譬如塑料。商家相应的使用传统皮革制造车座。

然而，不像现代的可塑性塑料能够很容易地变成任何形状，皮革的形态一开始是很硬的。为了让它变得更好加工，工人先把预先切好的座位模板泡在清水中两个小时。这有助于软化皮革，为下一工序做准备。

工人现在可以用锋利的刀片准确切割出真皮车座的形状。压力模具使它保持正确形状，工人顺着边缘切掉多余的材料。现在看上去已经像个自行车座了，但还不能支撑人的体重。加热潮湿的真皮会造成损坏，但在这里不会。车座在50摄氏度里静置2小时，有助于增加材料中的纤维强度。

现代生产方式简单易行并且更成本低廉，你可能会奇怪，为什么还有人想要一个硬邦邦的旧真皮车座，而不是一个舒适的橡胶车座？让你吃惊的是这种古老的车座在设计方面实际上比现代的橡胶车座对人更有益。在实验室条件下进行测试，可将现代的车座和手工皮革车座进行一番对比。

装在现代车座上的传感器显示出红色区域。红色表明过高的压力，天长日久可能导致组织损伤。当同样的压力传感器安装在140年的老设计车座上时，结果出人意料。老式的弹簧和车座形状能更好支撑车手。这意味着没有组织损伤和更舒适的骑行感。也许追逐时髦并非总是最好的？

英国的新潮设计已经很流行，因为它看上去效果更好。但也要考虑现代潮流的因素，复古造型正变得更加时尚。

不错，要保持车座的正确形状，需要一定的加固措施。为真皮车座加上铆钉和钢铁预制件。少了它们，车座就无法和支架连接。车座的支撑分为两部分。车座后面的弹簧提供悬挂，如同汽车上的减震器。让骑车的感觉更加舒适。安装在车座前面的框架提供支撑，平衡和力量，以确保它的经久耐用。这里的职工都为一个客户的故事而特别自豪。他的自行车上始终安着60年前买的车座，现在仍和刚买来的那天一样舒适。

手工装配所有的组件，是让这些车座以质量闻名的另一原因。每个零件都由手工调整，以确保正确装配。对车座每个细节的关注，让引为自豪的手工制作给客户带来持久的满意度。

组装车座的前部格外重要。客户希望昂贵的产品能经久耐用，所以车座的设计里包含了称手的小工具。前面部分有一个可调节的螺母。随着时间推移，皮革会有延伸，而这个螺母能使它再度收紧。

这个定制的真皮车座已接近完成。它花费90天的艰苦劳动才达到这样的水准，因此这些车座售价高达200英镑就不足为奇了。最后用高级的蜡抛光，进而改善车座的外观。它们可以出厂了。这个传统的设计，在你周日下午的自行车远足中焕发了青春。

1.20世纪20年代环法自行车赛选手使用这款车座
2.自行车车座的架子在这里生产
3.零件全部按传统设计手工制作
4.用传统方法制作老式弹簧
5.皮革车座板材泡在水中软化
6.压力模具使皮革保持正确形状

7. 车座在 50 摄氏度静置 2 小时
8. 车座加上铆钉和钢铁预制件
9. 开始装配车架弹簧

10. 车座前框架与车座连在一起
11. 高档自行车座架子装配完工
12. 用高级蜡给真皮车座抛光

Luxury Saddles

But first, the common steel spring. This spring is important because its part of luxury handmade leather bicycle saddle which has been keeping cyclists' behinds comfortable for more than 140 years.

These luxury saddles are made by a company called Brooks here in the UK. Although times have moved on, they still make some of their original designs that are over a hundred years old. The first thing that goes into this signature seat is the frame.

The design is so good that in the 1920's most Tour de France competitors used these high quality saddles. Times have moved on at the competitive level, but the construction process remains the same. The counters that keep track of what's been used are clockwork. There's no laser guided computer to malfunction here and the quality control is also performed the old-fashioned way. Keeping with tradition, these hand-made saddles shun the modern substances like plastic. Instead the manufacturers use traditional leather for the seat.

However, unlike modern malleable plastics which can be turned to any shape very easily, leather is very stiff in its original form. To get it into a more manageable state, the workers will bathe these pre-cut seat templates in plain water for two hours. This will help soften the material ready for the next stage.

With the sharpened blade the worker can now shape the leather seat precisely. Whilst the press holds it in the right form, he can cut away the extra material from round the edges. What emerges already looks like a bicycle saddle, but it wouldn't give you much support like this. Heating wet leather can damage it, but not here. The saddles are now cooked for 2 hours at 50 degrees Celsius, which helps strengthen the fibres in the material.

With modern production methods being readily available and far cheaper, you may be wondering why anyone would want a tough old leather seat instead of a comfy rubber one. Well it may surprise you to learn that this ancient seat design is actually better for you than modern rubber seats. Tested in laboratory conditions, the modern saddle was compared to the handmade leather variety.

Sensors on the modern saddle reveal red areas. This shows excessive pressure which can lead to tissue damage in the long run. When the same pressure sensitive equipment was tried on the 140 year old design, the results were unexpected. The old-fashioned springs and shape seem to support the rider far better. This would mean no tissue damage and a far more comfortable ride. Perhaps moving with the times, isn't always for the best?

扫描二维码，观看英文视频。

The innovative British design is popular because it seems to work better. But there are also modern trends to consider. Retro styling is becoming more fashionable again.

So, with the saddles in the right shape, they now need to be given some reinforcement. Rivets and the pre-made steel attachments are added to the leather seat. Without them there would be no way to join the seat to the framework. The seat's support is divided into two parts. The springs at the back of the saddle will provide suspension, acting like shock absorbers would on a car. This will make the cyclists' ride much more comfortable. The framework attached to the front of the seat provides support, balance and strength to help it survive plenty of use. Staff here are particularly proud of one customer's story. Apparently he's had the same saddle on his bike that he bought over 60 years ago and it's as comfortable now as the day he bought it.

Fitting all the components together is another part of the handmade process that gives these saddles their famous quality. Each specific part is adjusted by hand to ensure that it is properly assembled. The attention to detail that comes with a saddle that is proudly made by hand goes a long way to ensuring customer satisfaction.

This is particularly important when assembling the front of the seat. Customers expect expensive products to last so helpful little tools have been included in the saddle's design. Included in the front section is an adjustable nut. As the leather wears over time, it stretches and this nut can be used to tighten it back up once more.

The bespoke leather saddle is nearly complete. It can take up to 90 days of hard labour to reach this point so it's no surprise that these saddles can cost as much as 200 pounds. A final buffing with some high quality wax improves the final appearance and they're ready to go. this traditional design helps put the spring back into your Sunday afternoon cycle ride.

自行车打气筒

扫描二维码，观看中文视频。

当可靠的自行车打气筒不在身边时，车手的情绪就会像刺破的轮胎一样迅速瘪下去。打气筒由金属和塑料制成，开始时塑料是这样子的。将塑料颗粒送入注塑机，在那里被加热到熔化。液体塑料注入模具，形成不同的部件。

生产的第一个部件是打气筒本体，在这里是一根长管。它需要切成合适的大小，约 30 厘米左右。

其余的部件，如阀门、密封件和盖子由相同的基本原料制成，收集起来准备送入组装机。自行车打气筒是如何工作的呢？

随着活塞被拉回，打气筒内部空间充满了从顶部吸入的空气。当活塞不能进一步拉动时，它被推回。这将导致内部推杆顶端的密封垫膨胀。空气无法从来路逃逸，只能通过连接自行车轮胎的底部孔洞排出。

对于制作内部推杆来说，塑料不够结实，所以改用钢。钢料以大卷轴的形式送到工厂，使用之前必须先成形。一系列的辊压机把金属原料卷成管子，将接缝焊死。这家工厂每天制造超过 150000 个打气筒，因此需要大量的管子才能跟上生产进度。内部钢管做好后，切成合适的尺寸，安装到先前生产的塑料外管中。

所需要的最后一个零件是弹簧。通过改变打气筒的长度，就能适合各种自行车。虽然弹簧被成千上万地生产出来，但每个打气筒只需要一个，所以弹簧被分开，和其他零件一起装配。一个普通的自行车打气筒，需要产生 5 巴（1 巴 =10^5 帕 =10 牛顿 / 平方厘米）压力给轮胎充气。因此组装零件的机器需要精确。

现在骑自行车的人，也可以用二氧化碳气体罐给轮胎充气，不过尽管方便快捷，这些气罐只能使用一次。自行车打气筒可以反复使用，这家工厂 6 个零件一组进行装配以提高产量。零件被送入机器，然后装配机器人开始工作。内部钢制推杆定好方位，沿着生产线接受不同的零件，组装成打气筒。

首先安好最重要的阀，因为打气筒是从内到外组装起来的。橡胶套环安装到端部。确保严实地气密性配合，防止给轮胎打气时漏气。

下一步是安弹簧，接着在推杆顶部装上把手。然而在打气筒使用前，有个问题需要处理。打气筒内的阀是个精细零件。没有润滑将很快磨损。据估计，它甚至不能维持一个小时的正常使用。为了解决这个问题，整个装置涂上凡士林和油脂为打气动作提供润滑。涂了润滑剂的阀和推杆装进打气筒外壳，生产到此完毕。

最基本的工具让自行车手底气十足，那便是出色的自行车打气筒。

你知道吗？

自行车打气筒技术也能做出极好的浓咖啡。一项法国发明就是使用手动打气筒来冲泡新鲜的咖啡。

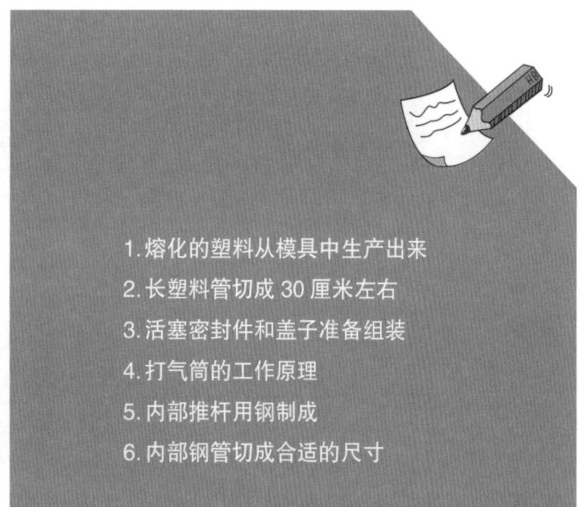

1. 熔化的塑料从模具中生产出来
2. 长塑料管切成 30 厘米左右
3. 活塞密封件和盖子准备组装
4. 打气筒的工作原理
5. 内部推杆用钢制成
6. 内部钢管切成合适的尺寸

7. 大量弹簧生产出来
8. 首先在钢管一端装上活塞
9. 每个钢管安装一个弹簧
10. 整个装置涂上凡士林和油脂
11. 活塞和推杆装进外壳
12. 手握打气筒底气十足

⑦

⑨

⑧

⑩

⑪

⑫

Bicycle Pumps

Like a punctured tyre a cyclists' mood would be swiftly deflated if the trusty bicycle pump wasn't around. Pumps are made out of both metal and plastic and the plastic starts out like this. The granules are fed into these injection moulding machines. Where they are heated until they melt. This liquid plastic is then injected into the moulds that will form the different parts.

The first part produced is the pump body, in this case; a long tube. But this needs to be cut down to size. 30 centimetres is about right.

The rest of the different pieces such as valves, seals and cap are all produced from the same basic raw material and collected up ready to be fed into the assembly machine. But how does a bicycle pump actually work?

As the piston is withdrawn, the inner space is filled with air sucked in through the top of the pump. When the piston can't go any further out, it's pushed it back in. This causes a seal on the end of the internal rod to expand. The air cannot escape the way it came in and can only escape through the nose or hole at the bottom attached to the bicycle tyre.

Plastic isn't strong enough to make the internal rod, so steel is used instead. It arrives at the factory on large reels but must be shaped before it will be any use. A series of rollers twists the raw metal into a tube and the seal is welded shut. Over 150,000 pumps are produced in this factory every day, so a lot of tubes are needed to keep up with production. Once the inner steel tubes are complete, they're cut down to fit inside the plastic outer tubes produced earlier.

And the final piece that's needed is a spring. By making the pump length variable, one pump can fit a wide variety of bikes. Although thousands are produced, each pump only needs one spring, so they're separated and fed through with all the other pieces. An ordinary bicycle pump needs to produce up to 5 bar of pressure in a tyre so the machinery that puts them together needs to be precise.

These days cyclists can also use cans of Carbon dioxide gas to inflate a flat tyre, but although they're quick and easy, these cans can only be used once. Bicycle pumps can be used again and again, and in the factory they are assembled in groups of 6 to speed production. The component parts are fed into the machine and then robotic assemblers get to work. The internal steel rods are orientated and passed along the line to receive the different pieces that make up the pump.

The all-important valve is first as the pump is built up from the inside out. Rubber collars are attached to the ends. This will ensure a snug airtight fit, preventing any air escaping as it's pumped into the tyre.

Next the springs are fitted, then the handles are clipped over the ends of the rods. However, before the pump is ready for use there's one problem that needs to be addressed. The valve inside the pump is a delicate component. Without lubrication it would wear out very quickly. It's estimated it wouldn't even last for an hour of normal use. To alleviate the problem the whole unit is coated with Vaseline and fat which lubricate the pump action. The coated valve and rod are then inserted into the pump housing and it's ready to go.

Essential kit that keeps cyclists spirits inflated, the simply brilliant bike pump.

Did you know?

Bicycle pump technology also makes an excellent espresso. A French invention uses a hand-pump to brew fresh coffee.

自行车轮胎

扫描二维码，观看中文视频。

如果你曾经骑自行车越过坚硬的马路牙子，那么可能你知道，是橡胶、尼龙和防弹凯夫拉纤维的结合，保护了你身体的敏感部位。轮胎是怎么制作的呢？轮胎是约翰·邓洛普1888年发明的，由几个重要部分组成。内部胎体为轮胎形成结构。在边缘的丝线被称为胎圈，使轮胎保持在轮毂上，而胎纹提供了抓地力。

直到第二次世界大战之前，自行车轮胎都是用有机材料制作的。橡胶来自树木，但供货不足迫使发明家寻找代用品。到1960年，合成橡胶已经成为规范，从那时起，现代自行车轮胎一直用合成橡胶制作。

制造轮胎橡胶就像是做蛋糕。必须按照配方来，才能得到完美的混合物。称出来适量的人造橡胶，通过重型滚筒送去与其他成分相混合。你可以从这个演示中看到发生了什么。轮胎通常是黑色的，但黑色并不是为了让你不用清洗。把炭黑加入白色橡胶里，可以提高抓地力。如果你想保持自行车直立行驶，这点非常重要。炭黑必须很好地混合起来，所以橡胶会几次通过辊子。准备就绪后，轮胎师傅可以添加其他成分了。矿物油、二氧化硅、氧化锌和硫都被加到组合里。这听起来相当有技术性，但基本上是为了改善轮胎的弹性、密度和耐久力。

在大辊子上进行完全相同的过程，只不过规模大得多。当混合完成后，新鲜的轮胎橡胶准备进行下一个阶段。

现在有了橡胶，但制作轮胎还需要很多其他东西。为了制作胎体，加入尼龙编织物使它的强度更高。大捆的尼龙被装载到机器上，准备与橡胶混合物结合。橡胶需要像糕点一样展开，所以工人们把它放回辊子。这会做出来一大片橡胶，然后就可以添加尼龙了。这种橡胶和尼龙的夹层是制作轮胎胎体的基本材料。

做轮胎时总是先从胎体开始。这个基础被精巧地缠绕在与轮胎尺寸一致的圆筒上，并用胶水固定到位。还记得凯夫拉吗？这就是的。这些线的制作材料，和卫士、警察的防弹衣相同。它强度很高，但很轻，可以帮助把轮胎固定在轮毂上。凯夫拉纤维缠绕在胎体上，一旦准确到位，圆筒膨胀将凯夫纤维拉压进橡胶。然后把胎体的两侧折叠，凯夫拉纤维能让轮胎防弹，当然只是假设轮胎的边缘被子弹击中。

这就是轮胎的结构，但说实话，如果你的自行车装上这样轮胎将不会很舒服，也没有多少抓地力。抓力来自胎纹。它由两层耐磨橡胶组合而成。一个专家连续测量两层的宽度，如果材料组成不对，轮胎不会抓地。

胎纹材料被送去加到轮胎胎体上。把轮胎包裹在圆筒周围，用力按下去，但它仍然还不能在轮毂上使用。首先它需要被“烹调”。这

你知道吗？

自行车陆地速度纪录达到每小时269千米，于1995年10月3日在犹他州邦纳维尔盐滩创造的。它被拖在高速赛车后面行驶。

个“烹调”过程被称为硫化。将混合的材料装进机器里，加热到180摄氏度停留3分钟。把各种组分都熔为一体，同时橡胶膨胀压进有胎纹的凹槽。

定时器时响了，机器打开，做好的轮胎被取走。接着，工人将从每批轮胎中随机选择一些样品进行严格的测试。不结实的轮胎碰上马路牙子会对你造成伤害，甚至会损坏你的车轮，轮胎在这台机器上测试三天。这就相当于连续50000次冲击马路牙子。爆胎非常讨厌，所以所有的轮胎都必须通过穿刺模拟测验。

如果你把过多的空气打进轮胎会怎么样呢？它们必须经受住至少每平方英寸（1平方英寸=6.4516平方厘米）8磅（1磅=0.4536千克）的压力。这个轮胎差不多达到了每平方英寸20磅。

下一次你碰到马路牙子时，记住你的防弹轮胎会承担压力，而你却不必紧张。

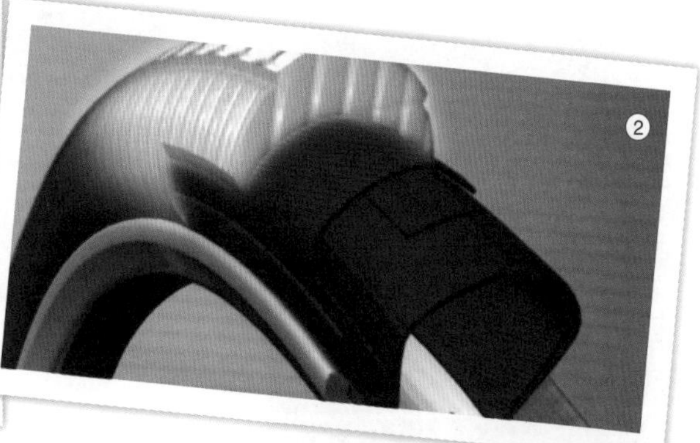

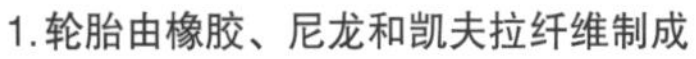

1.轮胎由橡胶、尼龙和凯夫拉纤维制成
2.轮胎的结构
3.自行车轮胎用合成橡胶制成
4.辊子把炭黑、橡胶及其他添加物混合
5.橡胶尼龙夹层是内部胎体的基础材料
6.胎体材料绕在机器圆桶上固定

7. 凯夫拉线缠绕在胎体上
8. 连续测量两层橡胶宽度
9. 胎纹材料加到轮胎胎体上
10. 硫化并压出胎纹
11. 样品进行冲撞测试
12. 穿刺模拟试验

Bicycle Tyres

But first, if you've ever ridden your bike over a really hard kerb, then you might like to know that it's a combination of rubber, nylon and bullet-proof Kevlar that's protecting your sensitive side. So how are tyres put together? Invented in 1888 by John Dunlop, the tyre is made up of several important parts. The inner carcass gives the tyre structure. Threads at the edges, known as beads, help keep it on the rim, and the tread provides the grip.

Until World War Ⅱ, bicycle tyres were organic. The rubber they were made out of came from trees, but poor supply forced inventors to come up with an alternative. By 1960 synthetic rubber became the standard and modern bicycle tyres have been made from it ever since.

Making tyre rubber is like making a cake. There's a recipe you have to follow to get the mix spot on. The right amount of the artificial rubber is weighed out and sent through heavy duty rollers to be mixed with the other ingredients. You can see what happens in this handy demonstration. Tyres are usually black, but that's not so you don't have to wash them. Black soot is added to the white rubber which improves the grip. This is pretty important if you want to keep your bike upright. It has to be mixed up really well so the rubber will be passed through the rollers a few times. Once it's ready, the tyre chef can add all the other ingredients. Things like mineral oil, silica, zinc oxide and sulphur are thrown into the mix. It sounds quite technical but basically they improve the tyres elasticity, density, and durability.

This whole process is the exactly the same on the larger rollers, just on a far bigger scale.

When the mix is complete, fresh tyre rubber emerges ready for the next stage. So, you've got your rubber, but there's far more to tyres than that. To create the carcass, nylon matting is added to make it stronger. Enormous rolls of nylon are threaded up onto these machines to be combined with the rubber mixture. It needs to be spread out like pastry, so the workers feed it back into the rollers. This will create a big sheet of rubber and the nylon can then be added in. This rubber nylon sandwich is what makes up the basic material for the tyre's carcass.

When you're building a tyre you always start with the carcass. This foundation is expertly wound round a tyre sized barrel and glued into place. Now remember the Kevlar, well that's this bit. These threads are made from the same bullet-proof material used to protect bodyguards and the police. It's very strong but light, and it helps hold the tyre on the rim. It's wound around the carcass and once it's in place, the barrel is expanded which actually forces the Kevlar into the rubber. The sides of the carcass are then folded over and the threads now make it completely bullet-proof assuming you get shot just on the very edge of your tyre.

And that's the tyres structure but to be honest, if you tried to ride a bike on a tyre like this it wouldn't be very comfortable or give you much grip. The grip comes from the tread. It's made from the combination of two layers of hard wearing rubber. An expert continually measures the width of the two layers otherwise the composition wouldn't be right and your tyre wouldn't grip.

The raw tread is sent off to be added to the tyre carcass. A layer is wrapped around the drum and firmly stuck down, but it's still not ready to be put on the rim just yet. First it needs to be cooked. This cooking process is known as vulcanisation. The combined elements are loaded into a machine that heats them up to 180 degrees Celsius for 3 minutes. This melts everything together, and at the same time it expands the rubber into the treaded grooves. When the timer goes ping, the machine opens up, and the finished tyre is removed.

Next, the workers will take a random selection of tyres from each batch and run some rigorous tests. Hitting a kerb with a weak tyre could hurt you and even damage your wheel, so the tyres spend three days on this machine. It's like hitting 50,000 kerbs in a row. And how annoying can a puncture be so all the tyres have to pass the puncture simulation test.

And what if you put too much air into the tyre? They all have to survive at least 8 pounds per square inch of pressure. This one manages almost 20.

So, next time you hit a kerb just remember your bullet-proof tyre is taking the strain so you don't have to.

Did you know?

The bicycle land-speed record stands at 269 km/h achieved on the 3rd October 1995. It was towed behind a dragster at Bonneville Salt Flats, Utah.

哈雷戴维森摩托车

扫描二维码，观看中文视频。

哈雷戴维森作为一种标志性的摩托车，让人们联想到著名的大胡子骑手。它们从好莱坞到摩托车族那里，都有深厚的历史传统。但这些高速公路的冒失鬼是怎么制造出来的呢？

这款世界著名的摩托车有许多秘密，但有一个经常被忽视的特点是车架。车轮、发动机和把手都需要一个稳固的构架来安装。这些钢管完美无缺，重量轻但且很坚牢。虽然哈雷戴维森公司在 1903 年投产，第一年只生产了 3 辆成品摩托车，现在这个工厂每天生产超过 500 个车架。机器人被用来加快生产进程，并帮助提高最终产品的质量。除了使用机器人，一些焊接活由手工完成，因为有些古怪的地方机器无法到达。做好的车架很重，因此要用起重机把它从工作台上卸下。

挡泥板是经典外观的一部分，这些都用钢板制成。一台巨大的压力机把钢材压成传统的复古风格曲线，它仍是这些摩托车受欢迎的一个特色。压力机塑造出挡泥板的标志性形状后，再压一次就切掉了多余的金属。研磨机磨掉粗糙的边缘，准备好进行下一步。

另一个由钢板压制的零件是油箱。油箱由 3 个不同的部分组成，从中间被分成两半，使它能安装在刚做好的车架上。激光切出半个油箱的形状。做出的边缘非常适合重新焊接。因为装的是汽油，油箱必须精心制造。有些摩托车生产商雇请其他公司做油箱，但哈雷戴维森公司自己生产油箱。工人小心翼翼用双手操作滚焊机，把两个部分背靠背焊在一起。把第三部分整齐安放在两边的半个油箱中间。再进行一些点焊，油箱就接近完成了。但重要的并不是仅是制造质量，工人的任务是确保新做好的油箱有一个完美无瑕的表面，不仅外形漂亮而且功能完好。接着工程师将成品油箱密封，并浸泡在水里。只要看到一个气泡上升到水面，这个油箱将被拒绝接受。

如果通过检查，油箱将被送到颜料室。首先将各个部分完全清洗。颜料只附着于一尘不染的金属表面。多种配色方案被使用。工人们用胶带把不同部分贴起来。这意味着一个部件可以有几种颜色，产生传统的双色调效果。特殊的复古外观，附加的彩色条纹，是由专人使用这些专用的模板和滚筒手工绘制的。最后是荣誉标志。工人使用激光校准器，确保贴纸每次都能恰好放在正确的地方。

闪闪发光的 V 型双缸引擎现在到达工厂。它们引吭狂吼的声音，是摩托车手追求的重要效果。

现在可以组装摩托车了。先是车架，然后把发动机整齐安放到位。接下来是汽油箱，它被包裹在玻璃纸里以保护定制的油漆图案。

成型的挡泥板加上去，但还缺少摩托车的

你知道吗？

有史以来出售的最昂贵的哈雷摩托车是“公路之王”。它在“今夜秀”上被拍卖，为抗击海啸募捐筹集了 *800100* 英镑。

一个主要部件。是的，车轮。现代摩托车轮通常用坚固的轻质合金制作。尽管清洁有些麻烦，但老式辐条轮圈仍然很受欢迎。当崭新的镀铬轮圈闪闪发光时，就知道这是为什么了。按照传统，轮圈配上白色侧面的轮胎，安装到位，各处紧固。摩托车完成了。

工程师需要对组装好的产品进行测试，但不会去道路上冒险。相反，他们开足油门尽情奔驰，却在一个封闭的房间。

这种设备被称为滚动路，能让工程师把摩托车开到每小时 130 千米的速度。让我们希望滚筒不停转动，否则他会很快冲出房间。

悬挂系统也很重要。它非常牢固，在二战期间，这些摩托车被用作越野车。超过 8 万辆在战争中被用作通讯摩托车。

当所有的测试完成后，它们被装箱运往世界各地的经销商。如同终结者和逍遥骑士在电影中与世长存，哈雷摩托车仍然是“公路之王”。

1. 哈雷戴维森是标志性摩托车
2. 机器人无法到达处由人工焊接
3. 起重机把车架从工作台上卸下
4. 切掉多余金属磨掉粗糙边缘
5. 第 3 个部分放在两半油箱中间
6. 油箱放置好后用手工点焊到位

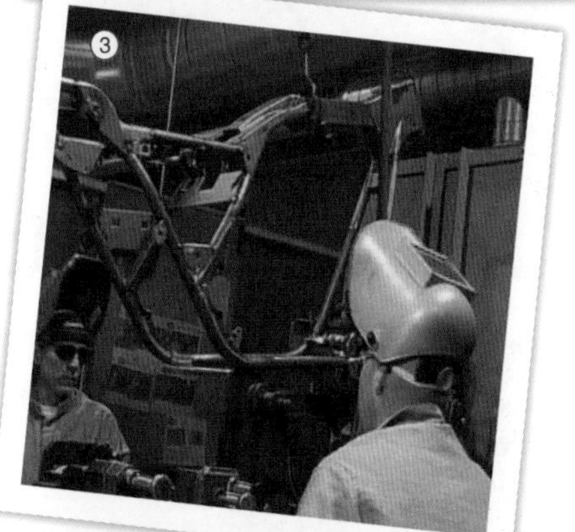

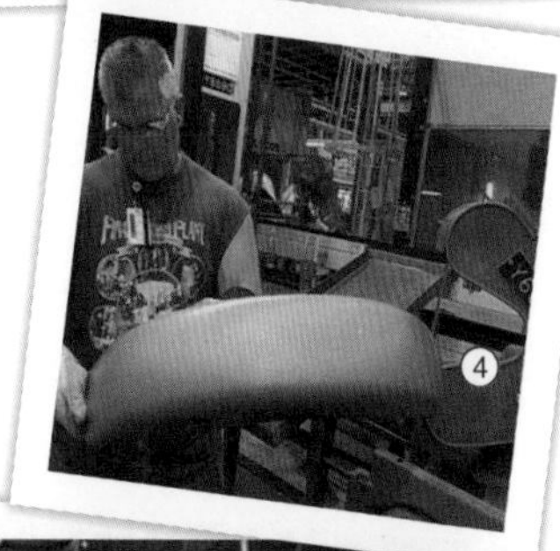

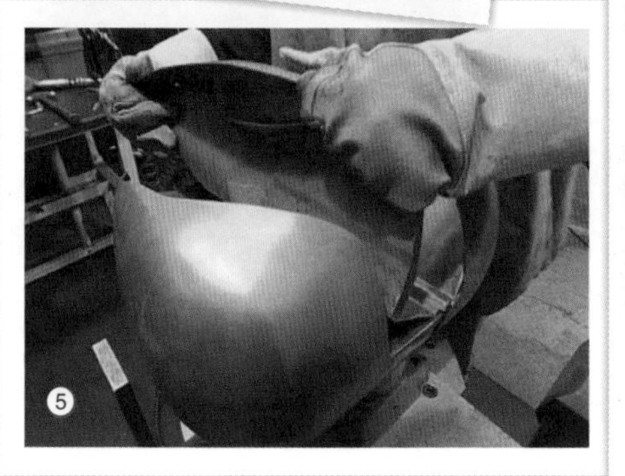

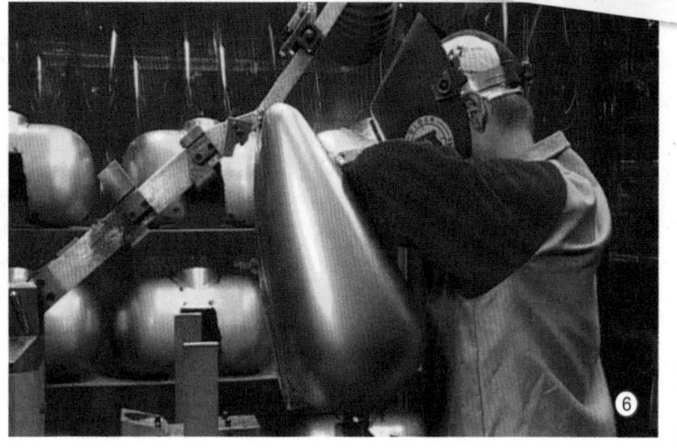

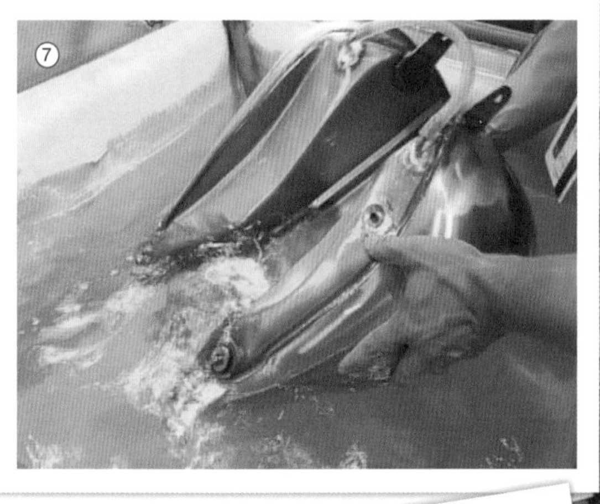

7. 油箱置入水中检查密封性
8. V 型双缸引擎安放到车架上
9. 安装仪表盘和汽油箱
10. 安装挡泥板
11. 紧固各处零件完成安装
12. 在高速滚动路上进行测试

Harley Davidson Motorbikes

But first, often associated with hairy bikers , the Harley Davidson is an iconic motorbike. they certainly have a famous heritage from Hollywood to biker gangs. But what goes into making this highway hellraiser?

There are many secrets to a world famous motorbike, but the one feature that is most often over-looked is the frame. The wheels, engine and handle bars need a solid structure to attach to, these steel tubes are perfect. Lightweight, but strong. Although the company started out in 1903, producing only 3 complete bikes in their first year, today over 500 frames are produced at this one factory everyday. Robots are used to speed up the process and they help produce a high quality finish. In addition to the use of robots, some of the welding is done by hand as the machines can't reach some awkward spots. The completed subframe is heavy so a crane is used to remove it from the workbench once it's done.

The classic look often includes mudguards, and these are made using sheet steel. An enormous press shapes the steel into the traditional retro-style curves that are still a popular feature on these bikes. After the first press has moulded the mudguard into its signature shape, another press cuts away the excess metal. The grinder removes the rough edges and its ready to go.

Another part of the bike pressed from sheet steel is the petrol tank. The tank is made of 3 different parts. 2 halves separated with a middle block so that it can sit over the subframe that has just been built. A laser cutter slices the tank shape in half. This gives it a perfect edge for when it gets re-welded later. Because they will be filled with petrol, the tanks must be built carefully. Some bike manufacturers hire other companies to do it for them, but at Harley Davidson, they produce their tanks themselves. Taking great care with his hands, the worker operates the roller welder which joins the sides back to back. They now have a space in the middle to take the third part which fits neatly between the two halves. A spot more welding and the completed tank is nearly done. However it's not just the build quality that's important. This workers job is to ensure the new tank has a flawless surface so it looks good as well as functions perfectly. Next an engineer will take the finished tank, seal it, and immerse it in this bath of water. If he sees a single bubble rise to the surface, the tank will be rejected.

However, If it passes inspection, it can be sent on to the paint room. First the different parts all have to be washed. The paint will only adhere to the metal if the surfaces are spotlessly clean. A variety of colour schemes are used. The workers will also tape up different parts. This means several colours can be used to paint one element giving them a traditional two-tone effect. For that extra special retro-look, additional coloured stripes are drawn on by hand using these special templates and paint rollers. And finally the badge of honour. Using a laser sight, the worker can line up the stickers so they are in exactly the right place every time.

Gleaming V-twin engines now arrive at the factory. Their throaty roar is an important part of the motorbike's appeal.

The bikes can now be built. The subframe is first and the engine is neatly fitted into place. Next comes the petrol tank, wrapped in cellophane to protect its custom paint job.

The shaped mudguards, or fenders, are added on, but there's one major component still missing. Ah yes, the wheels. Modern motorbike wheels are made of a solid lightweight alloy. But, despite being a fiddle to clean, old-fashioned wire rims are still popular. When the chrome is in this brand new sparkling condition, you can see why. Keeping with tradition, the rims are fitted with white wall tyres, and slotted into place.

Engineers need to test the finished products, but they don't venture on the open road. Instead, they open up the throttle and let rip in a closed room.

Known as a rolling road, this device lets the engineer drive the bike at up to 130 kilometres an hour. Let's hope the rollers keep turning or he'll be leaving the room very quickly.

The suspension is also very important. It's so tough that during World War II these bikes were used as off-road vehicles. More than 80,000 of them were used as courier bikes during the conflict.

When all the tests are completed, the new bikes are shipped out to dealers all over the world. Immortalised in movies like the Terminator and Easy Rider , the Harley is still 'King of the road'.

Did you know?

The most expensive Harley ever sold was a Road King. Auctioned on the "Tonight Show" , it raised £800,100 for the Tsunami Appeal.

冰球靴

花样滑冰是一种优美风雅而又规则严谨的技巧运动，几乎和横冲直撞的冰球世界毫无相似之处！但这些彪形大汉与花样滑冰运动员确实有超出你想象的更多共同点。他们都需要冰上的技巧和力量，就此而言，一双结实的冰鞋是基本条件。

冰球靴子需要足够强固来支撑球员，同时也需要足够柔韧能灵活自如。

一种几乎牢不可破的合成材料用于制作靴面。这种预先成形的塑料由手动的冲床从模板上压出来。虽然溜冰鞋其实只是一双底部装有冰刀的靴子，但制作高质量的冰球靴子需要54个部件。所有这些不同的部件都需要缝合在一起，使它们能承受住专业冰球比赛的强度。

缝入脚趾防护装置特别重要。冰球的击打速度能超过每小时160千米，足以严重伤害未加保护的脚趾。但不只是脚趾需要保护，脚踝同样容易受伤，因此靴子的这部分也需要加装密实的充填物。

现在准备好靴面成形，和运动员的脚相配。把它们套在这个特殊的模具上，技师插入脚下面的鞋垫。在技师把靴子放进这台机器前，鞋垫被暂时钉上去。机器按设计把所有部分闭合起来，形成靴子的模样。先把两部分粘起来，接着插入钉子让它们保持固定，直到胶水变干。所有这些强化措施，胶水和钉子的使用，看来似乎有点过分，但专业冰球比赛持续2.5小时，所以如果需要当此重任，这些靴子还面临一些重要的工序。

在添加靴底和冰刀之前多余的材料和钉头被打磨掉。这些材料如此结实，需要特殊的磨床来完成这项工作。当所有多余材料被切掉后，技师涂上一层热活化胶。

如果现在加上靴底将不会粘牢。要起作用，胶层必须加热，靴底也同时加热。当两个部分都达到70摄氏度左右，把它们放到一起。黏合剂现在已被激活，把各部分牢固地粘起来。最后用巨大的5吨压床结束整个过程。

但现在仍然缺少一个重要部件。冰球运动员要能在冰上以每小时40～45千米的高速度驰骋，需要安装额外的钢制锋利冰刀。

首先，刚安好的靴底需要钻洞。钻头打出8个孔，准备装上冰刀。记住，结实是座右铭，这些孔都安进金属螺线管。为螺钉提供一个能够抓牢的固体表面。

最后，是将钳在硬塑料夹具中的冰刀安装到靴子底部。八颗螺钉总共能承受约2.5吨的力。把冰刀固着在最健硕的球员脚上也绰绰有余。

但它们如何让球员在冰上保持直立？给它做出重要的刀刃，冰刀用强力机床磨快。锋利的金属可以让球员切入冰里，在比赛中提供至关重要的抓地力。但冰刀不能永远使用下去。每个冰刀在更换之前可以磨150次左右。

所以，虽然不像花样滑冰冠军那样优雅而曼妙，如果冰球运动员想要赢得致胜一球，也需要高品质的靴子承载着他们在冰上自如地驰骋。

你知道吗？

滑冰至少可以追溯到公元前1000年。在瑞典发现的一双古代溜冰鞋是用骨头制作的。

1. 冰球靴子需要足够强固和灵活
2. 靴面用手动冲床按模板压出
3. 缝入脚趾防护装置

4. 靴子加填充物以保护脚踝
5. 成形的靴面套在脚型模具上
6. 鞋底和鞋面黏合并用钉子加固

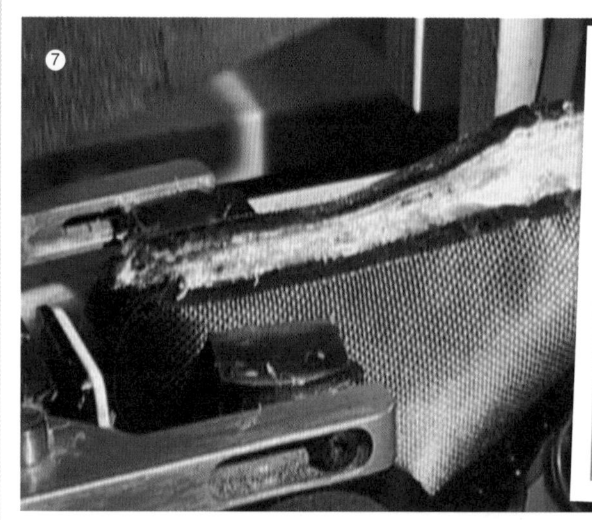

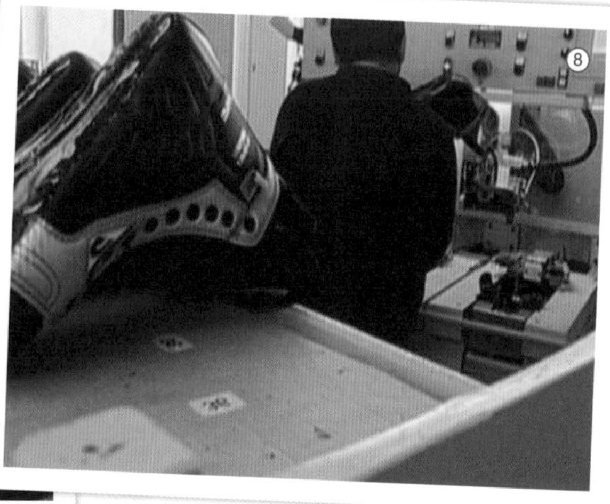

7. 特殊磨床磨掉多余材料和钉头
8. 黏合打磨后涂一层热活化胶
9. 胶层和靴底用 5 吨压床黏合
10. 靴底打孔并按上金属螺线管
11. 冰刀用机床磨快
12. 冰刀在更换前可磨 150 次

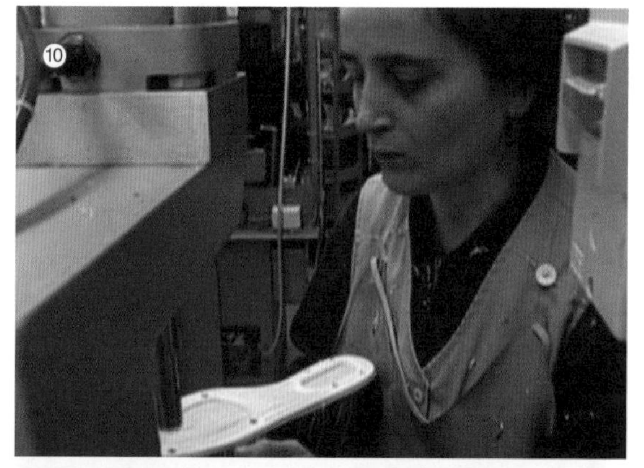

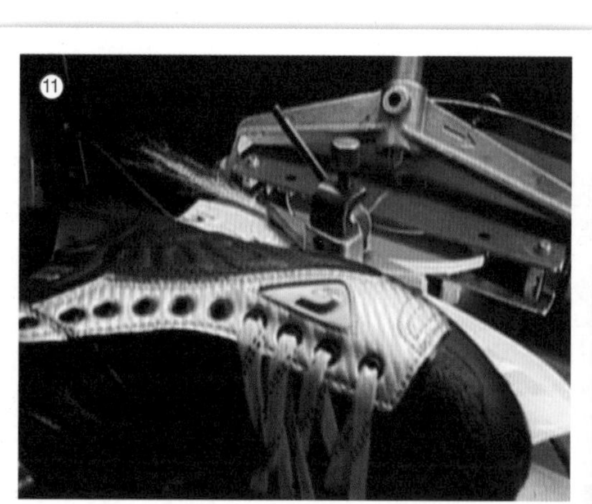

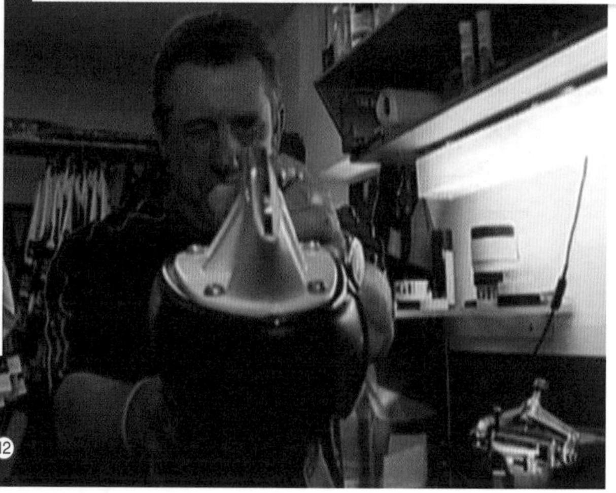

Ice Hockey Boots

But first…figure skating. A delicate highly disciplined skill that bears almost no resemblance to the hard hitting world of ice hockey! But these tough bruisers do have more in common with figure skaters than you might think. They too need skill and strength on the ice and for that, a solid pair of skates is essential.

Ice hockey boots needs to be tough enough to support a player, but flexible enough to allow full manoeuvrability.

A virtually unbreakable synthetic material is used for the boots upper. This pre-formed plastic is stamped out of the templates with a manually-operated press. Although skates are really just a pair of boots with a blade on the bottom, there are 54 parts that go into making quality ice hockey boots. All of these different parts need to be sewn together to give them the strength they'll need to survive a professional hockey match.

It's especially important to sew in the toe guard. Hockey pucks can be shot at over 160 kilometres an hour enough to cause serious damage to an unprotected toe. But it isn't just the toes that need protection. The ankles are just as vulnerable so that part of the boot is also fitted with solid padding.

The uppers are now ready to be shaped for the player's foot. They're fitted around this special mould, while the engineer inserts the bottom for the foot. It's stapled into place temporarily, before the engineer fits it to this machine. It's designed to close the whole unit up into its boot like form. First the two parts are glued then nails are inserted to hold them together until the glue dries. All the toughening, glue and nails may seem excessive, but a professional hockey match can last 2.5 hours, so if they're going to do the job these boots have got some serious work ahead of them.

Before the sole and blade are added, the excess material and nail heads are sanded off. The materials are so tough, special grinders are needed for this job. With all the surplus cut away, the engineer will now apply a layer of heat activated glue.

If the sole was added now it wouldn't stick. To work, this glue layer must be warmed up. The sole is also heated at the same time. When the two parts have reached around 70 degrees Celsius, they're brought together. The adhesive has now been activated and firmly sticks the parts into place. The whole process is given a final helping hand with a massive 5 ton press.

But there's one vital part that's still missing. For the hockey player to be able to speed round the ice at up to 40 or 45 kilometres an hour, extra sharp steel blades need to be fitted. First the soles that were just added need to be drilled out. The drill cut 8 holes ready for the blade to be attached to. Remembering that tough is the watch word, these holes are fitted with metal threads. This gives the screws a solid surface to grip.

Finally the blade itself, mounted in a hard plastic holder, is attached to the bottom of the boot. Together, the eight screws can withstand a force of around 2.5 tons. More than enough to keep the blade attached to the bottom of the heftiest player's foot. But how do they keep a player upright on the ice? To give them that all-important edge, the blades are sharpened with a heavy duty grinder. The sharpened metal helps players cut into the ice, giving them vital grip during a game. The blades don't last forever though. Each one can be sharpened around 150 times before they'll need to be replaced.

So although it isn't as gentle as the delicate form of the champion figure skater, ice hockey players need top quality boots to carry them skilfully over the ice if they want to score the winning shot.

Did you know?

Ice skating dates back to at least 1,000 BC. One ancient pair of skates found in Sweden were made out of bone.

有舵雪橇

一对高科技的冰刀，一个很小的手柄，和一个很大的决心。尽管听起来很简单，但要在高风险、高速度的长雪橇运动中取得成功，需要体力、技巧和勇气。即使是行家也需要最好的雪橇才能夺冠，否则垫底。

现代雪橇建在钢制框架上。但这个高科技制造过程中使用了很多不同的复合材料和金属。从基本框架到空气动力外壳和行进的滑板，雪橇由大约 150 个不同的部件组成。

雪橇会承受巨大的应力。为了在比赛中有竞争力，它需要达到大约每小时 145 千米的巡航速度。在转角时会承受高达 5 倍的重力加速度力，所以焊接一定要格外牢固。滑板也必须完美地进行研磨。即使最轻微的瑕疵也会产生阻力，减缓雪橇的速度，所以必须非常精心地打造。

现代有舵雪橇的理想宽度大约为 68 厘米。增加尺寸就会增加阻力，使雪橇速度减慢。在这种以毫秒计的运动中要想获胜，雪橇必须完美无缺。工程师建造雪橇结构的精度要在 1/10 毫米内，以尽可能达到高速度。

在完成雪橇结构时，另一组机械师制造它的外壳。它由复合玻璃纤维制成，涂上很厚一层聚酯树脂。树脂层逐渐增长，直到形成一个固体壳。然后静置晾干。

完全变硬后，机械师把外壳切割成最佳尺寸。精密的空气动力学，对确保雪橇达到最快的速度至关重要。他们用圆锯把新做的外壳切割成需要的形状，操作中要很小心地吸尘。玻璃纤维粉尘吸入肺部是非常危险的。

不同的部分完工后，一个竞赛级雪橇的基本外形就显现出来。但工程师又如何肯定他们的设计成功？在各种组装开始之前，早就用小比例的模型和风洞进行了大量测试。这样的技术水平意味着工程师能够测试新的设计，察看修改之处如何影响雪橇的性能。微小的细节，比如驾驶员头盔的形状，都会对雪橇的速度有很大的影响。

取得竞赛成功的最后关键要素是滑板。这些精密的刀片由哪些确切材料制作，是严加保守的机密，但会有包括铁、铜和磷的不同合金。使用普通的螺母和螺栓把它们安装到雪橇底部的可移动支架上。在选手把它带到赛道上之前，需要打磨滑板，去除任何缺陷。最轻微的划伤或磕碰也会增加阻力。

驾驶员如何控制雪橇呢？前面的人负责掌控。流线型外壳下面是一个由滑轮和手柄组成的简单设计。这就是转向装置。它直接连到前滑板，牵拉一侧或另一侧，就可以让雪橇向右或向左转。

整个雪橇完成后，剩下要做的就是去试试。雪橇运动取决于三个主要的因素。首先是车手的力量和竞争精神，第二是雪橇的质量，第三是沿赛道而下的驾驶技术。如果所有的制造、测试和培训都很到位，能否在雪橇比赛中成功赢得奖牌，最终就取决于选手了。

你知道吗？

《冰上轻驰》是关于有舵雪橇的电影之一，灵感来自牙买加第一支奥林匹克雪橇队的真实故事。

1. 雪橇有 150 个部件装在钢架上
2. 对滑板进行精心打磨
3. 雪橇制造精度为 0.1 毫米
4. 涂上多层聚酯树脂
5. 将静置晾干变硬的外壳取出
6. 雪橇基本外形显现

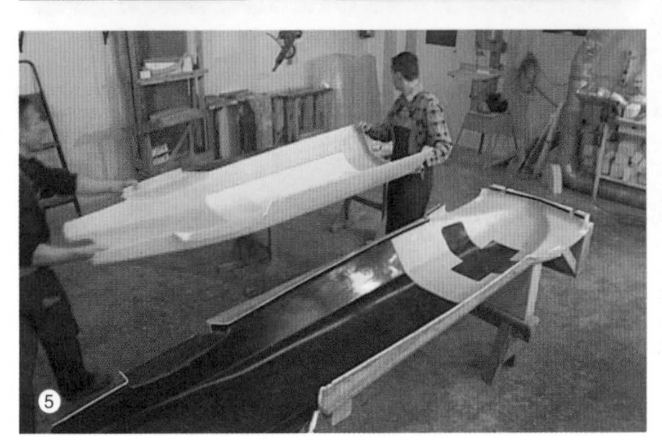

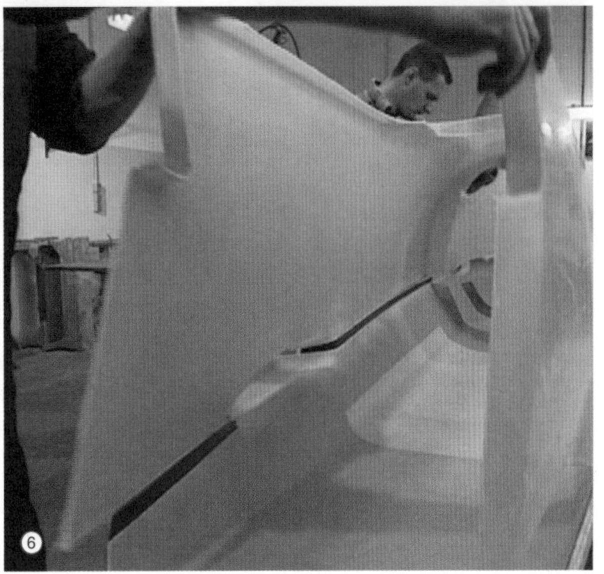

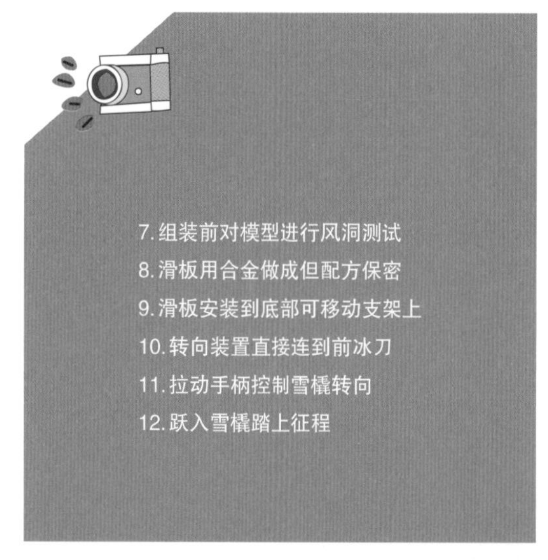

7. 组装前对模型进行风洞测试

8. 滑板用合金做成但配方保密

9. 滑板安装到底部可移动支架上

10. 转向装置直接连到前冰刀

11. 拉动手柄控制雪橇转向

12. 跃入雪橇踏上征程

Bob Sleds

A pair of high tech blades. A small handle. And a good deal of determination. It sounds simple, but anyone striving to make it in the high-stakes, high-speed sport of bobsledding needs stamina, skill and courage, but even the experts need the best possible sled to make it to the top or bottom.

The modern sled is built around a steel frame, but there are a lot of different compounds and metals used in this high-tech construction process. The sled is made using about 150 different pieces from the basic frame to the aerodynamic shell and the runners the sled will travel on.

It will be exposed to enormous stresses. It has to reach a cruising speed of around 145 kilometres an hour to be competitive. It will also have to withstand up to 5G of force as it corners, so the welds holding it together must be exceptionally strong. The runners too have to be perfectly milled. The slightest imperfection will cause drag and slow the sled down so they have to be very carefully crafted.

The ideal width for the modern bob sled is around 68 centimetres. Any larger and drag is increased, which again, will slow the vehicle down. In a sport where milliseconds count, a winning sled has to be perfect. The engineers construct the structure for the sled to within 1 tenth of a millimetre accuracy to ensure it's as fast as possible.

Whilst the structure is being built, another group of mechanics can produce the outer shell. It's made with a combination of fibreglass coated with a thick layer of polyester resin. The resin layers are built up gradually until a solid shell is completed. It's then left to dry.

Once it's completely solidified, the mechanics will cut the shell down to the perfect size. Precision aerodynamics are vital to ensure the finished sled will produce the fastest possible speeds. Using a circular saw, they cut the newly fabricated outer hull to the required shape, but they're very careful to vacuum as they go. Fibreglass dust can be very dangerous if it's inhaled into the lungs.

As the different pieces are completed the basic, outer shape of a competition grade bob sled begins to emerge.

But how can the engineers be sure they've got the design right? Intensive testing takes place using scale models and a wind tunnel long before any construction work even begins. This level of technology means the engineers can test new designs to see how their modifications affect performance. Tiny details such as the shape of the driver's helmet can have a big impact on the speed of the sled.

扫描二维码，观看英文视频。

The final element that is vital to a winning bob sled is the runners. The exact compounds that go into these precision blades are a closely guarded secret, but different alloys include iron, copper and phosphorous. They're attached to movable struts on the bottom of the sled using ordinary nuts and bolts. But before the riders take to the course, the runners are polished to remove any imperfections. The slightest scratch, or bump will increase drag.

But how does the driver steer his sled? The man at the front is in charge. Beneath the aerodynamic hull is a simple contraption that's made of pulleys and handles. This is the steering mechanism. It's connected directly to the front runners and pulling one side or the other turns the sled right or left.

Once the whole unit is completed, the only thing that remains is to try it out. Bobsledding is a sport that depends on three main factors. First, the power and aggression of the riders. Second, the quality of the sled; and third, how well it's driven down the track. If all the construction, testing and training have come together, it's finally down to the riders to make their sled a medal-wining success.

Did you know?

"Cool Runnings" , one of the few films about bobsledding, was inspired by the true story of the first ever Jamaican Olympic bobsled team.

滑雪镜

扫描二维码，观看中文视频。

滑雪很受欢迎，像所有的极限运动一样，安全装备至关重要。滑雪者跌倒，不仅是被同伴嘲笑的问题，还需要对眼睛很好地保护。每年有 150 万英国人去滑雪，所以很多人需要滑雪镜，而滑雪镜的外观非常重要。

多数人脑袋的大小基本相同，因此设计人员做出一个平均尺寸的聚苯乙烯模板。然后利用这个基本形状，确定镜片的位置。用一条蜡塑造出留给镜片的空间，加上几层腻子，建好模型的其余部分。他必须快速操作，因为腻子干得很快。当模型干燥后，蜡被除去，留在模型中就是镜片的框架。设计者现在可以打磨干燥的腻子，实现他所追寻的风格。

设计师还采用先进技术绘制出滑雪镜。在最后时刻进行调整，改善滑雪镜的外观和性能。这对生产的下一步——成形也是有益的。

滑雪镜框用一种被称为注塑法的工艺制成，这些金属块是成形模具。每个模具内有一个空腔，与新滑雪镜的设计形状吻合。熔化的塑料被压进模具，冷却后就会凝固成一副新的滑雪镜。

工人往机器里加进塑料颗粒，加热到 180 摄氏度，现在准备停当，可以注入模具了。机器被加热，使塑料保持液态。设计者希望塑料一旦注入模具中就能凝固。当塑料就位后，泵入冷水，使塑料迅速冷却和硬化。然后就能取出来一副新的滑雪镜框。

在山坡上滑雪的人非常讲究时尚，设计师必须让配色方案恰到好处，该给镜框涂油漆了。用手工把精细的油漆喷涂到每个镜框上。然后发送镜框到烘箱干燥。6 分钟以后油漆被固着到位。

下一个步骤是镜片。现代滑雪镜片由多层材料经过不同化学处理而成。内层来自有色塑料。颜色有助于人眼识别镜片的轮廓和边缘。同时将外层镜片做防雾处理。身体的热量和寒冷的室外温度相遇造成镜片起雾，是滑雪者的一大问题。如果不做这种处理，在寒冷天气滑雪时镜片会完全被雾遮挡，你将看不到任何东西。

滑雪者运动后很热，但仍希望看上去很酷。如果你戴了一副大滑雪镜就不能戴太阳镜，但滑雪时却常常阳光照耀，这是一个问题。因此，滑雪镜被喷涂上好几层硅和铬，可以反射阳光和保护眼睛。看起来也很酷。现在，所有的化学层已被添加到镜片上了，精密的切割机器进行塑形，让它和镜框完美配合。

还记得前面提到的带色内层吗？当镜片被处理和切割的时候，带色层被弯曲，以便和镜片相匹配。经过这台机器，在 60 摄氏度下待 5 分钟，一切都尽善尽美了。

现在我们已经备好所有的零件，剩下的就是把它们组装到一起。要让镜片结合起来，需

你知道吗？

为了在阳光下保护眼睛，古因纽特人用木头、骨头或象牙雕刻成了护目镜。

要把一层泡沫粘在外层镜片和带色内层之间。这样会产生类似于家中双层玻璃一样的热效果。

下一步是让它们看起来漂亮。闪光的零件和附件被添加上去，还有带子，让滑雪镜戴在你头上时不会掉下来。

镜片最终放入安身的镜框。从设计到制造，所有的辛勤工作汇集到了一起，最新的滑雪镜几乎要准备送往山坡了。

滑雪镜制作精良，看起来确实不错，但还要先进行一些测试。无论在你的口袋里还是滑雪摔倒时，镜片必须能抵抗划伤。镜片用一种能找到的最结实材料制成，甚至能经受百洁布的擦拭。

在严重事故的情况下，滑雪镜必须能承受强大的冲击。将轴承钢球以每小时高达 160 千米的速度射出来，如果镜片被打破了，滑雪镜就不合格。

你在哪里使用滑雪镜呢？在山地严寒的条件下。所以，它们被冷冻在 -20 摄氏度至少 2 小时。

无论你是初学滑雪的新手还是越野滑雪的老手，不管下雪还是晴天，这些滑雪镜都将保护着你。

1. 滑雪的安全装备很重要
2. 去掉蜡条按设计打磨模具腻子
3. 塑形机器被加热使塑料保持液态
4. 塑料注入模具凝固成一副新滑雪镜框
5. 镜框送到烘箱干燥
6. 外层镜片做防雾处理

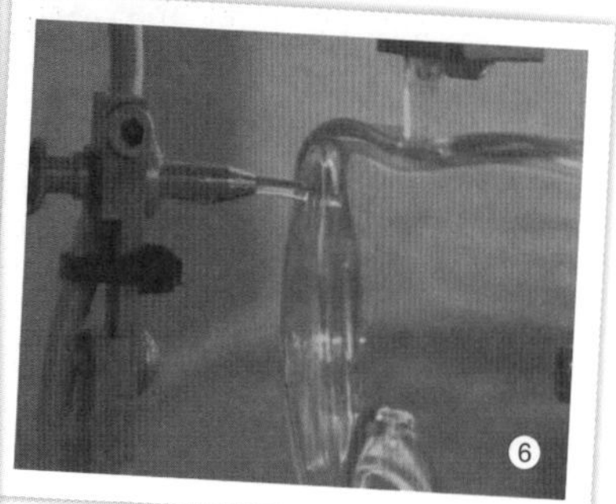

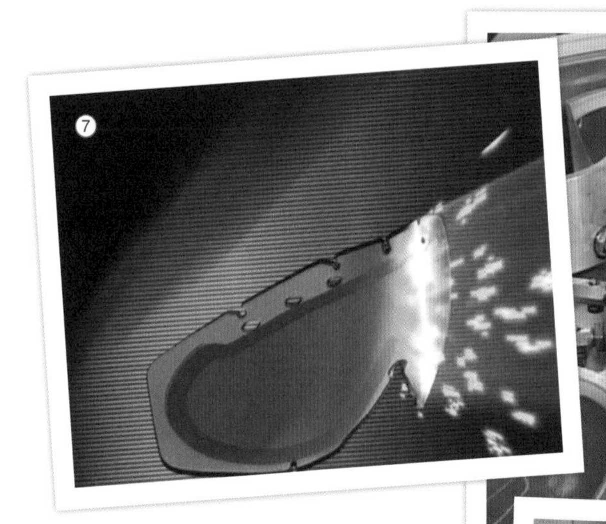

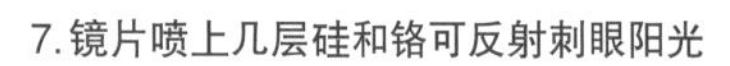

7. 镜片喷上几层硅和铬可反射刺眼阳光
8. 有色内层被弯曲以便和镜片相匹配
9. 把一条泡沫粘在外层镜片和内层之间

10. 开始组装并加上一些小配件
11. 滑雪镜经受钢珠的击打
12. 冷冻后测试不会断裂

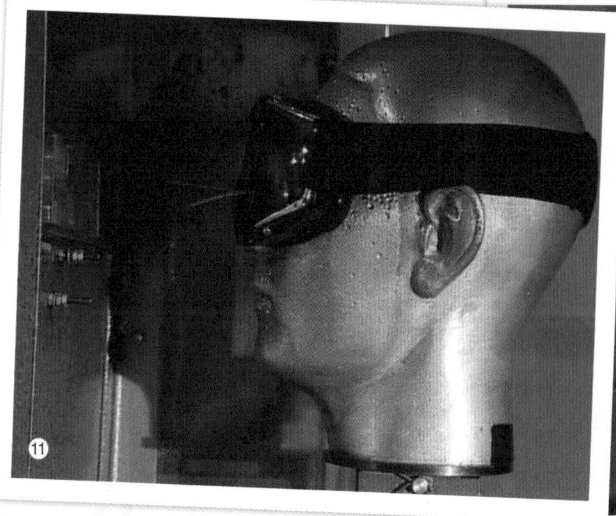

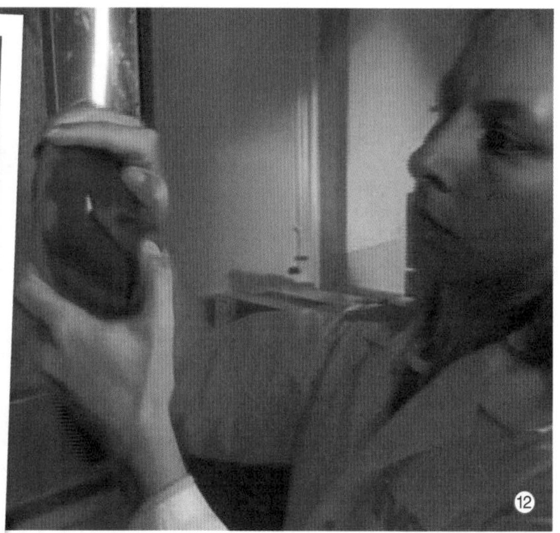

Ski Goggles

Skiing is popular and like all extreme sports, safety gear is vital. The skier that does this, will not only be laughed at by his mates, but will also need some really good protection for his eyes. Every year one and a half million Britons go skiing so there are a lot of people who need googles and getting the right look is important.

Now most people's heads are roughly the same size, so the designer makes up an average sized template from polystyrene. Then using this basic shape, he creates the space where the lenses would sit. Using a wax strip to shape the lens gap, he adds layers of putty to build up the rest of the model. He has to work quickly though because the putty dries fast. When it's dried the wax is removed leaving a frame for the lenses in the model. The designer can now sand down the dried putty to realise the final style he's looking for.

The designer also uses state of the art technology to map out the goggles. This will help him make any last minute adjustments to improve the look or the performance. It's also useful for the next stage of the production process, the moulding casts.

Goggle frames are made using a process called injection moulding and these blocks are the moulding casts. Inside each block there is a hollow space the same shape as the new goggle design. Molten plastic is forced in and when it cools it will solidify into a new pair of goggle.

Workers load this machine with granulated plastic which is superheated to 180 degrees Celsius. It's now ready to be injected into the cast. The machine is heated which keeps the plastic liquid. The designer only wants it to solidify once it's in the mould. Once its in place, cold water will be pumped in which rapidly cools and hardens the plastic. A new pair of goggle frames can then be pulled out.

With fashion on the slopes a high priority, the designer has to get the colour scheme just right so it's time for the frames to get a fresh coat of paint. Each set is air-brushed by hand using a micro-fine lacquer. The frames are then sent to the oven to dry. A 6 minute journey seals the paint into place.

The next stage is the lenses. Modern goggles are made out of a combination of several layers treated with different chemicals. The inner layer is cut from a sheet of tinted plastic. The colour helps the human eye to recognise contours and edges. The outer lens meanwhile is treated with an anti-fogging agent. The combination of body heat and cold outdoor temperatures mean lens fogging is a big problem for skiers. Without this treatment, a hard session in the cold would cause the lenses to fog up completely, and you wouldn't be able to see a thing.

So skiers get hot, but they also want to look cool. If you're wearing a large pair of goggles, you can't wear shades and as it's often sunny when you ski that's a problem. So the goggles are sprayed with several layers of silicon and chrome which reflects the sun and protects the eyes. It also looks pretty cool. Now that all of the chemical layers have been added to the lens, a precision cutter can shape it to fit perfectly into the frames.

Remember the coloured layer from earlier? Well, whilst the lens was being treated and cut, the coloured layer was being bent so it would match the lens. 5 minutes at 60 degrees in this machine and its just right. So now we've got all the bits that we need, all that's left is to put them together. To combine the lenses, a foam layer is stuck between the outer layer and the tinted inner layer. This creates a thermal effect similar to double glazing in a home.

The next stage is to make them look good. Any shiny bits or add-ons are stuck on here as well as the strap so they'll stay on your head.

The lens is finally put into the waiting frames. All the hard work from design to construction has come together and what's left is the latest ski goggles ready for the slopes…almost.

They're well made and they look really good, but there are a few tests to be run first. Whether in your pocket or during a crash, the lens has to resist scratching and because they've been made with the toughest materials possible, they can even resist a heavy duty scouring pad.

In the event of a serious accident the goggles must be able to withstand a heavy impact. Steel ball-bearings are fired at up to 160 kilometres per hour. If the lens breaks the goggles fail.

And where do you use goggles? In the freezing cold conditions you'd find on a mountain. So, they're stuck in a freezer at -20 degrees Celsius for at least 2 hours.

So whether you're a bunny slope beginner or an off-piste animal, these goggles will protect you come snow or shine.

Did you know?

To protect their eyes from the sun, ancient Inuit carved goggles from wood, bone or ivory.

滚轮垃圾桶

酒店床单清洗

滚筒毛巾机

有毒废物处理

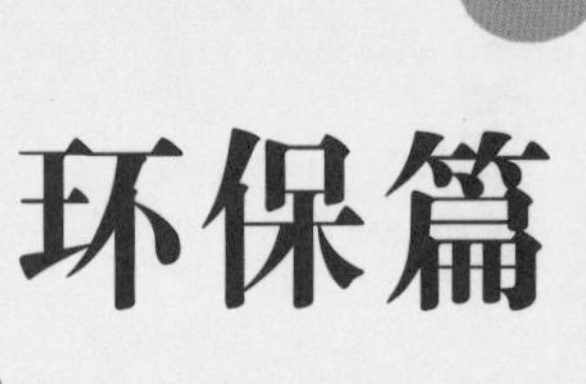

环保篇

汽车回收

废纸回收

电脑回收

轮胎回收

滚轮垃圾桶

扫描二维码，观看中文视频。

每年英国丢弃4.34亿吨之多的垃圾。从塑料包装到报纸，甚至花园中的废物。它们都需要被运到某个地方。第一站并不是垃圾场和垃圾车，而是滚轮垃圾桶。

取代旧式垃圾桶的现代升级版垃圾桶，是彩色的垃圾运送系统。一个关键的设计，是它与垃圾车的升降机相匹配。以前的垃圾桶由金属制成。它适合用于存储人们扔掉的各种垃圾，包括来自家庭壁炉和灶具的热灰。

然而，随着中央暖气和电饭锅的普及，使用壁炉的家庭越来越少，现代垃圾桶不再需要耐热了。不过它们必须很结实。

生产坚固的现代垃圾桶，从这样的工厂开始。这些容器装满了用于制造垃圾桶的原料，颗粒塑料。真空软管把塑料吸入注塑机，急剧加热到熔点。在260摄氏度下，塑料变成液体，并在短短的60秒内塑造成垃圾桶的形状。

英国每年用这种技术生产出约100万个滚轮垃圾桶。注塑成型在这个巨大的机器内运作，成品从另一端出来。正是这种喷射出来的热熔塑料，填充模具的空隙，变成新的垃圾桶。

为了方便回收，欧洲厂商给垃圾桶加上颜色，通过把色素颗粒添加到原料混合物里实现。但这是个渐进过程，产品需要仔细监视。颜色不正确的垃圾桶将被拒收。先要做出几次尝试，再把适当的颜料加进混合原料形成所需要的准确色彩。

最初生产出来的垃圾桶颜色不断加深，直到最终颜色正确并投入生产。不合格的垃圾桶被回收和碾碎，准备日后重新使用。

垃圾桶充分冷却后，可以添加为它们带来绰号的配件了。两个实心轮子安装在金属轴上，使垃圾桶具有机动性。安上盖子，高效的设计只需要用锤子敲击两下就能把它安装到位。盖子把气味封在里面，把狐狸挡在外面。

滚轮垃圾桶20世纪70年代首先在德国出现，直到20世纪80年代才传到英国。这种现代设计的一个有用功能，是便捷的升降系统。意味着清洁工人不再需要用手搬运沉重难闻的垃圾桶了。

滚轮垃圾桶可以有各种形状和大小。不论我们增加破烂回收，或是减少垃圾产生，英国各地的数以百万计的滚轮垃圾桶，会一直帮助我们将废品转移。

你知道吗？

按英国家庭和企业每年产垃圾的速度，只需2个小时就能填满阿尔伯特音乐厅。

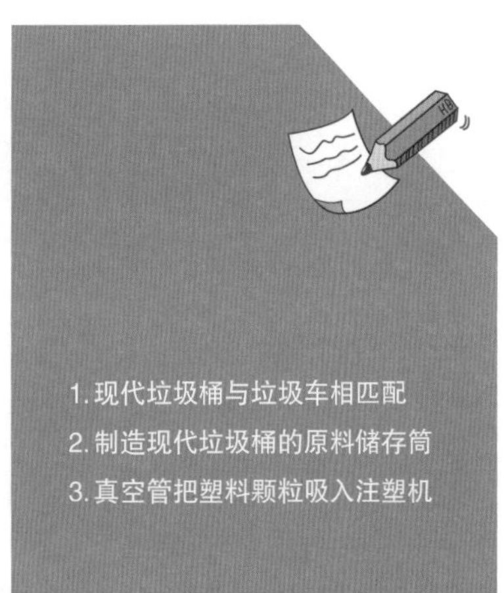

1. 现代垃圾桶与垃圾车相匹配
2. 制造现代垃圾桶的原料储存筒
3. 真空管把塑料颗粒吸入注塑机

4. 急剧加热到塑料的熔点
5. 塑料加热成液态塑造成方桶状
6. 垃圾桶在这里注塑成型

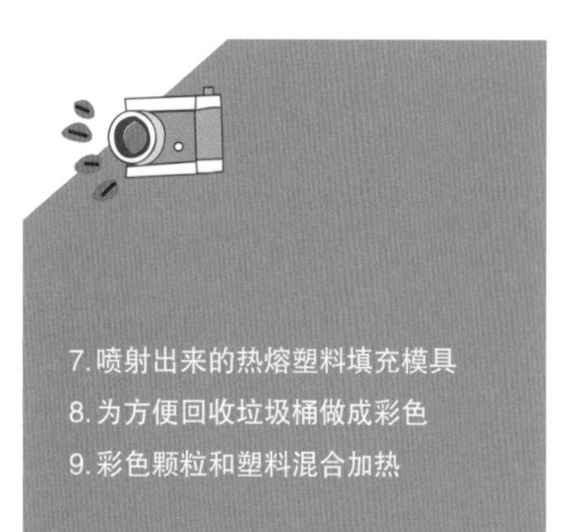

7. 喷射出来的热熔塑料填充模具
8. 为方便回收垃圾桶做成彩色
9. 彩色颗粒和塑料混合加热

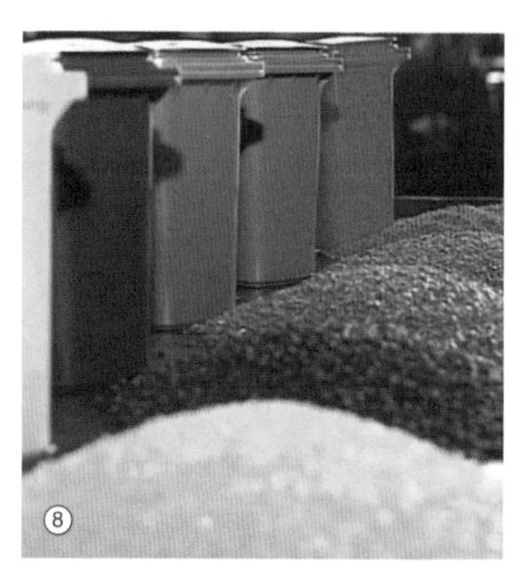

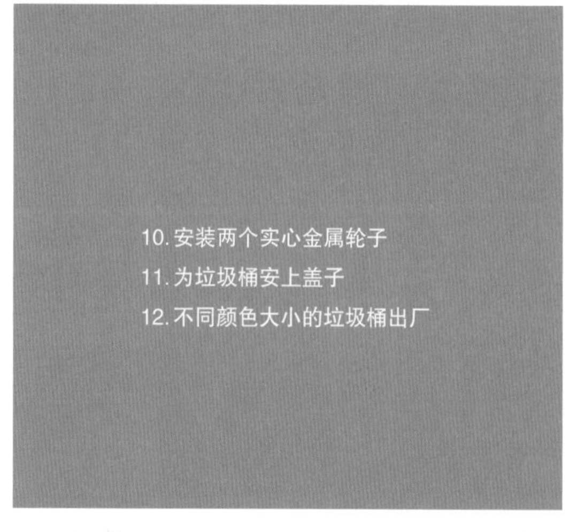

10. 安装两个实心金属轮子
11. 为垃圾桶安上盖子
12. 不同颜色大小的垃圾桶出厂

Wheelie Bins

Every year the UK throws out an enormous, 434 million tons of rubbish. From plastic wrapping, to newspapers and even garden waste. It all has to go somewhere. And the first port of call isn't the dump or the rubbish truck, it's a wheelie bin.

This modern replacement for the old ash can rubbish bin is a colourful upgrade to the rubbish removal system. A key feature to its design is it's compatibility with the rubbish truck's lifting mechanism. Previously rubbish bins were made out of metal. This was ideal for storing all the different rubbish that people would throw out, including hot ashes from home fires and cookers.

However, with central heating, electric cookers and fewer homes with fireplaces , modern bins don't need to be as heat resistant any more. They do however need to be robust.

The production of a modern, rugged rubbish bin starts out at a factory like this one. These containers are filled with the raw material that is used to make the bins; Granulated plastic. Vacuum hoses suck the plastic into the moulding machine where it is superheated to melting point. At 260 degrees Celsius, it's turned into liquid and can be moulded into the shape of the new bin in just 60 seconds.

About 1 million wheelie bins are made in the UK every year using this technique. The injection moulding takes place inside this enormous machine, and each new unit emerges from the other side. This hot jet of molten plastic is what fills each mould to become a new bin.

In order to help with recycling, some European manufacturers colour code their bins. This is done by adding coloured granulate to the basic mixture. However, it's a gradual process. Output is carefully monitored. Bins that are not the right colour will have to be rejected. It takes a few attempts before the coloured granules combine properly in the mix to create the exact shade required.

The first new units to emerge take on progressively darker shades until the right colour is finally produced. The rejected bins are recycled and ground down, ready to be re-used at a later date.

Once the bins have sufficiently cooled they can receive the accessories that give them their nickname. Two solid wheels are fitted on a metal axle to help make them more manoeuvrable. A handy lid is also fitted and an efficient design means it only takes two taps with a hammer to fit it firmly into place. The lid keeps the smell in, and helps keep foxes out.

Wheelie bins were first built in Germany in the 1970's but they didn't make it over to the UK until the 1980's. One useful feature of this modern design is this handy lifting system. It means bin men no longer need to lift heavy, smelly bins by hand.

Wheelie bins can come in all shapes and sizes. But whether we should recycle more, or just make less rubbish, the millions of wheelie bins throughout Britain will continue to help us shift that waste.

Did you know?

UK homes and businesses produce more than 434 million tons of waste every year. At that rate, it would take just 2 hours to fill the Albert Hall.

酒店床单清洗

扫描二维码，观看中文视频。

每天，在全世界的酒店里，服务员都会给客人换上干净的床单。浪漫早餐留在上面的草莓酱污渍，必须在24小时内消失。酒店是如何做到的？

你可能吃惊地获悉，酒店并不自己清洗所有的东西。当你每个月需要面对将近100吨的脏床单、毛巾、桌布和餐巾时，统统送到专业洗衣店去是个更明智的做法。

每天，大约70个酒店替换下来的各种纺织物被运到这里。首先被人工仔细分拣，将带颜色的挑出来。每次洗涤都确保满负荷运作。因为所有床单看起来都差不多，每份待洗织物都由电脑系统控制，以维持工厂顺利运转，并确保酒店能够按时拿回自己的干净床单。

传输带上的每份待洗织物都有一个独一无二的编号，这样就不会和其他酒店的搞混。随后，这份织物就加入队列，开始它的清洗旅程。

到达传输带的顶端后，粘着早餐草莓酱污渍的床单就掉入13个洗衣室中的第1个。第1个到第4个洗衣室进行预洗。后面的4个洗衣室完成主洗。再接着的4个洗衣室进行漂洗。最后1个洗衣室里加入织物柔化剂。从进入第1个洗衣室算起，大约32分钟以后，床单从第13个洗衣室被送出。接着，床单里大部分的水会被一个圆形机器压干。压出来的水被回收。重返洗衣室供下一轮织物使用。

干净床单经过圆形按压机器挤出水分后，看起来像一个巨大的阿司匹林药片。接下来，这个药片状的床单被放入一个巨大的甩干机里，旋转到将干未干的状态。留下的湿气能保证床单熨烫时不会粘到滚筒上。

最终，床单被熨平叠好。它们被推走，装上卡车，原路返回。

这项服务确保在24小时内完成，意味着酒店服务员能为新客人在刚洗干净的床单上，端去另一份浪漫早餐。

你知道吗？

英国第一家自助洗衣店于1949年5月9日在伦敦皇后大道开业。

1. 酒店每天更换床上用品

2. 早餐的草莓污渍须 24 小时内清除

3. 洗涤店工人把带颜色的织物拣出

4. 电脑系统控制流程运转

5. 带污渍的先掉入第 1 个洗衣室

6. 预洗、主洗、漂洗最后加柔顺剂

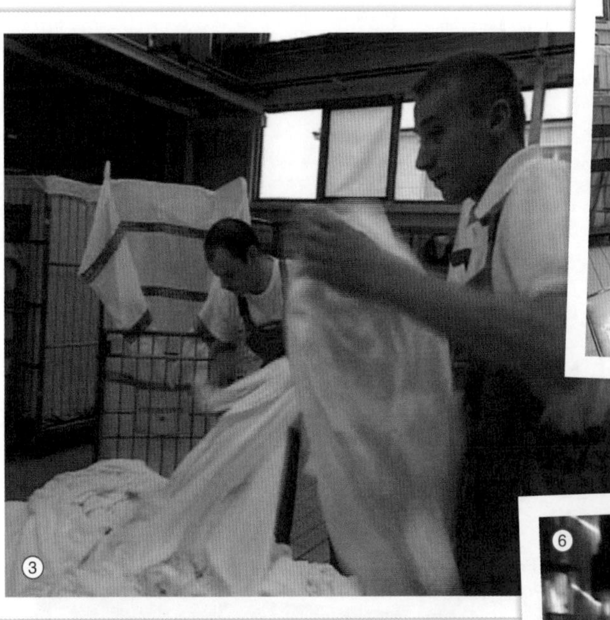

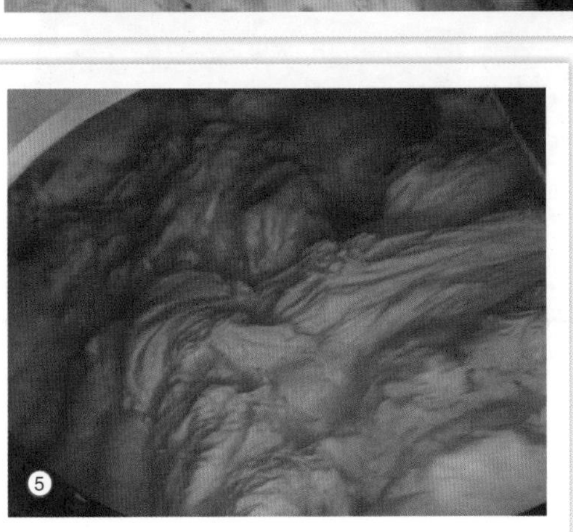

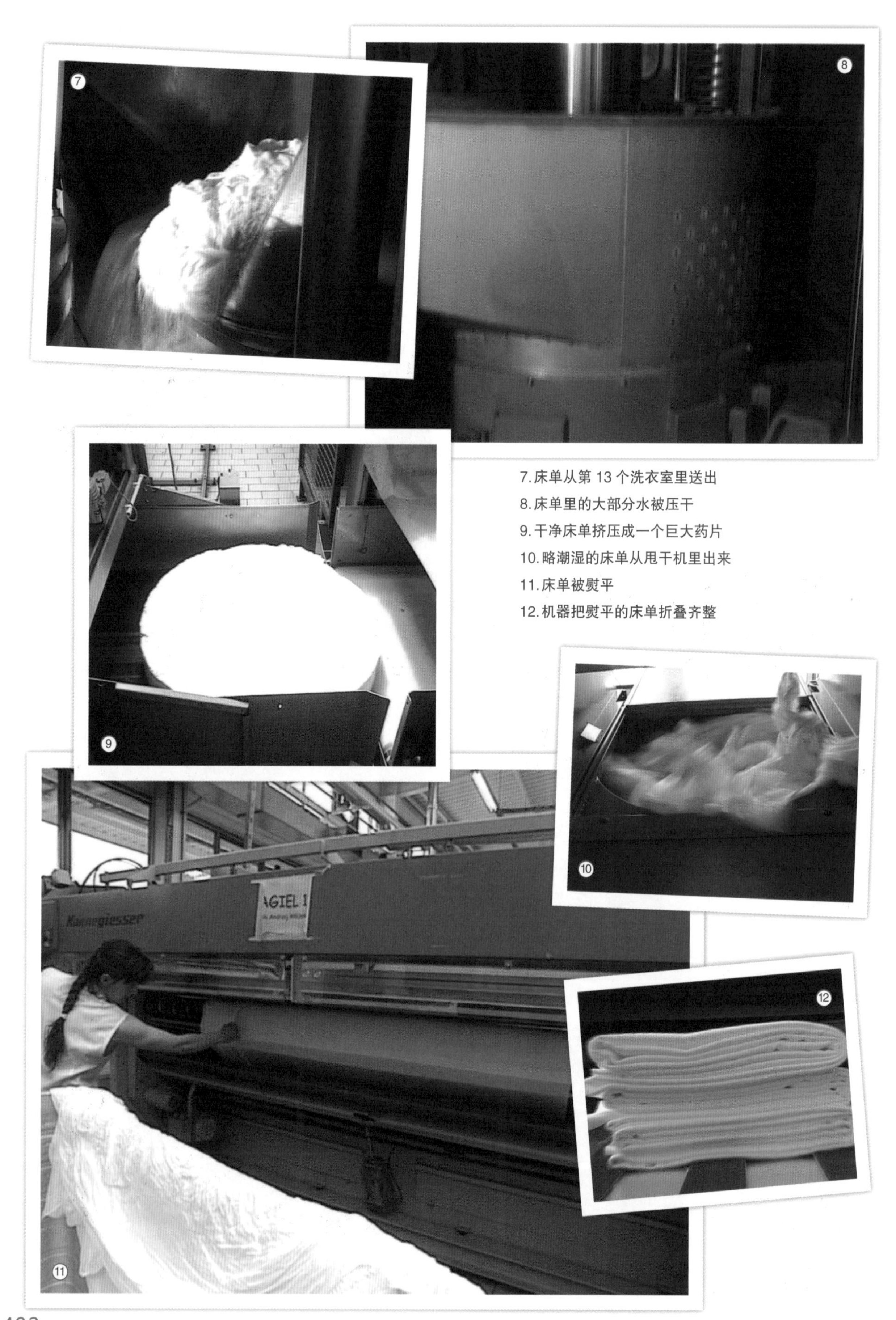

7.床单从第 13 个洗衣室里送出

8.床单里的大部分水被压干

9.干净床单挤压成一个巨大药片

10.略潮湿的床单从甩干机里出来

11.床单被熨平

12.机器把熨平的床单折叠齐整

Hotel Laundrette

Every day in hotels all over the world, maids have to change the sheets and provide clean linen for their guests. Strawberry jam stains from romantic breakfasts have to disappear in 24 hours. So how do they do it?

Well you may be surprised to learn that they don't clean all of this laundry in house. When you're dealing with nearly 100 tons of dirty sheets, towels, table cloths and napkins each month, it makes far more sense to send it all to a specialist laundrette.

Everyday the linen from around 70 hotels ends up here. First of all it's carefully sorted by hand to separate out any coloured items and to ensure a full load for each wash cycle. Because all the sheets look pretty much the same, the loads are managed by a computer system to keep the factory running smoothly and to ensure the hotels get their clean sheets back without delay.

Each load on the conveyor belt receives a unique number, so it doesn't get mixed up with another hotel's laundry. It then joins the queue to begin it's wash cycle.

When it reaches the top, the breakfast stained bed sheets fall into the first of thirteen chambers. Chambers one to four are the pre wash cycles. The next four are the main wash chambers. The following four are for rinsing and in the final chamber the fabric softener is added. Around 32 minutes after they went into the first chamber the sheets are ejected out of chamber thirteen. Most of the water is squeezed out by a circular press. It's recycled and flows back into the chambers to be used again on the next load.

The clean sheets emerge from the press looking like an enormous Aspirin. This pill of sheets is loaded into a giant tumble drier and spun until its nearly dry, but not completely. The bit of dampness that remains stops the sheets from sticking to the rollers while they're being ironed.

The sheets finally appear pressed and folded. They're wheeled away and loaded back into the truck for the return journey.

The service is guaranteed to take no more then twenty four hours, which means the hotel staff can service another romantic breakfast to the next guests on fresh clean sheets.

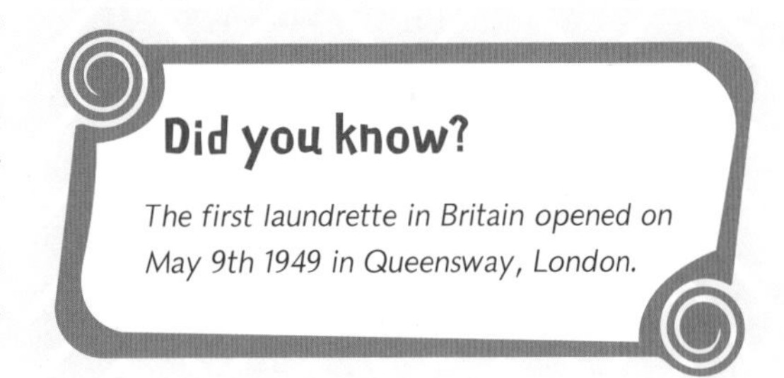

Did you know?

The first laundrette in Britain opened on May 9th 1949 in Queensway, London.

滚筒毛巾机

扫描二维码，观看中文视频。

当你在公共厕所洗手时，会面临一个年代久远的两难困境。如果你用毛巾擦手，它们可能不很干净。吹风机又似乎永远不胜任这件事。

最好的解决方案之一是滚筒毛巾机。每个人都能得到新的卫生毛巾，这确实解决了问题。但如果毛巾用完了呢？

这时，毛巾机公司的维护人员会来收取旧的脏毛巾，把它送回工厂。每天来自全国的15000条毛巾运到这样的巨大工厂。它们在这里被翻新。

首先第一件事，每一条脏毛巾仍然是卷着的，它们需要完全打开进行清洗。毛巾卷被拉开，然后排列整齐。这样肥皂就能洗到毛巾各处了。为了防止清洗时缠在一起，每条毛巾都被扎起来。它将和其他数百条准备好的毛巾一起，送到洗衣机里。

当下一台机器准备停当时，桶被打开。然而，就像不爱洗澡的孩子，毛巾似乎很勉强，但随着水的压力增大，最终把毛巾推了进去。在90摄氏度时，这台机器洗涤毛巾1千克，只使用4克洗衣粉，它每小时能清洗约1000千克的毛巾。比你家里的洗衣机更有效率。

现在烘干。首先挤压，除去所有的水。像大药片一样的毛巾块被挤干了，但也像石头般坚硬。接下来的机器会解决这个问题。大块干净毛巾被旋转并松开，互相再次分离。

工厂现在有了数百米干净但发皱的毛巾。该把它们再次卷起来了。捆绑的束带被剪断，把毛巾装到下一台机器上展开。接着毛巾被送进这个设备。这是最重要一步，因为它确保毛巾不会扭曲而堵塞机器。

用一根假想的线，这里以红色显示，计算机监视通过的毛巾。如果越过这条线，机器就会稍做移动让毛巾走直，这有助于避免意外的堵塞。

现在的毛巾可以用真空力量重新卷好。这个装置中的孔连接到抽气泵。当毛巾经过时被吸到孔上，随着金属棍的旋转，毛巾就绕着它卷起来。贴上一个标签告诉工作人员毛巾已经清洁完毕等待出厂。然后，毛巾被重新装入笼子里，准备装填全国各地的毛巾机。

每条毛巾大约有100次的洗涤寿命，但这一期间的任何时候毛巾都可能损坏。如果出现损坏，操作工将标记这条毛巾，并把它送到另一个部门。在那里，它将被修补或变成清洁布。

有了满满一笼新鲜毛巾，维护工把它装上面包车，按着自己的路线开出去装填毛巾机了。取掉翻新毛巾的标签，重新填入机器。现在不必在裤子上擦手了，有一条干净卫生的毛巾正在恭候你。

你知道吗？

据估计，全英国大约有18500千米长的毛巾被安置在厕所和洗手间里用于擦手。

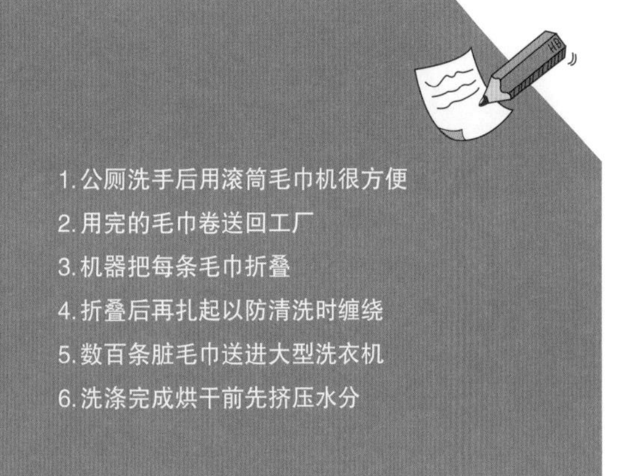

1. 公厕洗手后用滚筒毛巾机很方便
2. 用完的毛巾卷送回工厂
3. 机器把每条毛巾折叠
4. 折叠后再扎起以防清洗时缠绕
5. 数百条脏毛巾送进大型洗衣机
6. 洗涤完成烘干前先挤压水分

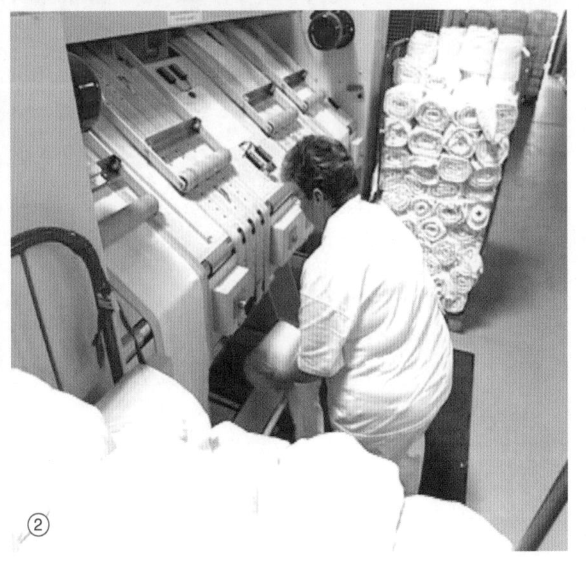

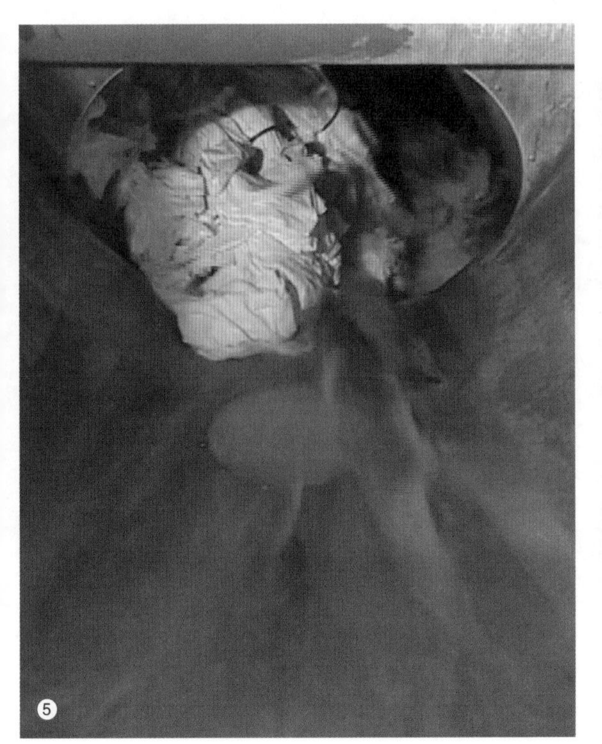

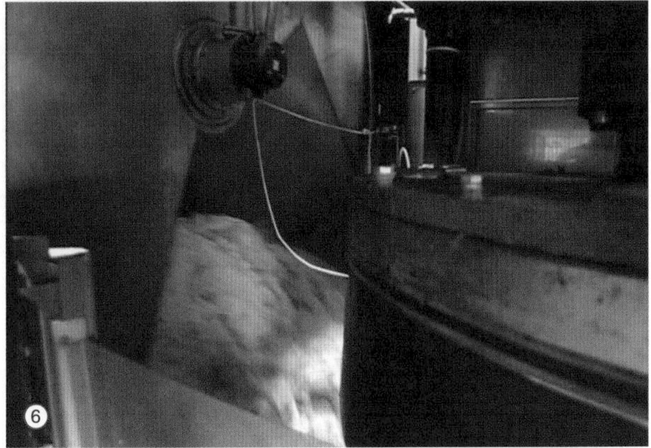

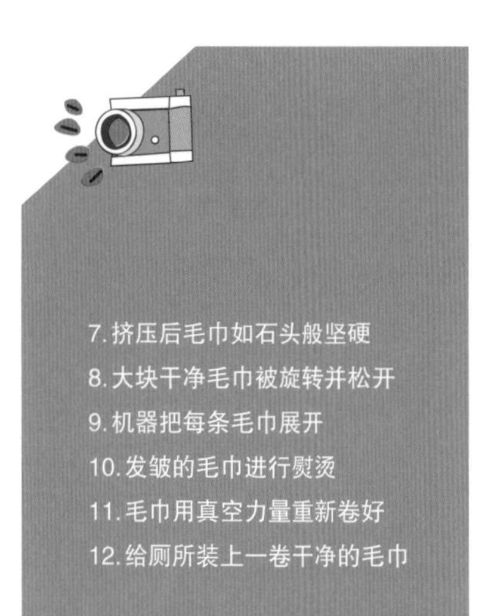

7. 挤压后毛巾如石头般坚硬
8. 大块干净毛巾被旋转并松开
9. 机器把每条毛巾展开
10. 发皱的毛巾进行熨烫
11. 毛巾用真空力量重新卷好
12. 给厕所装上一卷干净的毛巾

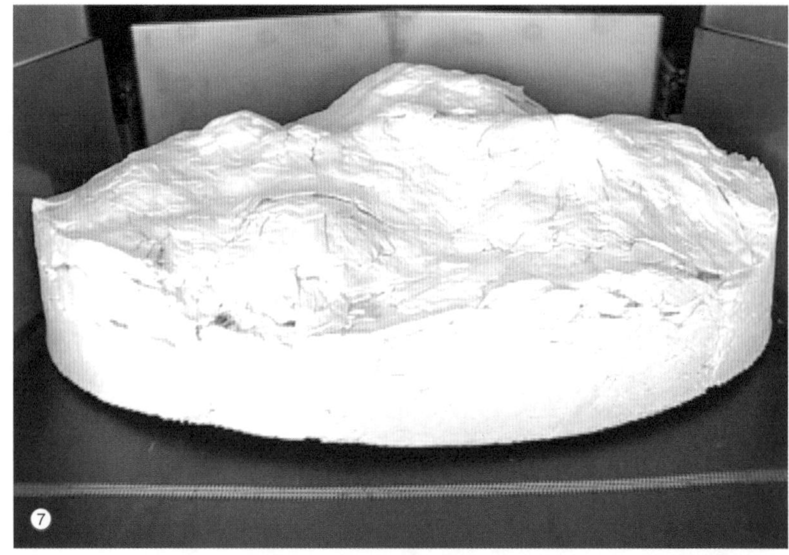

Towel Dispensers

It's the age old dilemma you face when you wash your hands in a public toilet. If you use the towel, it's probably not very clean. The electric dryer never seems to do the job.

But one of the best solutions is the roller towel dispenser. Fresh, hygienic towel for each new user and it does the job. But what if the towel runs out?

At that point the maintenance man from the Towel Dispenser company arrives and collects the old dirty towel to take it back to the plant. 15,000 towels are brought from all over the country to an enormous facility like this one every day. Here they will be refreshed.

First things first, each dirty towel is still rolled up and they need to be completely unravelled to be washed. The rolls are pulled out and then laid out neatly. This way the soap can reach all of the towel. To stop several getting tangled up in the wash, each one is now cable tied. It will join hundreds of others ready to be sent to the washing machines.

When the next machine is ready, the bucket is opened. However, like children who don't want a bath, they seem a bit reluctant to get in. But, the water pressure mounts up and eventually forces them inside. At 90 degrees Celsius, this machine uses just 4 grams of washing powder for every kilo of towels and it can rinse through about 1000 every hour. A bit more efficient than your machine at home.

Now for the drying. First they're crushed, this removes all of the water. The enormous pill-like block that emerges is dry but it's also solid as a rock. The next machine takes care of that though. The disc of clean towels is spun and loosened which separates them all once more.

The factory now has hundreds of meters of clean, but wrinkly towels. It's time to roll them back up again. The cable ties are cut and the towels fitted to the next machine which unravels them. Next they will be fitted to this device. It's vital for one important reason. It makes sure the towel isn't twisted which would jam the machines. Using an imaginary line, seen here in red, a computer watches the towel as it passes. If the towel crosses this line, the device tugs a bit which straightens it out and which helps to avoid unwanted jams.

Now the towels can be re-rolled using vacuum power. The holes in this device are connected to a pump which sucks air into them. When a towel passes over, it gets sucked onto the holes and as the bar spins, the towel is then rolled around it. A sticker is applied telling the staff that it's clean and ready to go. They are then reloaded into the cage ready to refill towel dispensers all over the country.

Each towel has a life-span of about 100 washes, but damage can occur at any time during that period. If a damaged one appears, the operator will mark this towel and it will be sent to a different department. Here it will be repaired or turned into cleaning cloths.

With a cage full of fresh towels, the maintenance man will load it back up into his van and head back out to refill the dispensers on his route. The sticker showing it's a fresh new towel is removed and it's reinserted into the machine. Now instead of wiping your hands dry on your trousers, you've got a clean hygienic towel ready to dry them for you.

Did you know?

It's estimated there are nearly 18,500 kilometres of towel used to dry hands in toilet and washrooms across the UK.

有毒废物处理

扫描二维码，观看中文视频。

这座电梯通往几乎 0.5 千米深的地下。门打开后，你将驾车驶过长长的通道。这个地下旅程的终点是什么呢？这里是存储有毒废料的地窖，由于太危险而不能建在地面上。这些工业垃圾场如果建在地面，将会污染土地和水。从重金属到有毒化学品都存储在这里。这个故事从德国南部萨克森安哈尔特州开始，那里有座生产钾盐的地下矿井。钾盐含钾，可以用作肥料。采矿有 2 个好处。首先是钾盐本身，其次是开采后留下的巨大洞穴。

钾盐矿在数百万年前形成，内陆海蒸发后留下矿物沉积层，今天在地下深处被发现。开采钾盐是一个简单的过程。钻孔后安放工业炸药，撤离矿井然后引爆炸药。当尘埃落定，矿工驾驶大铲车回到采矿面，铲起松散的钾盐，送到地面加工成肥料。采矿结束后，有毒废料运达。它经过处理后，被存储在地下洞穴中，像往日的钾盐一样。这批废料是砷，对人的健康十分有害，深埋之前要在实验室进行测试。科学家通过检查确定没有液体混在废料中。任何泄漏都将是灾难性的。接下来测试气体，打开袋子之前需要戴面具。废料中某些气体积聚会引起爆炸，这是储存有毒废料的另一个危险因素。检查无误后砷就能在地下深处垃圾场储存了。

但送出之前，还有最后一件事要做。科学家进行取样。到矿下就会明白这样做的原因了。将砷装在巨大的、清楚标记的容器中，准备运往地下，随着垃圾运往它们最后的归宿地，科学家把样品带到地下资料库。这是记录部门，但取代账本的是罐子，它们是这里存储的每批废料的样品。这个存储系统的功能像地图，科学家们能知道储存了什么废物，以及在哪里找到它。他把最近一批砷废料的样品添加到队列末尾。

同时垃圾场仔细准备接受新的废料。首先除去顶板上松散的留存物。石块掉下来会损坏专用的密封容器，任何泄漏都会招致灾难。接下来工人在顶板上钻孔，用螺栓拧紧特殊的强化钢筋，其作用像额外的保护层防止顶板坍塌。钢筋钻入坚硬的岩石层约 1.2 米。牢固地进抵到位后钢筋末端张开，如同膨胀螺丝抓紧坚硬的岩石，稳定松散层物质。这种额外的强化措施能在钢筋周围 80 厘米直径的面积内，承受住约 14 吨的岩石。整个矿井已经安装了几百万根这种有效的强化钢筋。

除了顶板，地面也需要准备。有专门的车辆把松散的碎石平整成道路，让有毒垃圾运输车安全往来。整个矿井有数百千米道路，如果你不认路就很可能会迷路。顶板固定完毕，道路准备停当，就可以运入废料了。由专门的拖拉机将袋子举起，使废料与人体的接触减少到最低限度。作为现代工业常见的副产品，有毒废料正成为一个日益严重的问题，解决方案之一是深埋在地下，如这里看到的，像沉睡万年的钾盐那样。

你知道吗？

科学家发现了消耗有毒废物的细菌，叫作 BAV1。它们吃氯乙烯——一种用来制造纸和塑料的化学物质。

1. 电梯通往地下 0.5 千米深的矿洞
2. 下了电梯驾车驶过长长的通道
3. 最终来到存储有毒废弃物的矿洞
4. 这是座生产钾盐的矿井
5. 在井壁上钻孔安放炸药
6. 对人体有害的砷废弃物运达

7. 砷装在有标记的容器里运往地下
8. 科学家带着样品到地下资料库
9. 装有样品的罐子取代了账本
10. 除去矿洞顶上松散的留存物
11. 整个矿井安装几百万根钢筋
12. 有毒废弃物被码放到位

Toxic Waste Disposal

扫描二维码，观看英文视频。

This lift goes almost half a kilometre underground. When the doors open, you have a long drive ahead of you. But what is at the end of this subterranean journey? This is where you'll find the vault that stores all the toxic waste that's just too dangerous to leave lying around on the surface. Dumps like these contain industrial waste that could contaminate land and water if left above ground. Everything from heavy metals to poisonous chemicals is stored here. This story begins in Saxony Anhalt, Southern Germany, where potash is extracted from underground mines. Potash contains potassium which is used as a fertiliser and the mining process has 2 benefits. Firstly the potash itself, and secondly the large holes it leaves behind.

The potash was laid down millions of years ago. As an inland sea evaporated it left layers of mineral deposits. Today these layers are found deep underground. Mining the potash is a simple process. Drills are used to prepare holes ready for industrial explosives to be inserted. The mine will then be evacuated and the explosives detonated. When the dust has settled, the miners will head back to the face with big tractors. They scoop up the loose potash, and it's sent above ground to be processed into fertiliser. When the mining's done the toxic waste arrives. It too will be processed ready to be stored far below ground in the hole where the potash used to be. This load is arsenic which can be very hazardous to people's health. Before it's buried, it's sent to the lab to be tested. Scientists check there is no liquid mixed in with the waste. Any leakage could be disastrous. The next test is for gases, so a protective mask is needed before opening the bag. The build up of certain gases in the waste could cause explosions, yet another danger factor in the storage of toxic waste products. Once checked and cleared, the arsenic is ready to be stored in the dump deep below ground.

But before it's sent off, there's one last thing to do. The scientist takes a sample. The reason why will become clear down in the mine itself. Loaded into large, clearly marked containers, the arsenic is ready to be taken underground. With the waste heading for its final resting place, the scientist takes his sample to a kind of underground library. This is the records department, but instead of books, these jars contain a sample of every waste product stored down here. The storage system acts like a map so the scientists know what kind of waste is stored and where to find it. He adds the sample from the latest batch of arsenic to the end of the line.

Meanwhile, the dump is being carefully prepared for its latest delivery. First any loose material is removed from the ceiling. Rock falls might rupture the specially sealed containers. Any leaks could be disastrous. Next the workers will drill holes in the ceiling and bolt in special reinforcing steel rods. This will act as a further layer of protection to prevent the roof collapsing. The rods penetrate about 1.2 metres into layers of solid rock. Once they're firmly fitted into place, the rod ends splay out like wall plugs, gripping into the solid rock layer to stabilize material in the loose layer. This extra reinforcement can help hold up around 14 tons of rock in an 80 centimetre diameter around the rod. The mine contains several million of these helpful reinforcements already.

As well as the ceiling, the floor is also being prepared. Special vehicles flatten the loose rubble into a stable road, so that the toxic waste transporter can come and go safely. There are hundreds of kilometres of roads throughout this mine and it is possible to get lost if you don't know your way around. With the ceiling stabilised and the road prepared, the waste can now be brought in. Special tractors pick up the bags, keeping human contact to a minimum. Often a by-product of modern industry, toxic waste is an increasing problem, and one solution is to bury it deep underground, in this case, where the potash used to be.

Did you know?

Scientists have found bacteria that consume toxic waste. Called BAV1, they eat vinyl chloride, a chemical used to make paper and plastic.

汽车回收

扫描二维码，观看中文视频。

人们很容易对旧汽车产生依恋，毕竟它是多年的伴侣。但是，当修车账单开始堆积起来的时候，该对旧车说再见了，为什么不让你的老朋友去做件有益的事呢？

这样的回收站为旧车支付现金。它们卸除零部件并回收碎金属。

每辆旧车里都会有油和工业液体，所以要进行钻孔，以便安全处置这些危害环境的液体。然后拆卸工开始摘除任何可以卖给其他司机的零件。

如果有人买这扇门，它就会把一辆3扇门的残骸变成一辆4门家庭轿车，和它从前一样。

对于想节省几英镑的车主来说，这些车场是阿拉丁的藏宝洞。不必购买昂贵的新部件，他们可以用几分之一的价钱买到旧货，并且为保护环境做出点贡献，因为回收节约了宝贵的能源和物力。

一旦回收了所有的可利用部分，旧车该回去见制造商了。汽车爱好者们现在最好不要看。

一台重型起重机把旧车放在破碎机里。所有的小男孩都喜欢砸坏玩具汽车，不过这个人能粉碎真汽车，最多一天50辆。

滑动铁墙用600吨的力量把汽车压缩成一块废金属。几分钟前刚吊走厢式车的起重机，现在把这堆碎金属吊起来放上火车，送它登上最后的旅程，去往回收工厂。这些车永远不会再开上乡间小路了，但也有好的方面，它们也永远不会再看到交通管理员了。

不管是阿斯顿·马丁跑车，或是奥斯汀·阿莱格罗系列，它们都受到同样的待遇。一个切刀把它们变成小块，以便放入粉碎机。这些曾经的汽车都被抬起来放上传送带，耐心等待着自己的生命被挤压出来。不管在辉煌时期曾经怎样精心打造，它们无论如何也顶不住2000马力的粉碎机。当你看到从粉碎机另一侧出来的是什么，就会知道为什么它被称为废金属了。

钢铁通过磁鼓从其他材料中分离出来。铝也被筛选出来，剩下的就是垃圾，它们将被送往填埋场。保留下来的金属聚成一堆。这是回收厂的宝贝。

巨大的拖拉机把废金属装到火车上，运到加工厂。在那里，它们会变成洗衣机、烤箱、桥梁，或者变回一辆车。

每年有大约两万辆汽车在英国被回收，每个车质量的大约80%将以另一种形式得到再使用。所以，当你忠实的朋友无法修复时，不要太伤心，这不是它生命的结束，而只是一个新生命的开始。

你知道吗？

英国汽车的平均寿命为13.5年。

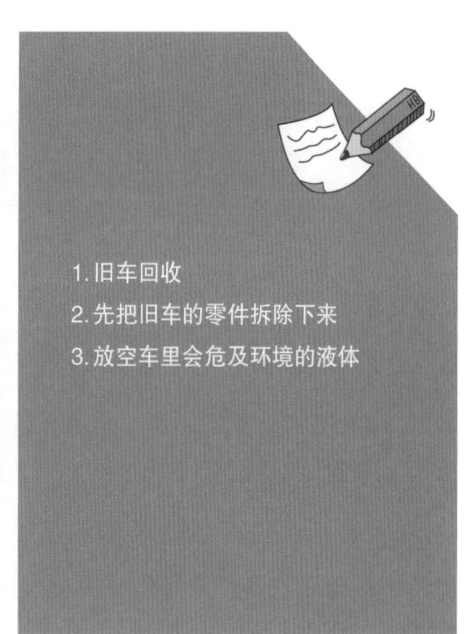

1. 旧车回收
2. 先把旧车的零件拆除下来
3. 放空车里会危及环境的液体

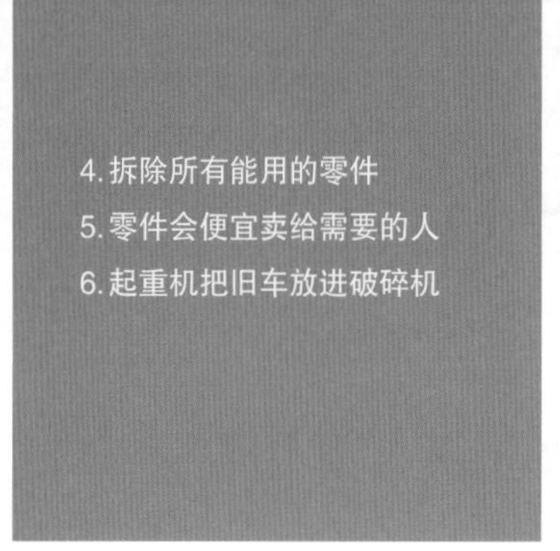

4. 拆除所有能用的零件
5. 零件会便宜卖给需要的人
6. 起重机把旧车放进破碎机

7. 滑动铁墙把车压缩成废金属块
8. 火车把废金属块运到回收厂
9. 切刀把它们切成小块

⑦

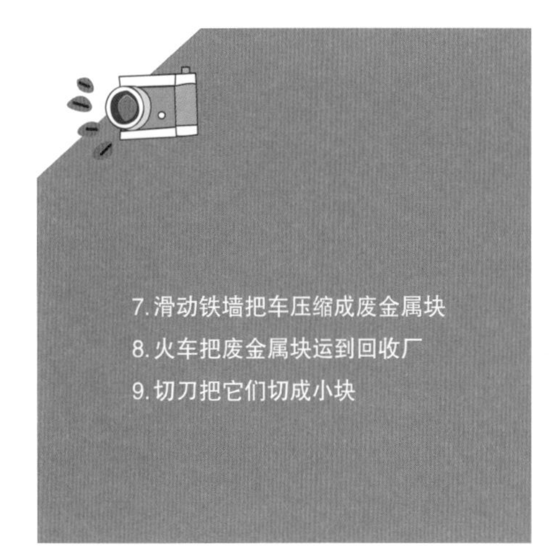

⑧

⑨

⑩

10. 金属碎块进入 2000 马力的粉碎机
11. 磁铁把钢铁分离出来
12. 车上的废金属将得到再利用

⑪

⑫

Car Recycling

It's easy to become attached to your old car, after all it's been your companion for many years. But when the garage bills start to pile up it's time to say goodbye so why not let your old friend go to a good cause.

Breakers yards like this one pay hard cash to strip away parts and recycle the strap metal.

Every used car contains oils and industrial fluids, so holes are drilled to safely dispose of the environmentally harmful liquids. Once that's done the breakers get on with stripping anything that can be sold on to another motorist.

When someone buys, this door it will transform their three door wreck into the four door family car it once was.

These yards are an Aladdin's cave for car owners who want to save a few quid. Instead of buying expensive new parts they can pick up old ones for a fraction of the cost and do their bit for the environment because this recycling saves valuable energy and resources.

Once they've salvaged all the parts they can it's time for the car to meet it's maker. Motor enthusiasts look away now.

A heavy duty crane places it into the jaws of the crusher. All little boys enjoy to smash up toy cars, but this guy gets to pulverize real ones, up to 50 times a day.

With a force of 600 ton the sliding walls reduce the vehicle to scrap metal. The crane that heaved up a hatch back minutes ago now lifts a heap of crushed metal onto a train for it's last journey, to the recycling yard. These cars will never speed down a country lane ever again but there is an up side, they'll never see a traffic warden again either.

No matter if it was an Aston Martin or an Austin Allegro, they all get the same treatment. A cutter chops them up into bite sized pieces so they'll fit in the shredder. The ex-cars are lifted up onto a conveyor belt where they patiently wait to have the life squeezed out of them. However well built they were in their heyday none of them are any match for the shredder with its 2000 horse power engine. When you see what comes out the other side of the shredder you can really see why it's called scrap metal.

Steel is separated from the other materials by this magnetic drum. After the aluminum has also been sifted out what's left is junk and this will be buried in a land fill site. All the metals that have been saved fall onto a vast pile. This is the treasure of the recycling plant.

Enormous tractors load the scrap metal onto trains and it's then taken off to processing plants. There it'll be turned into washing machines, toasters, bridges or maybe just back into a car.

Every year around two million cars are recycled in the UK and about eighty percent of the mass of each car is used in another form. So don't be too down hearted when your loyal friend is beyond repair, it's not the end of its life, it's just a new beginning.

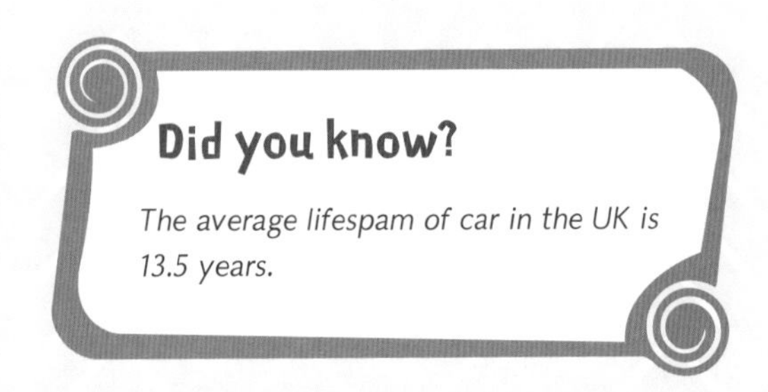

废纸回收

扫描二维码，观看中文视频。

废纸回收已经变得越来越流行。现代观念让更多人把废报纸、杂志和包装纸回收再利用。这也算为地球做一点事情。

英国平均每年使用超过 1100 万吨纸做包装和阅读材料，大约 40%最终来到这样的工厂。但并不在此回收利用。这里做的只是第一步，分类。

从夹层的信封到光洁的杂志都被送到回收站。很多不同类型的纸张让回收商大伤脑筋。他们需要归类，并把报纸增刊这样的东西分开，它们经常裹在塑料中，必须要除去。

夹层信封和硬纸板也堆在一起。要除掉它们，材料需要通过一系列滚筒。较轻的纸张落下，纸板、信封和包装留在上面。

对回收商来说不轻松的是，还要经过好几个分类阶段，才能获得纯粹的纸进行处理。最初分类后，一些彩色卡片仍会留下来。对颜色敏感的摄像头检查通过的东西，发现不对劲，就及时用气流把它从传送带上吹掉。这个分类设施使用了一些高科技设备，但归根结底，人力仍然是做这项工作的最佳选择。

一旦再生材料分类完毕后便来到这里。纯废纸可以准备回收了。利用废纸制造新纸的第一步是打纸浆。需要用一系列的化学物品把废纸粉碎。废物要再过滤一次，因为即使在这个阶段，仍然可能有杂质漏网。

混乱的东西被筛除后，工人得到灰色的纸浆。要想做成新的再生纸就必须进一步清洁。纸浆仍含有原来纸张上的印刷油墨，要加进化学品和肥皂除掉它。将空气泵进混合物中，油墨粘到肥皂泡上并浮到表面，在这里被除去。

这是一台把纸浆变成新纸的机器。它有 120 多米长，包括多种设备，用纸浆造出全新的 A4 纸。听起来很难相信，它只需要 20 秒就能让纸浆变成纸。纸浆进入机器时，含有 99%的水和仅 1%的纤维，因此必须去掉水。纸浆被摊开，通过有吸水布的滚筒。布吸收大部分的液体并在下面拧出。然后返回到机器再重复这一过程。造出的纸通过一系列加热辊。130 摄氏度可以很好地将它们烘干。

尽管纸浆此前已经清洗过，加工后的纸仍是灰色，所以要上色。两个涂有白色颜料的滚筒赋予它传统的整洁外观。

纸张再经过空气干燥器，就像你的工作上衣，它接下来要熨烫，抚平新纸张上的皱纹。最后，新做的纸张被卷到巨大的滚筒上。生产出的最大纸卷重达 30 吨。

强度是一个重要因素，决定新的纸张用于何处。书籍用的纸张级别远高于报纸。一本书可能被阅读很多次，而报纸只用两三次。

这台机器测试样品，以划定质量等级。它们被包装起来，在客户需要这些新库存时发送出去。但首先要将纸卷切成合适的尺寸。例如，图书出版商需要的规格与包装纸公司

你知道吗？

每吨再生纸能让 17 棵树免于砍伐和制浆。

的要求不同。

整齐切开的纸卷慢慢滚动着离开了铡刀，被用更多的纸包裹起来。这个保护层让新制造的回收纸保持干净。最后将纸卷存储起来，直到交付使用。

虽然废纸回收是环境保护的一大进步，但纸的回收至多五六次，此后就变得太脆了。

当回收公司把你的旧报纸变成全新的 A4 纸时，为环境保护所做的一切都功德无量。

1. 废纸首先运到回收站分类
2. 通过系列滚筒较轻的纸张落下
3. 摄像头发现要剔除的彩色卡片
4. 人工分类仍然是最佳选择
5. 加进化学品去除纸浆中的油墨
6. 纸浆通过 120 米长的造纸机器

7. 摊开的纸浆通过吸水布滚筒
8. 拧出吸水布的水重复使用
9. 造出的纸通过一系列加热辊烘干
10. 白色颜料滚筒为纸上色
11. 滚筒上的纸张抽样检查强度
12. 切好纸卷包装存储

Paper Recycling

But first, paper recycling has grown in popularity. Modern attitudes mean more of us are returning our wasted newspaper, magazines and packaging to be used again. Doing our bit for the planet sort of thing.

In an average year Britain uses over 11 million of tons of paper as packaging and reading material, and about 40% of it ends up at a facility like this. But, it's not recycled here. This is the first step where it's sorted.

Everything from padded envelopes to glossy magazines are sent to be recycled. That's a lot of different types of paper and this creates a big headache for the recyclers who need to sort and separate things like newspaper supplements. They're often wrapped in plastic. This has to go.

Padded envelopes and cardboard are also found in the pile. To remove this, the material is now passed over a series of rollers. Lighter paper falls through leaving the cardboard, envelopes and packaging on top.

Unfortunately for the recyclers there are still several more sorting stages to go through to get pure paper they can work with. After initial sorting some types of coloured card may have still be left. A colour sensitive camera checks what passes by and if it's not right a well-timed blast of air removes it from the conveyor. This sorting facility uses some quite high tech equipment, but at the end of the day, people power is still the best filter for the job.

Once all the recycled material had been sorted, it should end up here. Pure waste paper ready to be recycled. The first stage of making new paper from old is to pulp it. This is done by using a range of chemicals to break it all down. The waste is filtered once more because even at this stage impurities may have slipped through.

This clutter is sieved out and the workers end up with is a grey slop. This has to be cleaned if it's to be transformed into fresh recycled paper. The pulp still contains inks from the original paper print. To remove it, chemicals and soap are added. Air is pumped through the mix and the inks stick to the soap bubbles which are rising to the surface. Here they can be removed.

And this is the machine that will turn the pulp into fresh paper. It's over 120 meters long and contains a variety of devices to help turn sloppy pulp into pristine A4. Now, it may sound hard to believe, but it only takes 20 seconds to turn pulp into paper. When the pulp enters the machine, it's 99% water and just 1% fibre, so the water's got to go. The pulp is spread out passed through a roller with an absorbent cloth. The cloth captures most of the liquid which is wrung out below. It then returns to the machine to repeat the process. The sheet of emerging paper is now passed through a series of heated rollers. 130 degrees Celsius dries the paper perfectly.

Now, even though the pulp was "cleaned" earlier, the processed paper is still grey, so now it's painted. Two rollers layer on white ink giving it its traditional clean appearance.

The sheet then passes over air driers and, like your work shirt, it then gets a good iron. This smoothes out any wrinkles the new paper may have. Finally the fresh sheet is rolled onto enormous rollers. The biggest ones produced can weigh up to as much as 30 tons.

Strength is an important factor in deciding what the new paper will be used for. Books use a far higher grade of paper than newspapers. A book may be read many times where a newspaper is only used 2 or 3 times.

This machine tests samples so they can be assigned a quality grade. It can then be packed up to be sent to the right client when they need fresh stock. But first the roll will be cut down to the right size. Book publishers need a different size of paper from a company making wrapping paper for example.

The neatly sliced rolls tumble slowly off the guillotine and are sent off to be wrapped in yet more paper. This protective layer will keep the freshly recycled stock clean. Finally the rolls are stored until they are ready to be delivered.

Although paper recycling is a big step forward for the environment, paper fibres can only be recycled 5 or 6 times before they become too brittle.

But the savings for the environment make it all worthwhile as the recyclers turn your old newspaper into brand new A4.

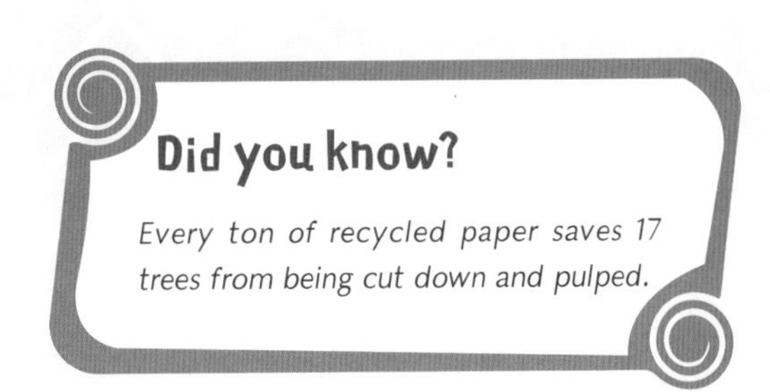

电脑回收

扫描二维码，观看中文视频。

很多人以为，当电脑不能工作后，唯一用途就是当个镇纸，让书桌杂乱。但他们大错特错了。有人正开创一项新的回收业——从旧硬盘里找到黄金。

你的旧电脑里隐藏着一个贵金属的宝库。电路板中含银和铜，微处理器含金，这些都是良导体，用于在电脑里传送电信号。旧电脑的贵金属被送到工厂回收。首先去掉电脑外壳。电缆被拔掉另收。剩下的就是芯片和处理器了。这些部件崭新的时候曾经很值钱。但是现在，如果被扔到垃圾堆，就一文不值了。

这家公司利润颇丰。一吨旧电脑可带来6500英镑的收益。微处理器最容易回收，因为它们所含的唯一金属是金。一种盐酸和硝酸的混合液，用来把金从陶瓷中分离出来。成百的微处理器放到一个大桶里同时进行溶解。剩下来的是纯金粉，它们被熔成金条——所有商人梦寐以求的东西。

电脑其他部件的回收要难得多，因为它们含有混合金属，但仍然值得花功夫。金属部分被剥离下来，放入炉子。熔化成金、银、铜的混合液态金属。再被倒出、冷却、压延，然后切成金属板。每个金属板经过浸泡，带上正电荷。如图中左侧所示，右侧是带负电荷的铜板，它会吸引从左侧混合金属板上游离的铜微粒，并在板上聚集。

在下一次浸泡里，钢板被用来吸引银，使它形成晶体。第二次浸泡后，金属板就只含金了，可以拿去熔化，将它做成不同形状，以满足多种用途。这些金箔能被用来制作金盐，出售给计算机行业。用金箔制造盐晶体，他们必须要用剧毒的氰化物，所以工程师要确保容器高度密封。这种盐不适合用在鱼和薯条上，但却是制造计算机芯片的绝佳材料。这桶金盐价值将近47000英镑。这里，金箔被切成小金属板，将用来做牙冠。曾经在计算机内存里传递信息的东西，马上就会在某个人嘴里安家。

这个工人正从炉中取出定制的金棒。它们被压缩和卷曲成盘旋状，然后变成计算机电路要用的细金属丝。另一个对黄金有巨大需求的领域是航天业。人造卫星和航天飞机等交通工具的引擎都需要用到黄金。每个人都知道黄金是一种贵金属，但谁又能想到，黄金还有这么多的工业用途。所有这一切，使回收旧电脑成为一项非常有利可图的生意。

你知道吗？

开，广泛用于测量黄金纯度的单位，是基于角豆种子的重量设定的。

1. 拆卸旧电脑
2. 留下含贵金属的芯片和处理器
3. 盐酸硝酸混合液把金分离出来
4. 得到金粉熔成金条
5. 其他金属被剥离下来投进火炉
6. 在炉中熔成混合液态金属

7. 铜微粒聚集在带负电的铜板上
8. 用钢板吸引金属板中的银
9. 第二次浸泡后金属板只剩下金
10. 用金箔制造盐晶体做芯片
11. 黄金线圈变成计算机电路
12. 航天业的引擎的某些部件都需要黄金

Computer Recycling

But first, many people believe that when their computer no longer works the only thing it's good for is a paperweight to clutter up their desks. But they couldn't be more wrong. These men work for a new breed of recycler who has discovered gold in old hard drives.

Hidden within your old computer is a treasure trove of precious metals. The circuit board contains silver and copper and the microprocessor is laced with gold, all of which are good conductors that carried electrical signals around your computer. The old computers are sent to a plant which salvages their precious metals. First the outer casings are taken off. Cables are unplugged and disposed of. And then all that remains are the chips and processors. All these components were worth a fortune when they were brand new but now they would be worth nothing if they were on a rubbish pile.

This company will make a handsome profit though. One ton of old computers could fetch them 6 and a half thousand pounds. These microprocessors are the easiest to recycle as the only metal contained in them is gold. A mix of hydrochloric acid and nitric acid is used to separate the gold from the ceramics. Hundreds are dissolved at a time in a large barrel. What remains is pure gold dust, this is melted into gold bullion. The stuff of dreams for any business.

Other parts of a computer's interior are a lot harder to recycle since they contain a mix of metals, but the effort is still well worth while. The metal parts are stripped down and put into a furnace. They melt into a liquid mix of gold, silver and copper which is poured out, cooled, pressed and then sliced into plates. Each plate is placed into a bath and given a positive charge. Here we see it on the left. On the right a negative charged copper plate attracts the copper particles which loosen from the mixed metal plate and accumulate there.

In the next bath a steel plate attracts the silver, which forms in crystals. After the second bath the plate contains nothing but gold and can be melted down. It will be shaped into different forms that have a variety of uses. These sheets can be used to make gold salt which will be sold to the computer industry. To create the salt crystals from the sheets they have to use cyanide which is highly poisonous so the engineer makes sure the container is sealed very tightly. This salt may not be so good for fish and chips but it is excellent for computer chips. This bucket of gold salt has a value of almost 47,000 Pounds.

Here the gold sheets are pinched into small plates that will become dental crowns. What once carried information round a computer's memory will shortly find a home in someone's mouth.

This worker is taking made-to-measure gold rods from the furnace. They're compressed and twisted into a coil which is then transformed into fine wire used in computer circuitry. Another industry that has a huge demand for gold is the space industry. Satellites and vehicles such as the Space Shuttle use gold in their engines. Everyone knows gold is a precious metal but who would have thought that it would have so many industrial uses. All of these combined to make recycling old computers a very lucrative business.

Did you know?

Carats, the unit widely used to measure the purity of gold are based on the weight of carob seeds.

轮胎回收

扫描二维码，观看中文视频。

烧胎，让轮胎冒烟的动作，是现代汽车拉力赛的乐趣之一。不过按这样的搞法，轮胎很快会被磨光。但无论是自然磨损或故意破坏，轮胎都可以回收。一个有趣的解决方案，是为牛棚做垫子，稍后我们会详加解释。

一年时间里，仅在英国报废的轮胎就超过40万条，而实际上只有10%被回收。它们首先集中到这样的先进设施里。

回收轮胎的第一步是检查状态。有些还没达到使用寿命。可以卖给经销商或运到别的国家。但如果轮胎上做了红色标记，那便是无法修复的且必须予以销毁。

分拣过程由这台机器完成，就像在邮局分拣邮包一样。但相机在这里扫描的是轮胎，而不是装满信件的袋子。如果看到任何红色标记，轮胎将被送到废品堆销毁。还没有完全磨光的轮胎可以被翻新或售出。它们去向如何，取决于磨损的严重程度。轮胎被分拣系统放进相应的容器里，等待自己的命运。

红色标记意味着只能回收，但却并非易事。回收厂不能只把轮胎撕开。现代轮胎由多种原料制成，包括橡胶、钢带和人造纤维。有许多不同的东西需要处理。

回收过程的第一工序是粉碎。轮胎破拆后能把金属从橡胶中分离出来。轮胎到了这一步，今后肯定不能再上路了。在轮胎的碎块里你能清楚看到金属和纤维，它们仍需要被分离出来。要做到这点，橡胶和金属块需要再磨上好几次直到变得更小。

这条传送带有一个强力磁铁。当通过细小的碎片时能把铁收集起来，送到金属公司卖上个好价钱。磁铁清除了大部分金属，但还会有些残留下来。后面的机器摇动剩余的混合物然后让它们通过更小的磁铁。剩余的碎铁在通过这些栅格时被捕获。这就是剩下来的东西——黑金。不是石油，而是纯粹的轮胎橡胶。这个设施每年能回收3万吨旧轮胎橡胶。

回收橡胶的一种巧妙用途，是给奶牛做地毯。为此将橡胶加热到50摄氏度左右。如果太热了就会粘在一起，太冷的话将不能充分混合。必须恰到好处。加进新的化学成分，将能产生一种坚固耐用的新橡胶。下一步用硅酮喷涂模具。防止再生橡胶粘在盘子上。这有点像做蛋糕时为饼模涂油。机器现在量出50千克橡胶混合物，放进每个模具，然后送入烤箱加热到150摄氏度。橡胶融化在一起，并形成新的形态。最后得到一张巨大的垫子，类似学校体育馆上课时用的那种。工人剥掉边缘，让垫子成为合适的形状。

但拿它们来做什么用呢？听起来似乎很不寻常，研究表明，舒适的环境能提高乳牛的产奶量。这些垫子将铺设在牛棚里，让乳牛的立足之地更柔软。

垫子需要有孔排水。这些孔是用特别刀具

你知道吗？

1989年，埋在威尔士垃圾场的1000万条轮胎起火，整整燃烧了9年！

切出来的。不是金属刀片，而是用每平方厘米近 4 吨力的高压水柱。它的强大难以置信，切开橡胶轻而易举。

垫子已经就绪，只需进行安装。如同放置地毯一样，农夫在牛进来之前先把它们铺好。希望能通过释放蹄子的压力，让牛群更轻松，产奶量更高。

显然，安逸舒适的奶牛能生产更多的牛奶。牛奶产量越高，农夫利润越多，大家都受益。这只是在废物利用道路上，旧轮胎转化的许多新用途之一。

1. 要报废的轮胎用红笔标出
2. 把金属从橡胶中分离出来
3. 传送带上强磁铁收集起小铁片
4. 残余金属通过栅格时被捕获
5. 得到纯粹的旧轮胎橡胶
6. 加进新化学成分使橡胶更坚固

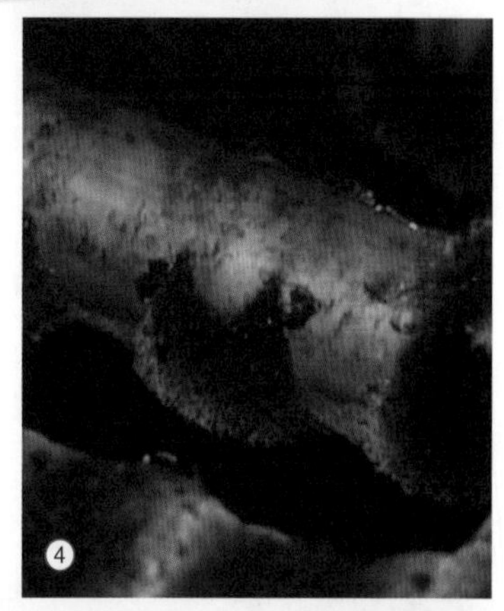

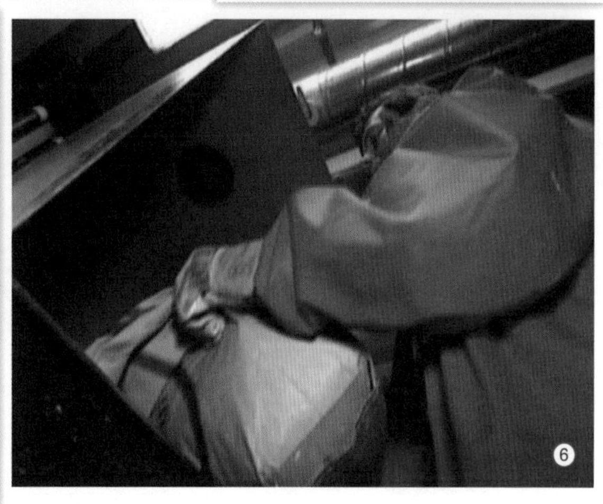

7. 用硅酮喷涂模具防止橡胶粘盘
8. 从烤箱出来的橡胶融为一体
9. 将垫子的边缘修整好
10. 垫子用高压水柱切出排水口
11. 牛棚地面铺上橡胶垫子
12. 垫子让牛群舒适产奶更多

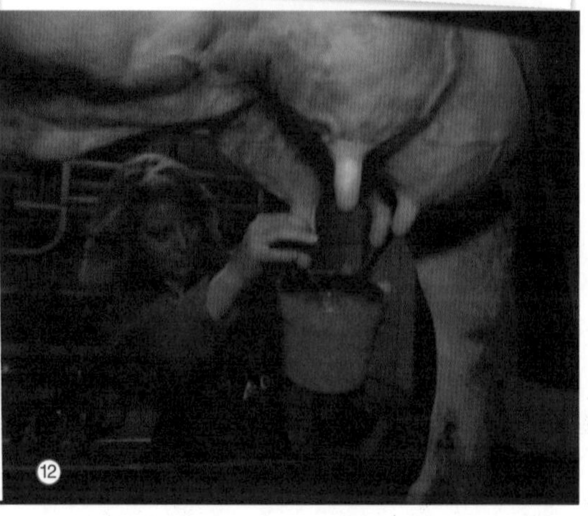

Tyre Recycling

But first, rubber roasting. Tyre-smoking action is all part the fun at a modern car rally, but at this rate the tyres will go bald pretty quickly. But whether its natural wear or deliberate destruction, you can still recycle them. One interesting solution is carpeting for cow sheds, but we'll find out more about that later.

In one year alone the British public scrapped over 40 million tyres but just over 10% of them were actually recycled. First they're collected at state of the art facilities like this one.

The first step when recycling a tyre is to check its condition. Some still have some life in them. These can be sold on to dealers or shipped to other countries. However if the tyre gets a red mark, it's beyond repair, and must be destroyed.

The sorting process is done by this machine which used to sort mail bags at the post office. Now, instead of bags of letters, the camera scans tyres. If it spots any with red marks, they are sent to the pile to be destroyed. Tyres with any wear left in them can be re-cut or sold-on. Where they will go depends on how worn they are. The sorting system identifies the right bin to put the tyre in and it's released to await its fate.

The red mark means recycle only, but this is no easy task. The recycling plant can't just rip them apart. Modern tyres are made up of many elements including rubber, steel bands and man made fibres. There are a lot of different things there to deal with.

The first stage of the recycling process is the shredder. Breaking the tyres down will help to separate the metal from the rubber. Tyres that end up in here certainly won't be rolling along the open road ever again. In the shredded pieces that emerge, you can clearly see the metal and fibres within the tyre which still need to be separated out. To do this the rubber and metal chunks must be ground down several more times until they're much, much smaller.

This conveyor belt contains a powerful magnet. As it passes over the finer shreds, the steel is collected. The metal industry will pay good money for this abundant source of scrap. The magnet has removed most of the metal but there's always some left over. The next machine shakes up the remaining mixture and then pours it across smaller magnets. The remaining fragments should be caught as the rubber passes through the grate. And this is what you're left with. Black gold. No, not oil, but pure tyre rubber. Every year this facility can recycle 30 thousand tons of old tyre rubber.

One of the more ingenious uses for this recycled rubber; carpeting for cows. To make it the rubber is heated to around 50 degrees Celsius. Too hot and it sticks together, too cold and it won't mix. It has to be right. New chemicals are mixed that will create a firm and durable new rubber. Next silicone is sprayed over the forms. This will stop the recycled rubber sticking to the pan. It's a bit like when you grease a cake tin. The machines will now measure out 50 kilos of the fresh rubber mix into each form. They're then fed into the ovens to cook at 150 degrees Celsius. The rubber melts together and takes on its new shape. What emerges is a gigantic mat, similar to the type you might find in a school gymnasium for PE class. The workers strip off the edges to get the mat into the right shape.

But what will they be used for? Well, it sounds unusual but research shows that a comfortable environment improves milk productivity. These mats will be laid in barns to soften the footing for dairy cattle.

The mats need holes to help with drainage. These are cut using a very unusual knife. It doesn't use a metal blade. Instead it uses water pressurised to almost 4 tons per square centimetre. It's incredibly powerful and slices through the fresh rubber with ease.

The mats are now ready and just need to be installed. Like fitting carpet, the farmer lays them down before letting the cows back in. It's hoped that by easing the pressure on their hooves, herds like these will be more comfortable and more productive.

Apparently comfy cows produce more milk. The more milk the cows produce, the more the farmer profits, so everyone benefits. Just one of the many new uses for old tyres transformed along the road of recycling.

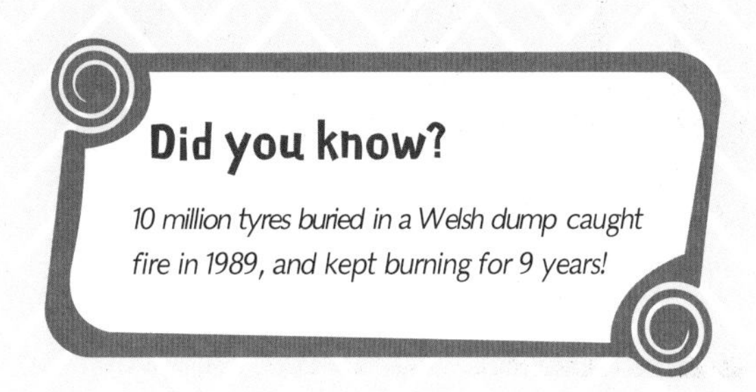

风力发电机

太阳能电池板

飞机跑道翻修

移动式起重机

工业篇

摩天大楼

预制房屋

锯鳐伐木机

高压电工

金矿

风力发电机

扫描二维码，观看中文视频。

世界各国政府都在致力于减少碳排放，推广可再生能源。风力涡轮发电机被认为是解决问题的关键手段。但这些庞大的设施是如何建造的呢？

起步工作之一就是准备地基。在这里他们使用的是钢筋混凝土。在钢筋框架中混凝土将被灌入。混凝土本身就可以承受住风塔的重量。但因为和任何高大建筑一样，风塔会遇到极端的大风天气，所以必须强化。如果没有钢筋加固，混凝土会逐渐退化并最终崩溃。钢筋框架给了混凝土足够的伸缩性来抵御冲击。地基上打进 40 米长的桩将风塔固定在地面。当 1200 立方米的混凝土浇筑后底座会重逾千吨。

风塔的所有部分都是丹麦的一家专业工厂生产的。这是风塔的底部。直径达 6 米，这个宽度足以承受整个结构其余部分的重量。起重机司机不能把这么巨大的截面精准降落到底座上，必须靠绳子和体力来安装到位。另外 3 个部分在接下来的 3 天内安装完成。第 5 个部分，即最后一部分的安装，才是真正的挑战。它将被安装到已经高达 120 米的塔架上。位于塔架顶端的机舱为发电室，是将风能转换为电能的地方。为了安装机舱，工人们必须沿梯子爬上去。这是一段 40 分钟的垂直跋涉，所以绝不是任何有身体缺陷或恐高的人能胜任的。

现在工人们已到达顶部，机舱可以吊到他们身边了。这是风力发电机的核心部分。风轮带动一个转子，转子连接发电机，让动能转化为可储存的电能。电流通过一个变压器将电压从 1 千伏提升至 2 万伏。此后被引到一个本地变电站进入主电网。

风轮将被安在这里。风轮的叶片在这个工厂生产。每年他们制造 1800 个，这些叶片是世界有史以来最大的。每片都有令人难以置信的长度，达 61.5 米。风轮叶片每次制造出半个，然后再拼装完成。塑料被玻璃纤维层加强，通过手工缝制在一起。这虽是个艰苦的过程，但使用这些轻质材料，让叶片非常高效。

为制作每半个叶片，他们用塑料外壳覆盖模具，抽出中间空气造成真空，然后泵入液态树脂。由于真空的缘故，液态树脂能迅速而均匀地延展开来。在接下来的几天里，树脂和玻璃纤维融合，变得干燥，形成一个非常坚实的结构。工人们再把做好的半个叶片拼装成一个完整的空心叶片。

为了模拟 20 年期间强风对叶片的影响，他们安装一个超过 50 吨的试验叶片，让它两边经受多达 500 万次摇晃。试验叶片通过了检测，现在 3 个相同的叶片可以送往工地了。要运输一个 61.5 米长的叶片，你需要一个非常非常长的运输工具。想象一下完成这个卡车的三点转向！

你知道吗？

苏格兰的凯恩戈姆山经受过英国有史以来最迅猛的风。时速 173 英里，是飓风平均速度的两倍多。

当3个叶片被安装到一起后，就能真正看到这个工程让人惊叹的规模了。他们夜以继日地工作，将叶片起吊到位。这些塔里的工人负责引导转子安装就绪。这是一项令人生畏的工作。他们对付如此巨大的物件，在很高的地方，又是深更半夜。当它们被插到引导螺栓上后，艰苦的工作才算结束。终于风力发电机准备开始发电了。一个这样的风塔所发出的电，足够5000户人家使用。

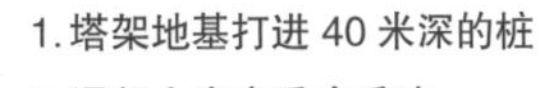

1. 塔架地基打进40米深的桩
2. 混凝土底座重逾千吨
3. 风塔底部直径达6米
4. 发电室机舱将被吊到塔架顶端
5. 机舱吊到塔架顶部准备安装
6. 每次制造半个发电机叶片

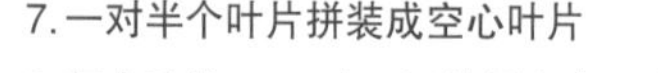

7. 一对半个叶片拼装成空心叶片
8. 每个叶片 61.5 米，运输是个大工程
9. 工地上 3 个叶片已安装到一起
10. 夜色中风轮吊到机舱旁
11. 塔里的工人引导转子安装
12. 一座风力发电机可为 5000 户人家供电

Wind Turbine

Governments across the world are trying to cut carbon emissions and renewable energy sources such as wind turbines are thought to be a key part of the solution. But how do they construct these massive structures?

One of the first steps is to prepare the foundations. Here they're using reinforced concrete. This is the steel framework that the concrete will be poured into. The concrete can bear the weight of the tower on it's own, but it has to be reinforced because like any tall structure it's going to face some stiff winds. Without the steel reinforcement the concrete would deteriorate and eventually give way. The steel frame gives the concrete enough flexibility to take the battering. The foundation also has 40 metre stakes driven through it, fixing it to the ground. When the 1200 cubic metres of concrete is added the base weighs over a thousand tons.

All of the sections for the tower are produced by a specialist factory in Denmark. This is the bottom section. At a massive 6 metres in diameter, it's wide enough to bear the weight of the rest of the structure. The crane driver can't drop such an enormous section on six pence it has to be dragged into place with ropes and brute force. Three more sections are fitted over the next three days. The fifth and final section is a real challenge. It's being fitted on to a tower that is already 120 metres tall. The nacelle is a power house that sits on top of the tower and this is where the wind energy will be converted into electricity. To attach it the workers have to climb up by ladder. It's a 40 minute vertical trek so it's definitely not a job for anyone that's out of shape or scared of heights.

Now the workers are at the top, the nacelle can be lifted up to them. This is the very heart of the wind turbine. The blades turn a rotor which is connected to a generator, This turns the kinetic energy into electrical energy which can then be stored. It passes through a transformer which bumps it up from 1, 000 to 20,000 volts. Then it can be diverted out to a local substation and onto the main grid. The rotor blades will be attached here.

They're made in this factory. Ever year they make about 1,800 , but these will be the biggest ones the world has ever seen. Each an incredible 61.5 metres long. The blades are made in halves that will be joined together later on. A plastic that's reinforced with layers of fiberglass and these are sewn together by hand. It's a painstaking process, but using these lightweight materials makes the blades very efficient.

扫描二维码，观看英文视频。

To finish each half of the blade they cover the mould with a plastic shell and suck out the air to make a vacuum, then they pump liquid resin in. Because of the vacuum the resin spreads quickly and evenly. Over the next few days the resin merges with the fibre glass and dries to make a very solid structure. Workers then join the halves together to make a single, hollow blade.

To simulate the effects of strong winds over a 20 year period they load a test blade with a weight of over 50 tons, then they shunt it from side to side an incredible 5 million times. It passes the test and now three identical blades can be taken to the site. To move a sixty one and a half metre blade you need a very very long vehicle. Imagine trying to do a three point turn in this truck!

When the three blades are put together you can really see the awesome scale of the project. They work around the clock to lift the blades into place. These men inside the tower are there to guide the rotor into place. It's a scary job. They're working with enormous objects at great heights, in the middle of the night. As it's slotted on to the guide bolts the hard work is over. Finally the wind turbine is ready to start producing electricity. This single structure will provide enough to supply five thousand homes.

Did you know?

The Cairngorms in Scotland endured the fastest wind ever recorded in the UK. At 173 mph it was over twice the average speed of a hurricane.

太阳能电池板

扫描二维码，观看中文视频。

当你想到阳光和沙滩时，通常映入脑海的是加勒比海度假，但你想到过太阳能电池板吗？使太阳能电池板工作的主要成分是硅或者沙子，来自像这类采石场中的岩石。岩石从矿场出来投进火中。第一阶段是从岩石中提取硅，这项工作在巨大的洪炉中进行。温度达到2000摄氏度以上。这是难以置信的高温，也相当危险。为了保护自己的眉毛，工人用长杆来拨动矿石。

从窗户玻璃到家用电脑处理器，硅用来做各种物品。离了它太阳能电池板就无法工作。用硅包在工厂内转运硅，没有别的东西能够承受硅的极高温度。硅在这种形式下是流动的熔融态，但要制作太阳能电池板，还需要添加特定的化学成分。

首先矿石需要结晶，这个过程由冷却机完成。硅的温度如此之高，足以烧穿机器，但专家拿出了一个解决方案：当它工作时在周围泵进冷水，如同水冷空调。

超热的结晶硅从机器中出现了，但这种物质怎样收集太阳的能量呢？

太阳能电池板用薄硅片做成，它有两个独特的材料层通过两种化学方法生产出来，在制作过程的不同时间对硅产生影响。第一阶段期间，第一种化学物质在这里添加，使硅更加导电。要使这两种物质充分结合，罐子被密封进一个大熔炉，放置2天。就像做一大锅煲汤，厨师始终密切地关注着他的配料。当蜂鸣器响起，烹饪完成。炉子被打开，硅块出现了。这个硅块太大，无法放在通常的太阳能电池板屋顶上。普通屋顶也无法承受，所以现在需要切小。

切割固体结晶硅不像是切黄油，有可能非常危险，所以要升起防护罩，以确保工人的安全。镶着金刚石的刀片切过硅块，被切成单个面板大小的硅块准备进入下一工序。硅块的大小现在合适了，但仍然太厚，不能把太阳能转换成电能，所以需要通过这台机器进一步减薄。机器中有几百条转动的金属线缓缓切过硅块，留下的晶体硅薄片正好制作太阳能电池板。切片必须分开清洗，但没有机器能敏感到不造成损坏。面板几乎没有指甲厚，并且难以置信地脆弱，必须由手工清洗。

每当人工成为生产过程的一部分时，就会产生污染，因为人会释放出百万计的微小颗粒。刚切好的硅片必须在第二种化学物质处理前彻底清洗。

一旦处理完成，面板被返回炉子进行烘烤，并做出两个隔离层。不，这不是一种新的城市伪装图案。它是制作巨大太阳能电池板的基本单元。它有2种不同的颜色，但不是为了时尚。加入蓝色使面板吸收更多阳光，产生更多能量。面板还会加上抗反射膜，由涂层烘烤固定而成。任何被面板反射的光线都是能量的浪费。

现在，面板干净了，处理停当了，并且具有合适的颜色，做好了发电准备。但如何把电能引导出来？高科技电脑把面板对齐，下个机

你知道吗？

仅6.6平方米的太阳能电池板就为星尘号飞船提供了动力，让它在7年间飞行了28.8亿英里。

器把一层灰色糊状物涂在上面。糊糊看起来像坏了的奶冻，但实际上是半液态金属，它将能源从面板导出输进电网，并通过一个插头来到你正使用的电器。

现在我们可以组建一个全尺寸的太阳能电池板了。每个晶片就像一个电池产生自己的电力，但为获得能量输出面板被组装成排由金属线路连接起来。整齐排列的面板连接完毕，机器就可以把所有零件组装到一起了。就像制作一个巨大的薯片三明治，只不过薯片是由硅制成的。一大块玻璃用来作基础，接着是胶纸层，然后是硅板，用更多黏合剂把它们固定到位。最后一层是框架把所有东西整合在一起。

尽管英国的天气并不总是阳光灿烂，太阳能电池板也可以在阴天工作，它们只需要清洗、安装和接上插头就能工作了。当英国的夏天越来越充满阳光，太阳能电池板将帮助我们减少有害的碳排放。

1. 太阳能电池板的原材料是沙子
2. 用硅包转移熔融状的硅
3. 在结晶硅中加第一种化学成分
4. 金刚石刀片把硅块切分开
5. 薄脆的硅片须经手工分开
6. 清洗干净后硅片返回炉子

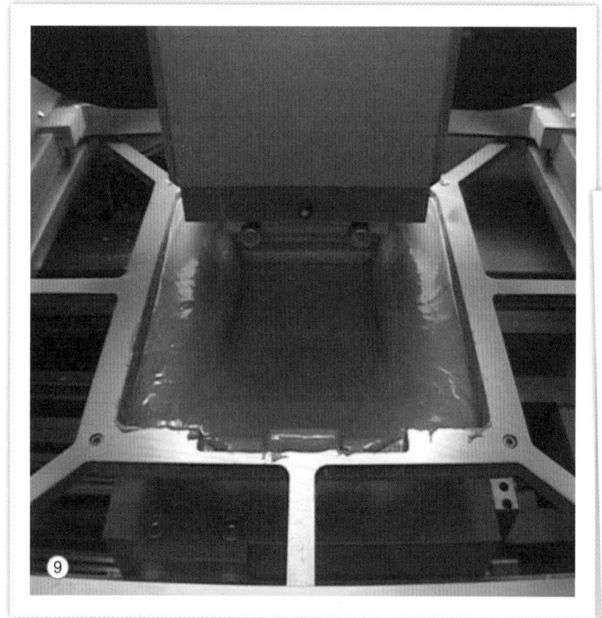

7. 硅加入蓝色可吸收更多阳光
8. 硅片加上抗反射膜
9. 硅片拼合好并涂一层半液态金属
10. 在玻璃上贴胶纸层
11. 框架把所有元件整合在一起
12. 太阳能电池板开始工作

Solar Panels

But first when you think of sun and sand, holidays in the Caribbean are usually what spring to mind, but have you ever thought about solar panels instead? The main ingredient that makes them work is silicon, or sand, which comes from rocks found in a quarry like this. It's out of the quarry and into the fire. The first stage is to get the silicon out of the rocks, and that's done in this enormous furnace. Temperatures reach well over 2,000 degrees Celsius. This is unbelievably hot and also quite dangerous. To keep their eyebrows safe, the men work the ore using long poles.

Silicon is used to make everything from window glass to the processors in your home computer and without it, a solar panel just wouldn't work. Huge barrels are used to carry it around the factory. Nothing else could withstand the immense heat involved. In this form it is liquid molten silicon but to make solar panels certain chemicals need to be added to it.

First the ore needs to be crystallized which occurs here in this cooling machine. The silicon is so hot, it could melt right through it, but the boffins have come up with a solution. Cold water is pumped round it as it works, which acts like watery air conditioning.

The super hot crystallized silicon emerges from the machine, but how will this help to harness the energy from the sun?

Solar panels are made using thin slices of silicon which have two unique layers in them. These are created using 2 chemical processes which affect the silicon at different times during production. Time for stage 1, this is where the first chemical is added. It will help make the silicon more conductive. To get the two combined properly, the tank is sealed into a large furnace and left for 2 days. It's like cooking a big stew and the chef keeps a close eye on his recipe. When the buzzer rings the cooking is complete. The furnace is opened and the silicon block emerges. This block would be far too big to put on your roof, which is where solar panels usually go. The average roof wouldn't support it so now it needs to be cut down to size.

Cutting solid crystal isn't like cutting butter. It can be very dangerous, so a protective shield is raised to keep workers safe. Diamond tipped blades then slice through it, cutting individual panel sized columns ready for the next stage. The columns are now the right size, but they're still far too thick to convert the sun's energy into electricity, so they need to be cut down even further by this machine. It has hundreds of spinning wires which slice their way slowly through each column. What they leave behind is wafer thin sheets of silicon crystal, perfect for solar panels. The slices must be separated to be cleaned but no machine exists that's sensitive enough to do it without breaking them. It must be done by hand because the panels are barely finger nail width and incredibly fragile!

Wherever humans are part of a process, contamination follows because we release millions of dust particles. The freshly cut slices must now be thoroughly washed before they can be treated with the second chemical.

Once the treatment is in place, the panels are returned to the furnace and baked to create the two separate layers. No, this isn't a pattern for a new kind of urban camouflage. It's the basic unit that makes up a larger solar panel. It comes in 2 different colours, but this isn't a fashion statement. The blue shade is added so the panel can absorb more light and make more energy. The panels are also treated with an anti-reflective coating which is painted on and then baked into place. Any light the panel reflects would simply be wasted energy.

Now the panels are clean, properly treated and wearing this season's colours. They're ready to make electricity. But how do you get the power out of them? High tech computers align the panels, and the next machine paints a layer of this grey gunge onto them. It looks like bad school custard. But it's actually semi-liquid metal. It helps conduct energy from the panel, into the electricity grid, and through a plug into whatever appliance you're using.

We can now assemble a full sized solar panel. Each wafer acts like a battery that generates its own electricity, but to get power out, the panels are assembled in rows connected by metallic pathways. With neat rows all hooked up, the machine can put all the pieces together. It's like making a gigantic chip butty only the chips are made of silicon. A large slice of glass makes the base, followed by layers of adhesive paper, the silicon panels and more adhesive to hold it in place. The final layer is the frame to hold it altogether.

Even though English weather isn't always sunny, solar panels can work in our cloudy conditions, they just need to be cleaned, installed and plugged in.

So as Britain's summers get sunnier, solar panels will work to help us cut down our harmful carbon emissions.

Did you know?

Solar panels measuring just 6.6 square metres powered the Stardust Spacecraft during a journey of 2.88 billion miles which lasted 7 years.

飞机跑道翻修

扫描二维码，观看中文视频。

德国法兰克福机场，是整个欧洲最繁忙的机场之一。每天有超过 500 架飞机在这条跑道着陆并造成磨损，因此 4 千米长的跑道需要定期更换。

一次性完成要花很长的时间，并且会中断航空运输，所以要一点一点来做。但怎样才能对付不停滴答作响的机场时钟呢？

这个建筑队只有 8 个小时来更换 15 米宽、60 米长的跑道，正好不耽误明天早上第一架航班降落。

首先，旧跑道由 4 个功能强大的挖掘机挖开。1500 吨的碎石被清除，货车把它们运走并回收利用。他们只有 1.5 小时完成这部分工作，主管要确保一切按时间表运行。

当 60 厘米厚的表面层被去除，爆破专家将登场，用传感器搜索地面，防止未发现的二战时期炸弹。当他给出“没有问题”的信号，下一群机械开进来为新跑道奠定基础。整个区域被挖到 70 厘米深并保持平整。跑道地基用回收的柏油材料建造。即使在时间的压力下，完美做好每一步仍至关重要。几小时后在此着陆的飞机的安危，掌握在这些工人手中。地基层必须要平，整个过程被密切监视。安放好管道——它们将放置跑道上的照明用电缆。坑的一端已经准备铺设沥青了，而另一端的混凝土仍然正在移除。沥青由卡车成吨地运来。当沥青到达时，用激光测量它的温度。沥青需要趁热铺设。首先将沥青集中铺设在管道周围，以保护管道不会被材料和机器的重量压坏。然后填充坑里的其余部分。

巨大的蒸汽压路机把路面压实压平。发挥团队协同效率，能在短短 1.5 小时完成这项工作。为了确保跑道承受大型喷气式客机降落的巨大负荷，这两台机器在地面添加另外两层沥青。蒸汽压路机再次将其压紧压平。

第一架飞机预计在 90 分钟后降落，建设仍在进行中。测量沥青的密度，必须达到标准，这样就不用在几个星期后回来返工了。像任何道路一样，跑道需要标记，以引导飞机和机场车辆。

最后，在逐渐冷却的跑道上铺一层细沙子，密封可能存在的孔洞。保洁车清扫刚铺好的跑道，保证没有碎片留下损坏飞机。然后该清理现场了。

总共只花了 8 个小时，工人就挖掉并更换了一段跑道，维护了这个大型机场的安全。

你知道吗？

乌鸦会将老鼠扔在机场跑道上来杀死它们。

1. 飞机跑道更换乘夜色完成
2. 4 个挖掘机同时开挖旧跑道
3. 1.5 小时清除 1500 吨碎石
4. 用探测器确定无战时留下的炸弹
5. 地基用回收的柏油材料铺设
6. 检测地面是否平整

7. 跑道另一端仍在移除混凝土块
8. 沥青要趁热铺设并用激光测温
9. 放置电缆的管道优先铺设沥青

10. 压平的跑道添加另两层沥青
11. 蒸汽压路机再次把跑道压平
12. 完成划线、细沙密封后竣工

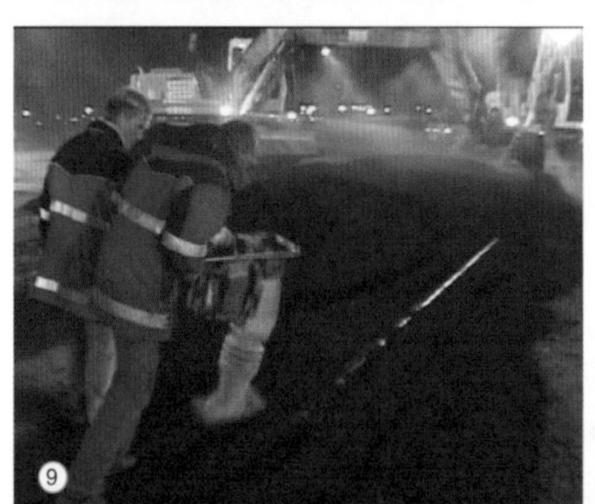

Runway Resurfacing

扫描二维码，观看英文视频。

Frankfurt in Germany has one of the busiest airports in the whole of Europe. Over 500 planes land on this runway every day and they all wear it down. So the 4 kilometres of tarmac needs to be regularly replaced.

Doing it in one go would take a long time and interrupt air traffic, so they do it bit by bit. But how do they beat the ever ticking airport clock?

This construction crew have just 8 hours to replace a 15 metre by 60 metre stretch in time for the first flights landing tomorrow morning. The heat is on.

First, the old tarmac is broken up by 4 powerful diggers. 1,500 tons of rubble is removed and lorries take it away for recycling. They only have an hour and a half for this part of the job so a supervisor makes sure that everything is running to schedule.

Once the 60 centimetre layer has been removed, an explosives expert moves in with a sensor to search the ground for undetected World War II bombs. When he gives the 'all clear', the next flock of machinery moves in to prepare the ground for a new tarmac. The whole area is flattened to depth of 70 centimetres. The runway's foundation is made from a layer of recycled tarmac. Even under the time pressure it is vital that every step is carried out perfectly. The safety of the planes that will land here in a few hours time is in these guys hands. The layer has to be even, so a close eye is kept on the depth throughout the whole process.

Tubes are laid out which will house the electric cables for the runway lighting. The pit is ready to be asphalted on one end whilst on the other end the last concrete is still being removed. The asphalt is trucked in by the ton. As it arrives the temperature is measured by a laser. It all has to be laid out while it's still hot. First the asphalt is first laid and condensed around the tubes to protect them from being crushed by the weight of material and machines. Then the rest of the pit is filled.

Enormous steam rollers condense and flatten it. Working as a team the crew can complete this section in just 1 and a half hours. To be sure the tarmac can take the enormous load of a jumbo jet coming into land these two machines add another two asphalt layers on top. Again steam rollers compress it into a level surface.

The first plane is scheduled to land in 90 minutes and construction is still in progress. The density of the asphalt is measured, it's got to be up to standard so they don't have to come back and redo it in a few weeks time. And like any road a runway needs markings to guide the planes and airport vehicles.

Finally a fine layer of sand is spread over the cooling tarmac to seal any pores. A vehicle sweeps the new stretch so no debris is left behind that may damage the aircraft, and then it's time to clear the site.

It's taken just 8 hours for the crew to strip and replace another section of the runway keeping this major airport working safely.

Did you know?

Crows drop rats on the tarmac of runways to kill them.

移动式起重机

扫描二维码，观看中文视频。

许多城市的天际线都被它们覆盖。它们被称为吊臂式起重机，离了它们就不能建造摩天大楼。当你建造一座拥有 90 米高的烟囱的发电厂时，便最能体现它们的重要了。

这是欧洲最大的移动式起重机，能够达到 200 米以上的高度。巨大的烟囱被卡车运到工地并准备就绪。它们先被放在一边，为巨大的起重机的到来留出足够空间。

起重机来了。这台起重机如此之大，安装它本身就是一个巨大的建设工程。第一步是先将基础打好。起重机下方的地下室结构必须加强，以承受起重机的重量。如果垮塌，后果将是灾难性的。

向导用喷漆做标记，让起重机排列在准确的位置上。单单搭建起重机就需要 3 天时间。次日早上液压支架到达，被小心地吊运到位，工人要确保支架放下时，他们的脚趾不在下面。

随着基础就绪并得到牢固支撑，竖立起重机塔身的工作就可以开始了。首先是压载物用来维持结构平衡，这个起重机用的 12 块配重平板总重量达 206 吨。

接下来是起重机巨大的框架结构。每个钢销重 36 千克，将分立的部件连在一起。让人熟识的威武身影慢慢开始出现了。这台起重机如此之大，要用普通起重机来支撑它高耸的框架。下一阶段需要格外谨慎。起吊烟囱必须达到足够的高度，起重机一定要相当长。随着不断升高，长臂伸向任何可用的空间……在这里便直接跨越到一条繁忙的公路上方。

工程必须限期完成，所以即使凌驾在高速公路上，起重机装配也必须继续进行。工人们一边关注着下面的车流，一边将每个部分安装到位。但谁又在这里照看着工人？此时起重机操作员没有太多事要做。等到起重机组装完毕后，他会对起重机位置做微小调整。而他的同事正在移动工作线上敲打 36 千克的钢销。框架完工后，添加一个简易风向标，为起重机操作人员指示风速。还有一个闪光灯标为飞行员提供辅助导航和障碍警示。

最后将滑轮和绞盘吊上来。单是这个巨大吊钩的钢缆就有长 900 米，重达 7 吨。这是个艰难的任务，如果手指不小心卡在机器里就会被压得粉碎。

最终轮到司机有所行动了。一切就绪以后他可以小心地抬起起重机，准备好明天大干一场了。为了把这个巨大的起重机安装起来，13 台卡车花费 3 整天运送设备和零部件。随着巨大的起重机建成，发电厂建设工程可以开始了。吊臂直指 200 米高空，这台起重机让周围每个起重机都成了侏儒。但这种规模的工程要靠团队的努力，较小的起重机仍然非常重要。

重达 90 吨，对于安全吊装来说，烟囱是一个大物件。因为它如此之长，又必须定位精

你知道吗？

世界上最大的起重机造价 630 万英镑，能够吊起达 3000 吨的物件。

确到毫米，风确实是个真正令人担心的因素。起重机操作员在工程中起关键作用。今天他来负责烟囱起吊。移动烟囱之前先做准备。操作员把它从临时支架提升到一定高度，工人还能利用最后机会修理或做必要改动。而当工长发出指令，巨大的起重机就投入工作。这不是一只鸟或飞机。这是一个发电厂的烟囱，耸立在100米高的空中。玩笑归玩笑，这是一项非常严肃的工作，安全压倒一切。巨大的烟囱下降到适当位置，起重机操作员明显松了一口气。如果没有精湛的技术和超级体量的起重机，完成这样巨大的工程是绝无可能的。

1. 高空作业离不开吊臂式起重机
2. 将起重机放置在准确的位置上
3. 液压支架被小心吊运到位
4. 12 块配重平板重量达 206 吨
5. 用压载物维持结构平衡
6. 起重机每个钢销重 36 千克

7. 用钢销将框架部件连接

8. 借普通起重机支撑它的长臂

9. 框架臂加风向标及闪光灯标

10. 吊钩的钢缆长 900 米重 7 吨

11. 起重机吊起 90 吨重的烟囱

12. 电厂烟囱耸立达到 100 米的空中

Mobile Crane

The skyline of many cities are covered in them. They're known as boom-type cranes, and you couldn't build a skyscraper without them.

Nowhere is this more so than when you're building a power plant with 90 meter high chimneys. This is Europe's biggest mobile crane, and it can reach over 200 meters high. The enormous chimney's are brought to the site by trucks and prepared. They're put to one side, creating a space big enough for the enormous crane's arrival.

And here it is. This crane is so big it's a major construction project in itself. The first step is getting the base into position. The cellar beneath the crane must be reinforced to take its weight. If it collapsed, the consequences would be disastrous.

The guides use spray paint markings to line the crane up perfectly. Constructing the crane alone takes 3 days and the next morning, the hydraulic supports arrive. They're carefully lifted into place and the workers make absolutely certain their toes aren't in the way when they're put down.

With the base now in place and firmly supported, work can begin to erect the crane itself. First comes the ballast to help balance the structure. The crane uses 12 slabs in total weighing a massive 206 tons.

Next, comes the massive crane's frame structure. Steel pins weighing 36 kilos each, hold the individual sections together. Slowly the recognisable shape of the boom begins to take form. The crane is so large, ordinary cranes are used hold up the towering framework. The next stage requires great care. Because it must reach high enough to manoeuvre the chimney, the crane must be very long. As it's raised one arm extends into any available space...In this case, directly over a busy road.

There's a deadline to be met so even though it's perched over a motorway, the crane construction must continue. Each section is put in place whilst workers watch out over the traffic below. But who is watching the workers up here? For now there's not much for the rig's operator to do. While he waits for the crane to be put together, he makes tiny adjustments to the crane's position, whilst his colleagues are hammering 36 kilo pins over moving buses. With the framework complete, a handy weather vane is included to indicate wind speed to the crane's operators. It also has a flashing light which acts as a navigation aid and obstacle warning for pilots.

Finally the pulley and winch mechanism are threaded up. The cable alone for this enormous hook is 900 meters long and weighs a massive 7 tons. It's a tough job and fingers could get crushed if accidentally trapped in the mechanism.

Finally the driver gets some action in his day. With everything in place, he can now carefully raise the crane, in readiness for tomorrow's big job. To put it all together this enormous crane has needed 13 truck loads of equipment and spare parts and 3 whole days. With the huge crane now constructed, work on the power plant project can now begin. Reaching over 200 meters in the air, it dwarves every crane around it. However jobs this size are a team effort and the smaller cranes are still important.

Weighing in at 90 tons, the chimney is a large object to manoeuvre safely. Because it's so long, and must be positioned with millimetre perfect precision; the wind factor is a real worry. The crane operator has a key role in the job. Today he is responsible for manoeuvring the chimney. Before it's moved it must be prepared. The driver raises it from its temporary stand to a height where workers can make any last minute repairs or necessary changes. And when the foreman gives the word, the massive crane is put to work. No, it's not a bird or a plane. This is a power-plant chimney over 100 meters up in the air. Joking aside, this is a very serious job and safety is paramount. The enormous smoke stack is lowered into position to the crane operator's obvious relief. Without his skill and the sheer scale of the crane, jobs as big as this just wouldn't be possible.

Did you know?

The largest cranes in the world can cost £6.3 million and can lift as much as 3,000 tons.

摩天大楼

扫描二维码，观看中文视频。

在世界上那些最大的城市里，摩天大楼构成了它们主要的天际线。建造一座摩天大楼是个浩大的工程。但慕尼黑人喜欢接受挑战，他们决定建造两座，每座的高度都超过 100 米，并且要在 18 个月内完工。

起重机操作员小心地把第一块材料吊到位。毫不奇怪当建造一个巨型高塔时，所用的各种建材也是庞大的。这块也不例外，它是一个巨型的 9 米长钢质支柱。

这些钢柱按一定角度倾斜安放和固定，使它比垂直放置能承受更多的负荷。每层楼的两端各有一对这样的钢柱来支撑两层楼的重量。这些钢柱用螺母和螺栓固定。但这些巨大而沉重的螺栓是不可能用扳手扭紧的。工人们使用电动螺栓起子来保证它们被牢牢拧上。

下一个需要起重机吊运到位的物件是支柱，也由钢制成，每根重量高达 24 吨。巨柱排成 8 列，一根建在另一根上面。使得高楼的重量从顶部被一直引导到地基，在那里被吸收。起重机操作员吊着柱子的一端让它沉降到位。下面的工人任务艰巨，负责引导柱子准确放到预定的位置。

大楼框架结构的最后一个材料，是一根巨大的加固钢质工字梁。业内人士简称它为工字梁，工字梁被安放在钢制支柱和巨柱的顶上。当它被焊接到框架上，惊人的力量能让大楼的整个结构连成一体。

每天多达 10 辆卡车满载混凝土来到工地。混凝土被吊上去用于浇铸和填充钢柱。它不仅让柱子更坚固，还使它们耐火。大楼里的地面也是混凝土材料。但在浇筑之前地面要先用钢筋网加固。

做完地面现在该安装窗户了。大楼每层都需要安装 125 个这样的大型窗格。工人们需要争分夺秒，但这不是一项可以仓促行事的工作。每个窗格都重达 600 千克。只要整个楼层的窗户安装就绪，室内装配大军就会进入岗位。每项任务都需要完成得迅速而准确。任何人的一点拖延就会让所有其他人等待，产生的连锁效应足以影响到整个工程。当室内装配工完成任务后就可以铺装地砖了。这项浩大工程进行到第 13 个月的时候，大楼所有的地板铺设完毕。

现在可以把玻璃电梯吊上去了。每个电梯重达 4.5 吨，需要运到 130 米高的电梯井顶端。把这些电梯安全运到顶端需要花费大量的时间。但它们开始运行后，速度就会快得多——时速超过 10 英里。升到楼顶只需 25 秒。

这两栋大楼看起来似乎全部完工了，但还需要连接到一起。一系列的桥梁连接不同的楼层，这样楼里的上班族就能方便地在两栋楼之间行动。起重机把一个桥吊上去，工人需要用引导绳把它安置到位，而不打碎一块玻璃。

在大楼的顶部一队工人正在固定钢架，用于支撑一面巨大玻璃，为建筑封顶。当最

你知道吗？

“摩天大楼”一词最早用于 200 多年前船上最高的帆。

后一根托梁即将放置到位时，对所有工人来说这是个特别的瞬间。他们用传统方式来纪念这一时刻。这是一项不朽的工程。在 18 个月的时间里，1500 名工人真正迎接了一场挑战。他们建造了两栋 61 层的高楼，总面积超过 6 万平方米。

1. 18 个月建造两座百米大楼
2. 9 米长的支撑杆按一定角度固定
3. 吊运 24 吨重的钢制支柱
4. 8 列巨柱一根建在另一根上面
5. 工字梁安放在支撑杆和支柱上
6. 混凝土浇筑和填充钢柱

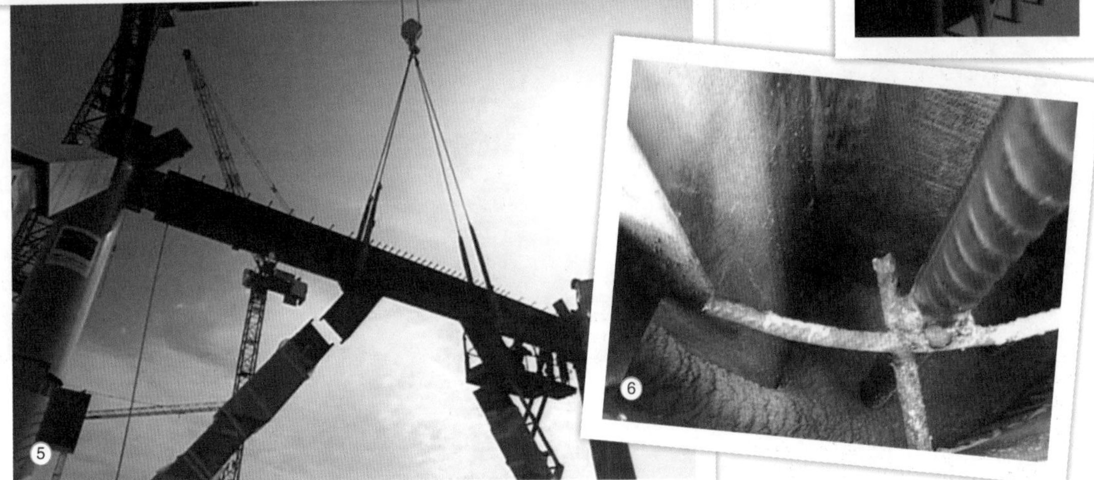

7. 地面先用钢筋网加固才能浇筑
8. 每楼层安装 125 个大型窗格
9. 室内装配完工成后铺设地砖
10. 电梯运到电梯井顶端
11. 两楼间架设一系列桥梁通道
12. 工人用传统方式纪念竣工

Skyscrapers

The skylines of the world's biggest cities are dominated by skyscrapers. Building one is a construction project on a massive scale. But they like a challenge here in Munich so they decided to build two, each over a 100 metres high, in just 18 months!

A crane driver carefully lifts the first elements into place. It comes as no surprise that when building a huge tower all of the parts that are used are also huge this is no exception it's a giant 9 metre steel strut.

The struts are fixed at an angle which allows them to bear more weight than if they were vertical. A pair will be fixed to both ends of each level and will support two floors of the tower. All that's used to hold them in position are nuts and bolts. But these massive bolts can't be tightened with a spanner! They use an electric drill to make sure they're firmly in place.

The next elements to be lifted into place are the pillars. These are made of steel too and weigh a whopping 24 tons each. They'll be fixed on top of each other in 8 columns. They allow the weight of the tower to be channelled from the top straight down to the foundations where it can be absorbed. A crane driver lowers a section of column into position and a worker below has the daunting task of guiding it into place.

The final piece of the framework is a colossal reinforced steel joist, known in the business as an RSJ. The RSJ's sit on top of the struts and columns and when they are welded to the rest of the frame their incredible strength allows them to hold the whole structure together.

Everyday up to 10 trucks full of concrete arrives at the site. It's hoisted up and then used to fill the steel columns. It doesn't only strengthen them it also makes them fire resistant. The tower's floors are made of concrete too but before it's added the floor is reinforced with steel mesh.

Done below it's time for the windows to go in. they need to fit one hundred and twenty five of these massive panes on each floor. They're up against the clock but it's not a job that can be rushed. Each one weighs six hundred kilos. As soon as a floor has its windows in place an army of interior fitters move in. Every job has to be done quickly and precisely. One delay from anyone of them will slow everyone down and could have a knock-on effect on the whole project. When all of their work is done, they can fit the floor tiles. Thirteen months into the mammoth project all of the towers floors are completed.

And they can hoist up the glass elevators. Each cabin weighs four and a half tons and has to be lifted right to the top of the130 metre lift shaft. It takes a considerable time to get them to the top safely but when they are operational they'll be moving a lot faster - over ten miles an hour. It'll take them just 25 seconds to ascend to the top of the tallest tower.

The two towers look like they're completely finished but they are going to be connected together. A series of bridges will link different floors so that the office workers will be able to move between them. A crane lifts one of the bridges up but they need a guide rope so they can get it into position without smashing all the windows.

At the top of the towers a team are fixing a steel frame it'll support an enormous fan of glass to top off the buildings. It's a special moment for all of the workers as they prepare for the final joist to be lifted into place and they commemorate it in traditional style. It was a monumental task, but over 18 months 15 hundred workers have truly risen to the challenge. They've created two towers with 61 floors between them and a combined floor space of over sixty thousand square metres.

Did you know?

The word 'Skyscraper' was first used over 200 years ago as a term for the highest sail on a ship.

预制房屋

扫描二维码，观看中文视频。

你也许会惊讶，仅2004年一年，英国预制构件房屋的销售额，就超过了16亿英镑。它们搭造起来也许很快，但要把所有的部件都组装完毕，还需要很多辛苦的劳动。

工作从木料开始。云杉木板的一面被涂上一种特制的乳胶，用以保护板材免受室外天气的侵蚀，并且防止变色。

将处理过的木板摞起来。接头处被安装在一起，形成外墙。一个厚木支柱切割成合适大小，沿着凹槽插入墙中用于锁定位置。镶着软木的纸板被固定在墙的内侧，用于阻挡穿堂风。不过整个墙都需要全面进行保温处理。他们用工厂里的废刨花来实现。

但首先这些刨花需要经过处理。它们在苏打和乳清里被剧烈摇晃。苏打能保护它们免受真菌侵蚀而导致腐烂，乳清则可使它们耐火。这是房子正面的上半部，将塞满刨花。刨花被撒进空腔，加压后大致成型。最终被压缩到初始体积的一半。这是被压好后的模样。

任何多余的刨花都会被吸走，进行再利用。接下来，他们给屋架安上一层硬纸板，用以封住保温层。硬纸板上面，再放上一块灰胶纸拍板，被切割成合适大小后，用一个形貌奇怪的工具——轮子上的装订机固定到位。

接下来，装上一排板条，留出空腔以容纳电缆。第二层灰胶纸拍板完成后，墙壁厚度达到30厘米。在房子运到工地前，窗户已被事先安装好，这样当房子被组装起来时，能够立即处于密封的状态，防止湿气进入。所有的墙壁用滑车搬运到一起，然后进行收缩包装，为下一步做准备。它们被装上货车运出，即将变成一座新房子。

一大清早，预制构件来到了工地。起重机很快把房子的整体底层从卡车上一举吊起。墙壁是按安装顺序堆放的，以节约时间和精力。

现在是整个安装过程中的关键时刻。当一个工人调好灰浆，准备将房子固定到水泥地基上时，第一面墙放置到位。如果稍有偏差，整个房子就会歪斜。

外墙就绪后，隔断墙就可以安放了。一面墙接一面墙，像三维拼图一样装配起来。当房子第一层的墙壁安装完毕后，工人们很快铺上天花板，没人愿意看到房子被灌进雨水。因为各个部分的重量都超过一吨，工人们必须和时间赛跑，在夜幕降临之前竣工。即使到了寂静的夜晚，工作仍在继续。下一步是把楼梯竖起来。楼梯到位后，就能往上搭建。他们分秒必争，连茶点也取消了。

在预制构件送达工地仅仅16小时之后，安装大功告成。钉上最后一根钉子，就只剩下给屋顶铺瓦了。

快速、高效又极其实用，预制构件房的确是一项惊人的成就。

你知道吗？

超过2200万美国人住在预制房屋里，这个数字超过了澳大利亚的人口总数。

1. 先在云杉木板一面涂上乳胶
2. 接头处安装在一起形成外墙
3. 一根支柱沿着凹槽插入木板
4. 房屋正面上半部的墙架完成
5. 为保温墙架塞满刨花
6. 装一排板条留出空腔容纳电缆

7. 加第二层灰胶石膏板，墙壁完成
8. 窗户提前安装在墙壁上
9. 房屋第一面墙将到位
10. 一层的墙壁装完后铺天花板
11. 在夜色中楼梯竖起来
12. 给屋顶铺上瓦，新房建成

Prefabricated House

Now it may come as a surprise that in 2004 alone more than £1.6bn worth of prefabricated houses were sold in Britain. They may be quick to put up but it takes a lot of hard work to bring it all together.

They begin with the wood, planks of spruce are painted on one side with a special emulsion that will protect them from the weather and prevent discoloration.

The treated planks are stacked on top of each other. The joints fit together to form the exterior wall. After a thick wooden strut is cut-to-size, it is slotted into the wall to lock it into place. Cork-studded cardboard is fixed to the inside of the wall to help prevent drafts. But all of the walls have to be fully insulated too. They use wood shavings gathered from factory waste.

But first the shavings have to be pretreated. They are tossed about in soda and a milk serum. The soda protects them from the fungus that can cause rotting and the milk serum makes them fire-resistant. This is the top half of the front of the house. It's going to be stuffed with the shavings. They're sprinkled inside the cavity and then roughly squashed into shape. Finally they are compressed to half of their original volume. And this is how it looks afterwards, any surplus is hoovered up to be re-used.

Next they fix a layer of cardboard on to the frame to seal in the layer of insulation. A plaster board is placed on top of that and after it's cut down to size, it's fixed into place by a very odd looking gadget, a stapler on wheels.

Next a row of slats is attached which creates a cavity to house the electrical cables. A second plaster board completes the wall which is now 30 centimeters thick. Before the House goes to the site the windows are fitted so when it's put together it can be sealed instantly and no humidity can get in. All of the walls are brought together with the help of a pulley and then shrink wrapped for the journey ahead. They are loaded on to a lorry and then sent off to become a new home.

Early in the morning the prefab arrives on site.

A crane makes short work of hoisting the whole ground floor off the truck in one foul swoop. The walls have been stacked in the order which they will be erected to save time and energy.

It's time for a key moment in the whole process, while a worker prepares some mortar to secure the house to its concrete foundations the first wall is lifted into position. If it is even slightly out the whole house will be skewed.

Once the exterior walls are in position the partitions can be slotted into. Wall by wall it slots together like a three dimensional jigsaw. With the ground floor walls in place the fitters are quick to add the ceiling. The last thing that anyone wants is for the house to fill with rain. As each section weighs more then a ton it's a race against time to get the job finished before night fall. Even in the dead of night the work continues. The next step is to fix the staircase into position. With that in place they can build upwards, which they do without even stopping for a tea break.

Just 16 hours after the prefab arrived on site the assembly is completed. The last nail is banged in and all that remains is for the roof to be tiled.

Fast, efficient, and incredibly practical the prefab truly is a spectacular achievement.

Did you know?

Over 22 million Americans live in prefabricated houses– more than the entire population of Australia.

锯鳐伐木机

扫描二维码，观看中文视频。

这是完全不同的伐木。加拿大最重要的自然资源之一，是它广袤的森林。多年来，数以百万英亩（1 英亩 =0.4047 公顷）长满树木的原野，成为木材行业和环保人士的战场。

但是，有家公司可能已经找到了解决方案的端倪。不是砍伐活着的树木，而是砍伐死掉的树木。通过锯鳐伐木机。世界各地大约有 2.6 亿棵树木竖立在水库和水坝的底部。当它们被淹没后，树死掉了，但木材被淹死它们的水保存下来。

穿着潜水设备操作链锯是一件非常危险的事，所以一家加拿大公司开发了完成这项工作的理想工具。这是一台遥控车，可以潜到湖泊的底部，并把死树用链锯锯倒。链锯超过 4 英尺（1 英尺 =0.3048 米）长，可以在几秒内切断死树。

37 个气囊的每一个，都被固定到一棵树干上，然后浮到水面。两个螺旋桨可朝任一方向旋转。它们引导锯鳐伐木机在水下的树丛之间穿行。

这个 3 吨重的机器被小心放到幽暗的深水处。当它的空气室被水充满时，锯鳐伐木机沉下去。上面安装八台摄像机，还有声呐。所以待在平台上的团队，可以看见它在湖底所遇到的东西。

控制室对它进行安全遥控。驾驶员用几个操纵杆引导它靠近一棵死树，准备砍倒。当锯鳐伐木机到位后，用一对夹子抓住树干，把自身固定，然后在充气之前把浮标钻进树干。

接着就该行动了。一旦锯鳐伐木机截断树干，气囊就会把巨大的树木托举到水面。如果没有这些漂浮装置，木头会沉到水底并丢失。

锯鳐伐木机不会停歇，它带有 37 个气囊，并打算把它们全都用上。这是一台非常高效的机器。从抓住树身，安上气囊，到锯断树干，一切都在 30 秒之内。使命完成了，锯鳐伐木机被起重机吊出水面。

靠着钉在树干的浮标，不需要很长时间就可以找到树木，并把它们拖回来。将原木固定在平台上，气囊被解开以便反复使用，不断地把珍贵的木材送到水面。因为这些树木经过多年的缓慢生长，木材密度很大。这一优势加上环保认证，使它们的价格比普通木材高 3 倍。

世界各地大约有 2.6 亿棵淹在水下的树木，锯鳐伐木机的牙齿在未来很多年都不会闲下来。

你知道吗？

锯鳐长着锯齿状的鼻子，用来探测和杀死那些掩埋在地下的猎物。

1. 世界各地水下约有 2.6 亿棵树
2. 遥控锯鳐伐木机能潜到水底锯断死树
3. 每个气囊都将被固定在一棵树上

4. 两个螺旋桨可朝任一方向旋转
5. 锯鳐伐木机上有摄像机和声呐
6. 水面上的驾驶员用操纵杆控制伐木

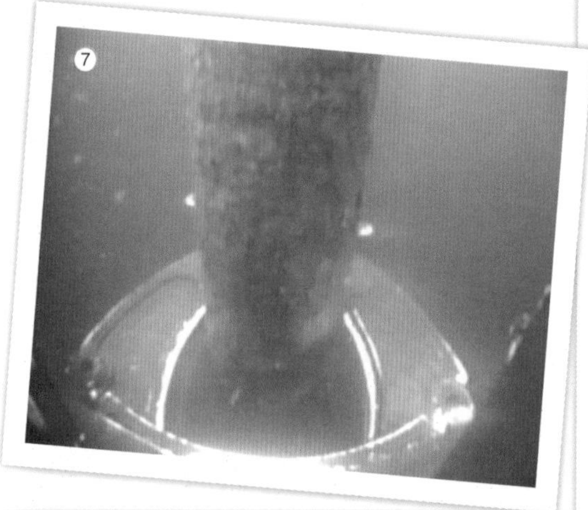

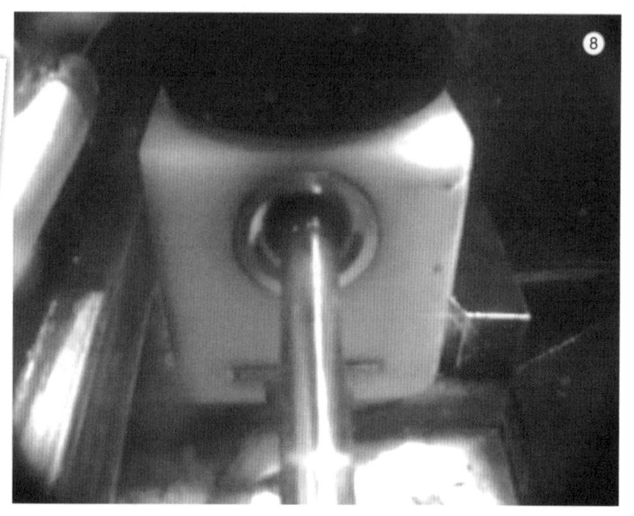

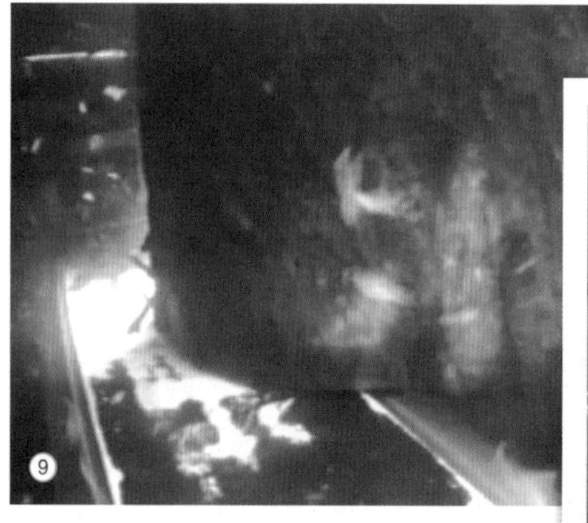

7. 伐木机夹子抓住树干固定自身
8. 把浮标钻到树干上后充气
9. 用 4 英尺长的链锯把树锯断
10. 气囊把巨大树木托出水面
11. 把木头固定在平台并收回气囊
12. 水下木头密度大又利于环保

Sawfish Harvester

Logging with a big difference. One of Canada's most important natural resources is it's vast forests. Over the years the millions of aches filled with trees have been something of a battle ground, between the timber industry and environmentalists.

But one company may have found a small part of the solution. Instead of chopping down living trees they chop down dead ones. With sawfish harvester. Around the world, there are about two hundred and sixty million trees that stand on the bottom of reservoirs and dams. When they were flooded the trees died but their wood has been preserved by the water that killed them.

It would be a very risky business to try and use a chain saw while you're wearing a scuba tank so a Canadian company developed the ultimate tool for the job. It's a remote-control vehicle which plunges to the bottom of lakes and cuts down the dead trees with a chain saw. The chain saw is over four feet long and can cut through the dead wood in seconds.

Each of these 37 airbags will be screwed into a trunk and were lifted up to the surface. There are two propellers which can spin in either direction. These steer the saw fish in and out the underwater trees.

The three ton machine is carefully lowered into the murky depths. As it's air chambers fill with water the saw fish sinks. The saw fish has eight cameras and sonar on board. So the team on the platform can see what it sees at the bottom of the lake.

It's remotely operated from the safety of a control room. A driver uses a couple of joysticks to steer it towards a dead trunk which is about to get the chop. When the saw fish is in position it deploys a pair of pinches to grasp the trunk and hold itself in place. Then it drills one of the floats into the trees trunk before inflating it.

Then it's time to get physical. As soon as the saw fish cuts through the trunk the airbag heaves the mighty tree to the waters surface. Without these floatation devices logs would sink to the bottom and be lost.

The saw fish doesn't hang about it's got thirty seven airbags and intends to use them all. It's a very efficient machine. It can attach itself to a tree, secure a float and saw through the trunk all in just 30 seconds. It's mission accomplished and the saw fish is hoisted out of the water by a crane.

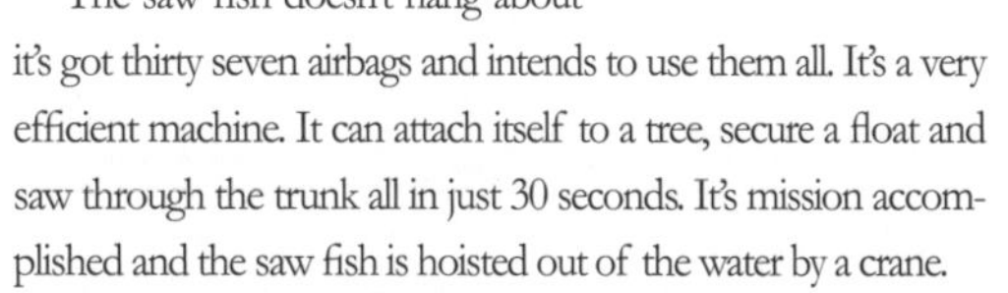

With the floats attached to the trunks, it doesn't take long to track them down and haul them back. The trucks have been secured to the platform and now the floats can be detached these will be used over and over again to keep bringing the valuable timber up to the surface. Because these trees grow slowly over many years the wood is very dense. This fact coupled with the eco-credentials will help the timber to fetch up to three times the price of ordinary trees.

With around two hundred and sixty million more emerged trees around the world, the saw fish will be getting it's teeth suck in for many years to come.

Did you know?

A sawfish is a fish with a saw-like snout which it uses to detect and kill buried prey.

高压电工

在高压电缆上工作是件危险的事。操作者要和带有345000伏特高压电的电缆打交道，把自己捆在直升机上。

他们可不能穿着牛仔裤和T恤衫干这种活。他们穿的所有衣服直到袜子，都布满特殊的金属丝网。其中的原理是这样的。电流总是选择最容易走的路径。由于金属的电阻远远低于肉体和骨骼，因此电只会安全地绕过身体而不是通过身体流动。

在美国，绝大多数这类工作是把维修工绑在直升机附加的平台上完成的。他们飞近和输电塔相连的高压线。因为直升机仅仅触及电缆，电流只在其周围环绕而不会产生到其他地方。

此时他在更换一个绝缘子，阻隔电顺着支撑杆向下泄漏。维修人员的手显然需要非常稳，驾驶员同样如此。众所周知，直升机即使没有人悬挂在边缘上也很难操纵。如果失去控制，毫无疑问会让两人会一起殒命。为了减少飞行负荷，小组的一个成员留在地面协调各种事务。飞行员在悬停的时候心越静越好。

有些工作需要数小时完成，这意味着飞机要几次往返加油。由于直升机外加了一个平台，因此不能全程加满油箱，否则总重量会破坏平衡，驾驶员也无法保持直升机的稳定。

如果工作点临近高压线塔，直升机就必须把维修员投放下来，让他孤身作战。这不仅需要消防队员的勇气，还要有走钢丝的绝技。如果你有恐高症，那就肯定不需要申请这份工作了。电流和风引起电线震荡，天长日久会造成螺栓松动。因此需要人工维护，有时还要更换器件。

当又一件折磨人的工作完成后，他终于能从高压线上走回来，停在一个能被直升机安全接走的地方。

如果还嫌这项工作不够繁难和惊险，此处，他只能在2根电线上行走。

这个职业的危险程度难以置信，即使给再多的钱，一般人也不愿意涉足。但对于电力公司来说，每到月底付出巨额的薪酬是非常划算的，因为这些伙计们每天为他们节省1600万美元。

坐在相对安全的直升机悬挂平台上，维修工要飞回基地休息了。他知道自己的英雄行为确保了电力流向千家万户。

你知道吗？

一束雷电能达到300万伏特的高压。

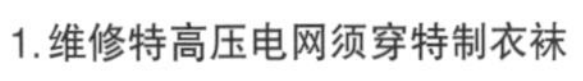

1. 维修特高压电网须穿特制衣袜
2. 维修工被固定在直升机附加平台上
3. 飞近和输电塔相连的高压线
4. 维修人员在更换一个绝缘子
5. 地面人员协调飞机的各种操作
6. 数小时工作飞机几次往返加油

7.外挂平台使直升机不可加满油

8.临近高压线塔工人只能独行

9.在高空行走需要勇气和绝技

10.完成高压线塔零件维修更换

11.在两根电缆上爬过和维修

12.完成任务等候直升机

High Voltage Workers

Working on high voltage cables is a risky business. The people who do it are handling cables that carry 345,000 volts whilst strapped to a helicopter.

They can't just wear jeans and a T-shirt when they're doing this job all of their clothes are lined with a special metal mesh right down to their socks. Here's how it works. Electricity flows along the easiest path. As metal has a much lower resistance then flesh and bone, the electricity is conducted safely around the body rather then through it.

Here in America most of the jobs are done with the repair man strapped onto a platform which is welded to a helicopter. They edge up to the power line which is attached to the tower. Since the helicopter is only touching the cable the electricity flows round it in a loop, and doesn't build up in anyone place.

Here he is changing an isolator, that stops the electricity flowing down the supporting pole. The repair man obviously needs to have steady hands and so does the pilot. Helicopters are notoriously hard to fly even without someone hanging off the edge. If it were to lose control now it would almost certainly be the end of both of them. To ease the pilots load a member of the crew remain on the ground to coordinate everything. The less the pilot has to think about whilst he's hovering the better.

With some jobs taking hours it can mean several refuelling trips. Because of the platform attached to the helicopter they can't fill the tank all the way up or the total weight will upset the balance and the pilot wouldn't be able to keep the helicopter steady.

扫描二维码，观看英文视频。

If a job is next to pylon then the helicopter has to drop the repair man off and leave him to make his own way. Not only does he need the courage of a fire fighter now he also needs the skills of tightrope walker. If you're scared of heights, then there's definitely no need to apply for this job. The electricity and winds make the lines vibrate which can cause the bolts to loosen over time. This means that they have to be serviced and occasionally replaced by hand.

With another exhausting job complete he can shuffle back down the line to a place where it's safe to be picked up.

As if that job wasn't tough enough, here he's only got two lines to move along.

This is an unbelievably dangerous job and most people wouldn't do it for all the money in the world but for the power companies it's well worth the hefty pay packets they hand out at the end of the month. These guys save them about 16 million dollars a day.

Back on the relative safety of the platform suspended beneath the helicopter the repair man can head back to base to put his feet up knowing that his heroics have kept power flowing to thousands of homes.

Did you know?

A bolt of light can contain up to 3 million volts of energy.

金矿

扫描二维码，观看中文视频。

南非大约因盛产钻石而闻名于世，但在这里人们正忙着挖金子。随着约翰内斯堡附近的金矿每天在炸药的轰鸣声中向地球深处掘进，岩石便奉献出了珍贵的黄金。每天大约 150 个矿工沉降到世界最深的矿井中，这里距地面达 1300 米。爆破是在每次换班之间实施的。因此，上班后第一项工作常常是清除昨天炸出的大堆矿石。

井下作业环境十分恶劣，温度高达 35 摄氏度，湿度几乎是 100%。矿工用高压水枪将岩石松动。当他们钻进地层深处时，要把木材截成合适的尺寸作柱子，支撑起矿井顶板的万钧重量。昨天的爆破清理完毕后，一位专家便会标记出新的爆破点，此时工人们用强力电钻在岩石上打出 200 个孔，每个深 1 米。这些炮眼备好后把炸药安上导火线，然后推送进炮眼中。一天的工作接近尾声，撤出矿井的时间到了。当矿工安全到达地面后，井下便开始了起爆。

第 2 天，9000 吨碎矿石被传送带运到地面，再被火车拉到加工厂。另有传送带把卸下的矿石运往车间，黄金将在这里进行提炼。把水和矿石混合，再用巨大的滚筒把它们碾碎，处理后的原料便被泵进一个广口的大罐中。加进氰化物，使金属和其他物质分离。

接下来是清水淘洗，留下的金沙被送进熔炉。融化的金沙倒进模具后迅速固化，这些产物看上去像纯金但却不是，其中仍然含有银和其他金属。在接下来的步骤中它将进一步被提炼。把得到的金属块再次放到 1064 摄氏度高温下熔炼，此时加进氯。由于氯气有剧毒，防毒面具和护目镜是必不可少的安全设备。这道工序使得其他金属上浮到表面，并被工人撇除。

这时候，熔融状态的金属几乎完全是纯金了。工人们将它倒进模具后变成金块。但在冷却和凝固之前，还要用煤气喷灯烧掉表面残留的杂质。一旦处理完毕，金条跃然而出。接着只用几秒的时间便从 1000 多度高温冷却到常温。每块金砖重 12.5 千克。价值超过 12 万英镑。这些岩石曾经亿万年沉睡地下，一旦从矿井下挖出来，送到地面上精心提炼，就变成了世界上最昂贵的金属之一——赤金。

你知道吗？

黄金如此稀有，世界上开采过的所有黄金能被容纳进 20 立方米的体积中。

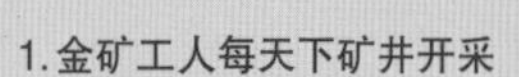

1. 金矿工人每天下矿井开采
2. 到井下先清理昨天炸出的矿石
3. 用原木支撑顶板的压力
4. 按专家标记打出 200 个炮眼
5. 炸药推送进 1 米深的炮眼里
6. 前一天爆炸的碎矿石被送到地面

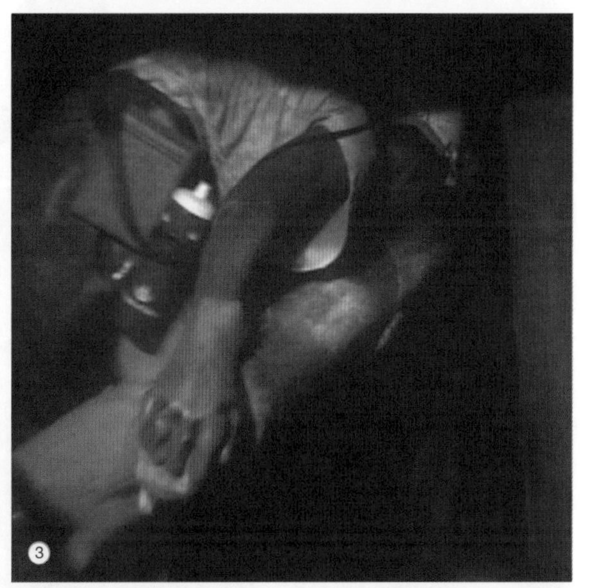

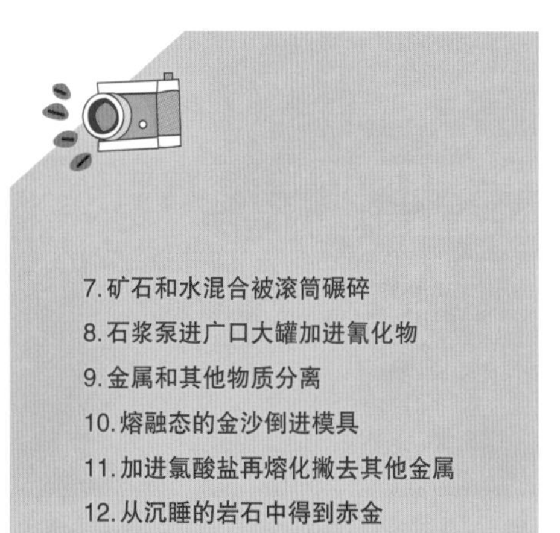

7.矿石和水混合被滚筒碾碎

8.石浆泵进广口大罐加进氰化物

9.金属和其他物质分离

10.熔融态的金沙倒进模具

11.加进氯酸盐再熔化撇去其他金属

12.从沉睡的岩石中得到赤金

⑦

⑧

⑨

⑩

⑪

⑫

Goldmine

South Africa might be famous for its diamonds, but here they're digging for gold. These mines near Johannesburg get deeper every day as a series of explosions blast away the rock to reveal the precious metal. Here around 150 men descend every day into the deepest mine in the world - thirteen hundred metres below the surface. Explosions take place in between each shift so the first job is to always clear out the rubble from yesterday's blast.

It's gruelling work down there with temperatures around thirty five degrees Celsius and humidity approaching 100 percent! To loosen the rock, they pound it with high pressure water hoses. As they burrow deep into the earth, they cut timber joists to size and use them as makeshift columns to support the massive weight of the mine's ceiling. Once the site of yesterday's explosion is cleared a specialist marks out the position for the new charges. Then the crew uses a powerful drill to create two hundred holes in the rock face which are each a metre deep. When the holes are dug they attach the explosives to fuses and slide them into the holes ready for action. At the end of the miner's working day it's time for them to evacuate the mine, and once they're safely above ground the explosives are detonated.

The Next day, nine thousand tons of rubble is dropped onto a conveyor belt that carries it to the surface and from there a train takes it to this processing plant. The rubble's offloaded onto another conveyor belt and it heads into the plant where the gold will be extracted. Water is added. And then huge barrels grind up the mix. The resulting slag is pumped into these open tanks. Cyanide is added which separates metals from the other materials.

扫描二维码，观看英文视频。

After it's rinsed with water, the sand that is left is shovelled into furnaces. The molten sand is poured into moulds where it rapidly solidifies. These bars may look like pure gold but they're not - they still contain silver and other metals - these will be extracted in the next step. The bars are melted at a thousand and sixty four degrees Celsius and then they can add chlorine. A gas mask and goggles is an essential piece of safety equipment as the chlorine gives off highly poisonous fumes. It causes the other metals to rise to the surface so they can be skimmed off the top.

Now the molten liquid is almost pure gold they pour it out into moulds where it will set into bars. But before it cools and solidifies they blast off any remaining impurities with a gas burner. Once set, the gold bullion is flipped out and cooled from over a thousand degrees Celsius to room temperature in just a few seconds. Each gold bar weighs twelve and a half kilograms and is worth more than one hundred and twenty thousand pounds. What was once rock blasted out of a mine has been brought to the surface and refined into one of the most precious metals in the world – solid gold.

Did you know?

Gold is so rare that all the gold ever mined in the world fit into a 20 metre cube.

加油站

防撞护栏

交通标志

安全带

交通篇

雨刮器

加长豪华轿车

旅行拖车

液化石油气汽车改装

修复老爷车

加油站

扫描二维码，观看中文视频。

给汽车加油是一桩日常生活事务。加油站到处都有，但你是否想过它们是怎么来的？不错，它们从地上的一个大坑开始。根据事先订好的计划，工程师大约需要 3.5 个月时间，把一大片中间挖了坑的空地变成全新的、功能齐全的加油站。

工程第一步是储油罐。将它们放置在地下，占用空间少并能受到保护，避开不靠谱的司机。当储油罐运到后，工程师首先要把它从卡车上卸下来并进行检查。如果要在其中储存易燃的汽油，储油罐就必须安全。

使用电压测试表，机械师检测油罐是否完全密封。暴露的金属会与仪表作用，产生电火花，警告机械师需要维修。使用喷灯和沥青石膏，机械师封住暴露金属，并再次测试以确保可靠。当机械师确信储油罐密封完好，便将它们沉降到预先挖好的洞中并放置到位。储油罐被放稳后要进行测量，它们必须是水平的，以便安装管道。大家都满意后，挖掘机开始用沙子填充储油罐周围的空隙，支撑它保持在正确位置。如果建造合格，每个储油罐至少能使用 20 年。

储油罐就绪后，下一步要修建检修孔，包括其中的管道连接。管道系统把燃料从地下储存罐送到地面油泵，从这里给汽车加油。维修孔包括对管道连接处的操作，这使维护工作更容易进行。这些管道输送易燃物，任何泄漏都是灾难性的，因此管道焊接得非常牢固。

地面的工作现在可以开始了。先铺设水泥地面，工人接下来建造熟悉的商店和重要的油泵。用预制板来建造店铺的墙壁。

接下来机械师取出来数百个这样的螺栓。做什么用呢？它们拿来固定加油站前常见的顶棚，像一把大伞，下雨时确保顾客不被淋湿。初具规模后，其他工人把商店内部也装修起来。从放热咖啡、小吃和地图的货架，到监控加油的电脑，都安装齐备并进行全面测试。

最后把这些非常重要的油泵安装完毕。它们需要在管道系统中精确定位，然后与地下的储油罐连接起来。

只有在万无一失后，油罐车才会开来，将数千升无铅、超级无铅和柴油燃料充满储油罐。终于，现场管理人员测试油泵。这是为了校准。如果油泵显示供应了 10 升，那么顾客就应该准确得到 10 升汽油。

安装在新店面的最后一件东西，是公司名牌和那些可怕的燃料价格牌。虽然看来油价天天上涨，但一个新的加油站从无到有只需要 3.5 个月时间。

你知道吗？

英国第一个加油站是什鲁斯伯里的渡鸦车库。它建于 1913 年。

1. 加油站的建设是从地面上的一个大坑开始的
2. 用电压测试表检测储油罐是否密封
3. 用喷灯和沥青石膏封住暴露金属
4. 密封好的储油罐沉降到位
5. 水平仪测量储油罐是否放平
6. 在储油罐周边填充沙子

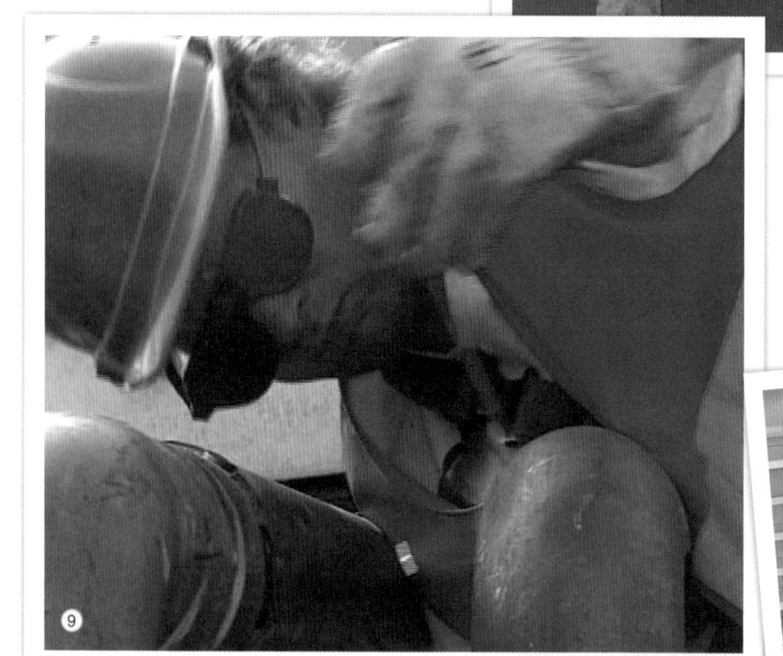

7. 修建检修口包括管道系统
8. 管道连接处的维修孔
9. 管道焊接得非常牢固
10. 店铺货架及监控加油的电脑备齐
11. 油泵与地下储油罐对接
12. 地下储油罐注满油，新的加油站建成

Petrol Stations

Filling your car with petrol is a fact of life. Petrol stations are everywhere, but have you ever wondered where they come from? Well, they all start with a big hole. Following well drawn up plans, engineers need about 3 and a half months to turn a big empty space with a hole in the middle of it, into a brand new fully functioning petrol station.

The first step of the process is the storage tanks. They're housed underground because they take up less space that way and they are also protected from erratic drivers. The first task for the engineers when the tanks arrive is to get them off the trucks and inspect them. If you're going to store flammable petrol in them they need to be safe.

Using a voltage test meter, the mechanic tests whether the tanks are perfectly sealed. Exposed metal will react with the meter and produce sparks warning the mechanics that repairs are needed. Using a blow torch and a tarmac plaster, the mechanics seal up the exposed metal and test it once more just to be sure. When the mechanics are satisfied that the tanks are completely sealed, they will be lowered into the pre-dug hole and put in position. The tanks are stabilised so that they can be measured. They must be level for the pipes to fit. And when everyone's happy the diggers will start filling the surrounding space with sand to hold it in the right spot. If it's properly built each tank should be serviceable for at least 20 years.

With the tanks in place the manholes where the pipes are connected are next. The network of pipes will then carry the fuel from the underground storage to the pumps above ground where you fill your vehicle. Access to the connections is included in the manhole which make maintenance work easier to carry out. With all the flammable fuel these pipes will be carrying, any leakage could be disastrous, so the pipes are welded together securely.

Work can now begin above ground. A cement floor is laid and the workers can now build the familiar shop and the all-important pumps. Prefabricated sections are used to erect the walls of the shop.

Next the mechanics count out hundreds of these bolts. Their job? They will hold up the cover that is traditionally built over the station fore-court. It makes an excellent umbrella if its raining which helps keep customers dry. As it takes shape, other workers are kitting out the inside of the shop. Everything from the shelving for the hot coffee, snacks and handy maps, to the computers that monitor your petrol purchase are fitted, and tested thoroughly.

And finally those all-important pumps are fitted into place. They need to be positioned perfectly over the pipe network. They will then be joined up with the tanks below the ground.

Only when they're fully in place, does the tanker arrive and fill the storage containers with thousands of litres of Unleaded, Super-Unleaded and Diesel fuels. Finally, the site managers test the pumps. This is to calibrate them. If the pump says it's providing 10 litres, that's exactly how much the customer has to get.

And the last piece to be fitted to the new forecourt is the sign with the company name and those dreaded fuel prices. Although it seems like those prices rise everyday, it only takes three and a half months to raise a new petrol station from scratch.

Did you know?

The first petrol station built in the United Kingdom was the Raven Garage in Shrewsbury. It was built in 1913.

防撞护栏

扫描二维码，观看中文视频。

每年都有数以百万计的英国人沿高速公路来往旅行。也许只有很少人停下来，看看路中间令人惊叹的钢制安全屏障。这是一种挽救了许多生命的创新。

最初的设计使用的是混凝土，但技术进步让生产对象转向钢铁制品。这两种材料的性能，在遭遇碰撞时表现不同。

高速行驶的车辆，在前进方向有很大动能。当它撞上完全不动的水泥护栏时，能量无处可去，汽车毫不含糊地被撞回去。混凝土不吸收车辆的任何能量，而只会让它转向。

混凝土的替代品是预制钢护栏。它看起来比混凝土更脆弱，更有柔性，但这也正是防撞护栏能拯救生命的特性。

当车辆撞向护栏时，钢铁缓和冲击。动能被吸收，减缓汽车向前的运动。同时，像水泥护栏一样，钢铁护栏也能防止汽车冲入对面的逆向车道。

这种重要的安全保护设施是如何建造的？

成卷的原料钢安放到这台机器上。每卷重约 8 吨，足够制作约 170 个护栏。

钢料送进这个巨大的压床，钢板被压平后经铡刀裁剪到合适长度。每块 4.3 米。

扁平的金属相对较弱。一个巨大物体高速度撞击时，会很容易弯曲，它不能吸收冲击力并防止重大事故。

压床解决了这一问题。用惊人的 800 吨压力，将钢板压成波纹状，使金属强度大增。褶曲强化的钢结构，使得这些栅栏可以吸收 7 吨以上的冲击力。

下一个步骤是准备将褶曲栅栏安装到路边。每块都送进这台压床，内有定做的模板。它在金属上冲出准确的螺栓孔。

如果想要护栏在开放的道路上使用多年，就需要在自然环境中进行有效保护。钢不是特别顽强的金属，很容易生锈，所以护栏要镀锌。

镀锌将形成坚韧的外层，具有保护作用，能耐受 30 年的自然天气而不会生锈。

这些护栏是用来保护德国高速公路的。高速公路的某些部分没有速度限制，使得中线的保护更为重要。

护栏被螺栓固定到位，然后用机械拧紧。

成千上万公里的护栏，保护着欧洲的司机。单单在英国，沿主要道路就装有超过 10000 千米护栏。

单株树木也能造成对司机的威胁，一种创新设计可以节省材料和人力。

在德国，人们提出了一个量身定做的解决方案，来隔离路边特殊的障碍物，比如树或者灯柱。

把塑料塞在两端来减缓车辆的冲击速度，弯曲的护栏也有助于阻止事故车辆翻越护栏。

护栏根据所保护的障碍物而弯曲，然后焊

你知道吗？

在英国，每年有 40 人死于车辆越界事故——驾驶员将车辆越过中央分道区撞上迎面而来的车辆。

在一起，增加强度。装好充填塑料，唯一剩下要做的是碰撞测试，看它是否能够胜任工作。测试过程要拍摄归档，并准许科学家研究结果。这些信息有助于改进设计方案，从而拯救更多生命。

当尘埃落定后，专家检查车辆碰撞的现场，看起来护栏成功地发挥了作用。无论车辆和护栏的损坏都显著减少了。

所以，你下次在公路上拥堵时，凝视一下这些护栏吧。请记住，这一条钢带让驾驶更加安全。

1. 汽车撞上混凝土护栏会被撞回
2. 车辆撞向钢铁护栏会缓和冲击
3. 钢铁护栏能防止车辆冲入对面车道
4. 制作钢铁护栏先把钢材压平
5. 把钢板裁剪成合适长度
6. 钢板压成了波纹状增加强度

7. 将波纹钢板冲出螺栓孔
8. 钢板要镀锌防锈
9. 为不限速的公路安装中线护栏
10. 螺栓固定护栏后用机械拧紧
11. 遇特殊障碍物护栏须弯曲
12. 钢铁护栏冲撞试验现场

Crash Barriers

Every year millions of Britain's travel up and down the motorways. It's maybe just as well very few stop to look at the amazing steel safety barrier along the middle. It's an innovation that has saved many lives.

Original designs used concrete but technological advances saw production turn towards the steel variety. The two substances have different effects on qualities when struck.

A speeding vehicle has a lot of kinetic energy in the direction it's travelling. When it hits the completely immovable concrete barrier, the energy has nowhere to go so the car or vehicle literally bounces off it. The concrete doesn't absorb any of the vehicle's energy, but instead deflects it.

The current alternative to concrete is to use formed steel barriers. It may seem weaker and more flexible than concrete, but these are also life saving qualities for a crash barrier.

As the vehicle strikes the barrier, the steel cushions the impact. The kinetic force is absorbed, slowing the cars forward motion. And, like the concrete barriers, the steel stops the vehicle from passing into lanes of oncoming traffic.

But how are these vital safety measures put together?

Rolls of raw steel are fitted to this machinery. Each roll weighs about 8 tons, that's enough metal for about 170 barriers.

The steel is fed into this giant press where it's flattened out and a guillotine trims the roll down to the right length. These ones are 4.3 meters long.

Flat metal is relatively weak. A large object striking it at speed would bend it too easily and it wouldn't absorb the force needed to prevent a major accident.

However, this press takes care of that problem. Using a staggering 800 tons of pressure, the sheet steel is corrugated and this is what gives the metal its added strength. The folds reinforce the steel's structure to such an extent that these barriers can now absorb an impact of over 7 tons.

The next step is to prepare the freshly folded barriers so they can be installed at the road side. Each one is passed into this press which has a custom-made template inside. This stamps out the correct bolt holes in the metal work.

Now if they're going to survive for years by the open road, they're going to need strong protection from the elements. Steel is not a particularly hardy metal and is prone to rust, so the barriers will now be coated with molten zinc. This coating, or galvanizing, creates a tough outer layer of protection and they should be able to withstand 30 years of weather without rusting.

扫描二维码，观看英文视频。

These barriers are destined to protect the roads of the German autobahns. Some parts of the autobahn have no speed limit, making that central protection even more important.

The barriers are bolted into place, then and machined closed.

Thousands of kilometres of them protect drivers all over Europe. In England alone there are well over 10,000 kilometres bolted along major roads.

Individual trees can also pose a risk to drivers, but an innovative design is saving material and manpower.

In Germany, a tailor-made solution has been developed to isolate specific roadside obstacles like trees or lamp-posts.

The plastic inserts at the ends will absorb some speed of an impacting car, and the barriers' curvature will also help to stop a vehicle from crashing through.

The barriers are curved according to the obstacle they're guarding and then welded together for extra strength. The plastic inserts are put into place and the only thing left to do is run a crash test to see if it's up to the job. The test is filmed for the record and to allow scientists to study the results. The information helps improve designs which might save even more lives.

When the dust has settled and the experts examine the crash scene, it appears the barrier has worked successfully. Damage to both car and barrier has been significantly reduced.

So the next time you find yourself in a traffic jam, staring at the barriers, remember that this band of steel makes driving safer.

Did you know?

Crossovers–incidents where drivers pass over the central reservation into oncoming traffic–result in up to 40 death in the UK every year.

交通标志

扫描二维码，观看中文视频。

无论你在上班高峰时期开车经过伦敦，或者是在欧洲度假，你总能依靠路牌的指引。但这些标志是如何制作出来的，让每个人包括游客都能够看懂？这个德国办公室制作用途广泛的各种路牌，设计师的工作是要确保它们容易被所有人理解。

标志用于指引各种方向。它们能告诉你在哪条街上，或者限速是多少。形状也很重要，圆形标志告诉行路者他们应该做的事，而三角形警告前面有危险。甚至颜色也是重要的。在英国，有蓝色边框的圆形符号告诉行路者他们必须做什么，而红色边框表示他们不可以做什么。

标志本身由多层压板和反射材料等组成。一旦设计师们满意了，打印机和切割机就开始工作。各层材料要结合严实，这点很重要，气泡会产生不寻常的反射，在夜间分散司机的注意力。这台压床确保基底层平滑地贴在空白金属圆盘上。

生产团队员工确保材料切割整齐，具有正确的形状。对于标准的路牌，德国制造商使用两层金属箔。这些箔层含有玻璃球，白天能吸收一些环境光，或夜晚前车灯的光束，这意味着减少反射给司机的光。

选取的第三类箔片用于重要的警示标志。它不含有玻璃球层，所以日光或前车灯光以最大强度反射给司机。这种材料是用来提醒司机改变限速的，或者前方道路有危险。演示表明，这种材料的反射能力有多强。用高反射率的第三类箔层覆盖，这个限速标志放置在100多米长的大厅另一端。即使在这个距离，一个低功率的手电筒也足以把标识照亮，即使最微弱的光源也能被清楚地反射回来。

据估计，仅英国的道路系统就有超过250万个标志和交通信号。虽然有些可能被忽视，所有的高速公路标志每年会清洗一次，帮助司机提高或降低车速。

要生产一个新标志，印刷工拿一张设计方案的彩色底片，把它和新鲜的颜料一起送进印刷机里。把准备好的带有反射箔的空白标志放置到位，印刷机开始工作。

这个生产设施每天可以制作约1000个标识，足够一个中型城镇使用。老板对产品做常规检验。他的公司所做的工作能帮助挽救人们的生命，因此所有标识都要符合高标准。

除了用机器打印，一些标识仍然用手工制作。这是一个比较耗时和费钱的过程，工人必须知道他们在做什么。

拿这个限速标志的生产为例。技术人员的工作是把数字完美地排放在圆盘中心。她只有一次机会把它做好。完成的标志不允许有气泡。经验和技能让她精确无误。

标志的另一个重要部分是如何向司机展示。

你知道吗？

交通标志收藏家大卫·摩根的藏品保持着最独特标志的世界纪录。他收藏有137块有代表性的交通标志。

大部分标志安装在路边的杆子上，所以需要有安螺栓的孔洞。有些标志使用更复杂的安装系统。

首先是框架，用一条不生锈的铝条弯曲成形。然后将标志安装到框架中，并铆接在一起。最后，一个专门的压床用来弯曲标志周围的框架。这额外的紧固，使标志更经久耐用，能抵御大风或过往车辆产生的气流。

从不同寻常的旅游信息到重要的警告标志，都保障着你的旅行安全。迷路应该不是大问题，只要跟着交通标志走就行了。

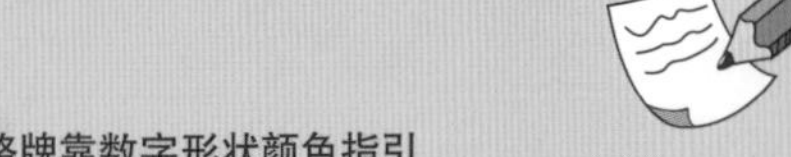

1. 路牌靠数字形状颜色指引
2. 设计路牌要容易理解
3. 标志由多层压板反射材料制成
4. 让基底层平滑贴在金属盘上
5. 确保切割整齐形状正确
6. 标志用含有细玻璃球的金属箔

7. 警示标志反射强度大
8. 印刷生产新的标志
9. 一些标识要用手工制作

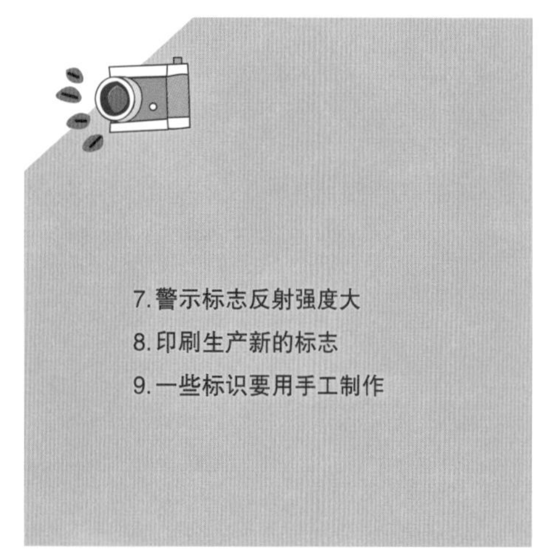

7

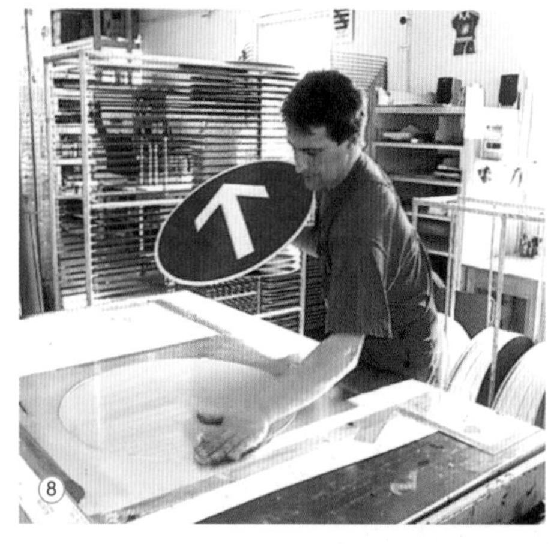

8

10

9

10. 完成的标志不允许有气泡
11. 标志安装到框架中紧固
12. 为标志牌打孔便于悬挂

11

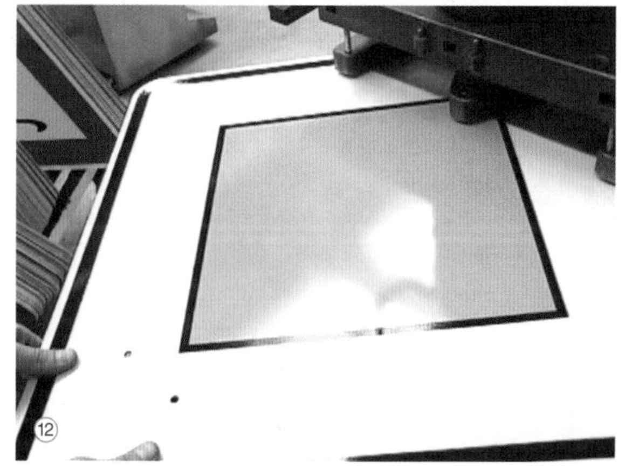

12

Traffic Signs

Whether you're driving through London in rush hour, or you're on holiday in Europe, you can always rely on the street signs for guidance. But how are these signs put together so that they can be understood by everyone, including tourists? At this German office, road signs are produced for a wide variety of uses and it's the designer's job to make sure they are easily understood by all.

Signs are used to deliver all sorts of directions. They can tell you which street you're on or what the speed limit is. The shape is important too, round signs are for informing the road user of what they're expected to do, whilst triangular shapes warn of hazards ahead. Even colour is important. In Britain a circular sign with a blue border tells the road user what they must do, whilst a red border indicates what they must NOT do.

The signs themselves are made up of multiple layers of laminate and reflective material. Once the designers are satisfied the printers and cutters get to work. It's important the layers are applied properly as bubbles could cause unusual reflections that might distract drivers at night. This press ensures the base layer is applied smoothly to the blank metal disc.

A member of the production team then makes sure it's neatly cut to the right shape. For standard road signs, the German manufacturers will use two foil layers. These foils contain glass balls which absorb some of the ambient light during the day, or headlight beams at night, this means less light is reflected back to the driver.

An alternative third type of foil is used for important warning signs. There is no layer of balls here so the full strength of daylight or headlights is mirrored back to the driver. This material is used to alert the driver to changing speed limits, or dangers in the road ahead. A demonstration shows just how powerfully reflective this material is. Covered with the highly reflective third type of foil, this speed limit sign is placed at one end of hall that is over 100 meters long. Even at this distance, a low-powered, hand-held torch is enough to illuminate the sign that clearly reflects even the most feeble light source.

It's estimated that there are over 2 and a half million signs and traffic signals on the UK road system alone. Although some may appear neglected, all motorway signs are washed once a year, helping to keep motorists up or down to speed.

To produce a new sign, the printers will take a colour negative of the design and add it to the printing press along with some fresh paint. A ready prepared blank that already includes the reflective foil is put into place and the press gets to work.

This production facility can produce about 1000 signs a day, which would be enough for a medium sized town. The boss makes regular checks of the output. The work his company does could help to save lives so it's important that all signs are made to a high standard.

As well as machine printing, some signs are still produced by hand. It's a more time-consuming and costly procedure and the staff have to know what they're doing.

Take the production of this speed limit sign for example. The technician's job is to line up the numbers perfectly in the centre of the disc. She only gets one shot at getting it right and no bubbles are allowed in the finished work. Her experience and expertise help her to get it spot on.

Another important part of the sign is how it's displayed to the driver. Most are mounted on poles by the side of the road so holes are needed for the bolts that they will be attached with. Some signs are produced with a more sophisticated mounting system.

First, the frame, made out of a strip of rust-free aluminium, is bent into shape. The sign can then be fitted into the frame which is then riveted together. And finally a special press is used to bend the frame around the sign. This extra grip makes the sign more durable against strong winds or turbulence from passing vehicles.

From unusual tourist information to vital warnings to keep you safe on your travels, getting lost shouldn't be too much of a problem; just follow the signs.

Did you know?

Traffic cone collector David Morgan holds the world record for the most unique cones in a collection. He owns 137 of the iconic traffic guides.

安全带

自从1983年以来，英国司机在汽车里被强制性系安全带。据估计，一年大约拯救2000条生命。

根据法律，轿车和面包车必须配备安全带，保护所有乘客和司机。但如何制造出这种单薄的黑色带子，让它坚韧到足以在高速撞击中拯救生命？

这是聚酯纤维，它是用于编织安全带的基本材料。为了能承受极大的力，安全带需要多层编织。

制造多层安全带的第一步骤，是制成线材的平行结构，称为经纱。接着把强化的线材编织其中，称为纬纱。

织针每分钟超过2500次上下运动，把线材来回通过基础结构。没有慢动作，快速编织看起来一片模糊。要看清它是如何工作的，我们需要停住机器，一步步观察。你能看到这块金属板在纬线对齐后将其收紧，但首先它们是如何编织到位的？

红线是一根纬纱。它穿过经纱。绿线设计成固定纬纱的锁。它被勾住和带回，环绕纬纱循环。然后金属板关闭，将织物收紧。红线再次被拉过来，这个程序不断重复，生产出超强的结构，当你被迫急刹车或出车祸时也不会崩断。

做出的安全带是一条连续的长带子，需要截断以适合每辆汽车，裁剪过程在楼下进行。而在裁剪成合适尺寸之前，它需要通过烤炉。聚酯安全带被加热到230摄氏度并进行拉伸，使分子结构得到加强。

这家工厂每天能生产40000多条安全带。事关安全性，一定要坚持高标准，新的安全带必须通过防火测试。样品被点燃并计时，要通过这项测试，火焰一分钟内的蔓延速度须少于10厘米。

不用说，安全带必须足够结实以抵御破坏。如果在受力下断掉了，就没多大用处，所以要进行常规测试。

这些安全带必须能经受28千牛的力。大约等于在短短的1/10秒内，阻止70千克重的人以每小时95英里速度行进所需要的力。

编织完毕后，50道激光束对安全带扫描寻找缺陷，但彻底检查还需要人工触摸。测试员的指尖非常敏感，能发现机器错过的瑕疵。如果机器和人都对安全带的强度认可，它将被裁到3.2米长。

而工人必须很小心。剪开织带会减损强度，厂家不能让它散开，所以额外补上72针，以确保安全。接下来是加上重要的扣环。安全带穿过用来连接座位插槽的锁扣，并缝纫固定。同时加上标识，以便厂家了解这条特定的安全带装配于何时何地。

接下来是安全带的工作机制，既能救命，也能让你爬进爬出毫无麻烦。

你知道吗？

一根普通安全带可以安全地承受一个重达2.8吨的物体，大约相当于一头母象的平均体重。

白色塑料棒安装到装有弹簧的转轴上。转轴里隐藏着安全带工作的秘密。正常运动情况下，左侧的银色金属球仅稍微移动，不会激发锁定机构，可以让司机或乘员自由拉动。但是，任何的突然运动将使金属球把臂杆推向连在安全带上的白色塑料管。由此带来齿轮被锁定和停转，从而阻止安全带运动，能把使用者牢牢固定。

回到生产线上，每个锁扣都与插槽样板进行测试，以确保良好的配合。生产到此完毕。

按照保守估计，过去25年中，在以汽车为主要运输方式的国家里，安全带拯救了超过30万人的生命。仅在英国，据信事故死亡率就降低了惊人的20%。

1. 轿车和面包车必须配备安全带
2. 聚酯纤维是基本编织材料
3. 制造多层安全带首先要制成经纱

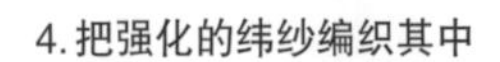

4. 把强化的纬纱编织其中
5. 编织好的长带进入楼下烤箱
6. 聚酯带被加热并进行拉伸

7. 样品进行防火测试

8. 安全带必须经受 28 千牛的力

9. 为剪开的安全带额外补针

10. 联动设计会让齿轮锁定

11. 锁扣都与插槽进行测试

12. 安全带大大减少交通伤亡

Safety Belts

But first…Since 1983 it been compulsory in the UK for drivers to wear a seat belt in a car, and it's estimated they save around 2000 lives every single year.

By law cars and vans must have safety belts fitted for the protection of all passengers and drivers. But how do you make a thin black strap strong enough to save lives in a high speed smash?

This is polyester fibre and it is the basic material that is used to weave safety belts from. To make them as tough enough to withstand extreme forces, seat belts need more than a single layer of weave.

The first step in making a multi-layer belt is to create a parallel structure of threads. This is known as the warp. The strengthening threads will then be woven through this and those threads are called the weft.

Needles race up and down over 2,500 times a minute passing the threads back and forth through the basic structure. Without slow-motion this swift stitching becomes a blur. To see exactly how it works we need to stop the machine for a step-by-step look. The bar you can see here tightens the weft threads once they are properly aligned but how are they woven into position in the first place?

The red thread is one of the weft threads. It's passed through the warp. The green thread is a lock designed to hold the weft in place. It will now be caught and carried back looping it around the weft. The bar then closes, which tightens the weave. The red thread is then carried through again and the process continues over and over creating the super-tough structure that won't snap if you are forced to brake hard or if you have an accident.

The emerging belt is one long continuous strip which needs to be cut down to be fitted to each vehicle, but the trimming process takes place downstairs. However, before its cut to size, it's passed through this oven. By heating the polyester band to 230 degrees Celsius and stretching it, the molecular structure is strengthened.

This one factory can produce over 40,000 individual seat belts every single day. Where safety is concerned, standards have to be kept high and the new belts must now pass a fire test. Samples are set alight and timed. The flames may not travel more than 10 centimetres in a minute for the belt to pass this test.

And it goes without saying that a belt must be strong enough to resist breaking. If it snapped under pressure, it wouldn't be much use, so regular tests are carried out.

These belts must survive up to 28-kilo Newtons of force. This is about the same as the force created when trying to stop a 70 Kilogram person travelling at 95 miles an hour in just 1 tenth of a second.

Once they're fully woven, 50 laser beams scan the belts for imperfections, but a thorough check requires a human touch. The tester's fingertips are so sensitive they can find flaws that the machinery may have missed. When both machines and humans are satisfied the belt is strong enough, it is cut down to 3.2 metre lengths.

However the workers have to be careful. Cutting the woven belt weakens it, but the manufacturers don't want them unravelling, so an extra 72 stitches are put in place to keep them secure. The next step is to add the all-important buckles. The belt is looped through the tongue attachment that will fit into the clasp in the car and sewn into place. An identification tag is also included so the manufacturers know where and when this particular belt was assembled.

Next, the working mechanism of the safety belt that both keeps you alive and lets you climb in and out of your car without any hassle is built.

The white plastic bar is fitted into the spring mounted roller. This roller holds the secret to how the safety belt works. Normal movement causes the silver ball on the left to move only a little. This doesn't activate the locking mechanism and drivers or passengers can move freely. But, any sudden movement will cause the ball to push the arm against the white plastic tube attached to the belt. This engages a lock on the gear wheel, the gear wheel stops revolving, which stops the belt and holds the wearer firmly in place.

Back on the production line, each tongue is tested in a sample of the clasps to ensure a good fit, and the belt is finished.

Conservative estimates suggest that over the past 25 years in countries where cars are the main form of transport, seat belts have saved over 300,000 lives. In Britain alone, they are believed to have reduced accident fatalities by a staggering 20%.

Did you know?

An ordinary seat belt can safely support an object weighing up to 2.8 ton, about the average weight of a female elephant.

雨刮器

扫描二维码，观看中文视频。

雨天下班驾车回家总让人受罪，雨刮器的发明让开车人能看清前面的路。现代雨刮器由两个关键部分组成，一是固定雨刮片的臂杆，二是柔性的橡胶雨刮片本身。雨刮片生产始于这种外观奇特的材料。这种古怪而粗陋的东西是原始形态的橡胶。要用合成橡胶也能行，但这家工厂使用天然橡胶，由马来西亚橡胶树所收获的树液制成。

生橡胶首先需要滚压。目的是彻底去除其中所有气泡，这需要花点时间。约 30 分钟后，它变成这个样子，更光滑，更有弹性。接下来加进雨刮片橡胶中最重要的成分——炭黑粉末。这使橡胶具有传统的黑色，并增加强度以抵抗外界损害。橡胶已经铺开，现在从传送带输进搅拌机，炭黑将被添加进去。从搅拌机出来的混合物，看起来更像我们熟知的传统橡胶。

我们已经得到了雨刮片的橡胶，但制作雨刮器远远不止于此。下一步是制造臂杆，用来固定和带动雨刮片在风挡上来回运动。它由这种钢板制成。钢板被送入压床，它非常精准，极少浪费材料。这台压床每天能生产超过 90 万个雨刮器臂杆。成型的雨刮器臂杆仍然附着在原材料钢板上。它们暂时这样待着，以便在生产过程的下几个阶段进行运输。

雨刮器的任务是清理挡风玻璃。这意味着它们将暴露在最恶劣的天气里，所以需要良好的防锈蚀保护。这个装置是用来给雨刮臂杆上涂料的。涂料颗粒在 80000 伏电压下带上静电。有助于涂料吸附在金属表面。水幕使喷涂区保持隔离状态，从空气中滤掉多余的涂料。

雨刮器臂杆被送去烘干。现在开始把新鲜橡胶制成雨刮片。首先将原料送到挤压机中，迫使它变为正确的形状。一根连续的橡胶条切成两半，可以做成两个单独的雨刮片。如果保持这个长度将会做出很长的雨刮片，但现在仅仅是生产过程的开始。接下来长橡胶条通过热的盐水浴变得更加结实。现在橡胶条被切成适当尺寸。这家工厂为多种汽车生产雨刮片，需要不同型号风挡的雨刮器。雨刮片被切成合适的长度，现在可以分成两半了。旋转的铡刀把橡胶条切成完美的两半，产生的结果是新雨刮片。

为确保橡胶的高质量，每批产品都要取样，并通过严格测试。在这里将脏水喷在模拟挡风玻璃上，然后监控雨刮器的清除能力。

这个喷嘴在零下 30 摄氏度的风洞里向汽车吹雪。每个雨刮都要在这种条件下生存 24 小时，并不停顿地工作。如果它们不能清除积雪，或出于某种原因被卡住，就算测试失败。工程师利用这段时间来评估雨刮器臂杆的设计以及整体的性能。如果雨刮片和臂杆能胜任工作，工程师就开始将它们组装起来。

这在雨刮器工厂是一项重要而重复的工

你知道吗?

有毛病的风挡雨刮器，会让你付出多达 *1000* 英镑和驾照扣 *2* 分的代价!

作。她的任务是把新雨刮片送入机器，使其与臂杆结合在一起。机器人将抓手滑动到雨刮片上方，并将它们夹紧到位。

新做好的雨刮器通过质量控制计算机的摄像头。人的眼睛也用来检查从机器里出来的每个新雨刮。除了清理挡风玻璃上喷洒的尘土和淤泥，雨刮器也必须应付像秋季枯叶一类大点的东西。还有泥泞的长途旅行后产生的大量污垢。制造商建议你时常清理雨刮器，这将延长它的使用寿命。再加上玻璃洗涤液，无论天气如何，你都会有清晰的视野。

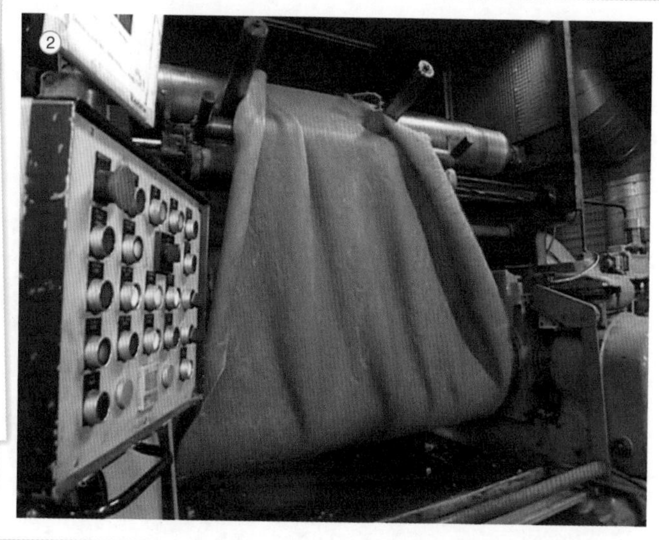

1. 雨刮由臂杆和橡胶刮片构成
2. 滚压使橡胶变得光滑有弹性
3. 准备制作雨刮器的臂杆
4. 压床把臂杆从钢板上冲压出来
5. 臂杆要进行防锈蚀处理
6. 涂料带静电并吸附在金属表面

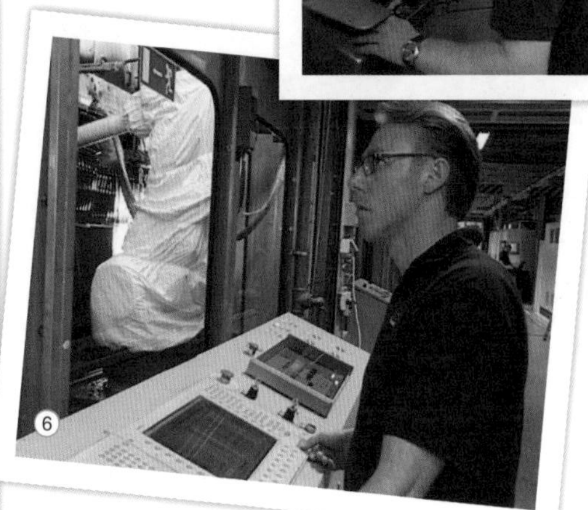

7. 新鲜橡胶送进挤压机
8. 橡胶条分成两个雨刮片
9. 在风洞中向汽车喷雪
10. 测试雨刮器在极寒天气的性能
11. 新雨刮片与臂杆结合
12. 雨刮器让你视野清楚

Windscreen Wipers

But first, the drive home after work always miserable when it's raining, but the invention of the windscreen wiper has helped provide a clear view ahead for many car drivers. The modern windscreen wiper is a made of two key parts. First, the arm that holds the wiper blade and second, the flexible rubber blade itself. Wiper blade production begins with this bizarre-looking material. This strange and lumpy substance is rubber in its raw form. Synthetic rubber is available but this factory uses the natural variety, made with sap harvested from rubber trees grown in Malaysia.

First the raw rubber needs to be rolled out. The aim is to completely remove all the air bubbles from substance which can take a little time. After about 30 minutes, it looks more like this. Smoother, and well, more rubbery. Next one of the most important elements of wiper blade rubber can be added. Carbon black powder. This gives the rubber its traditional black hue, and strengthens it to help it resist the elements. The rubber has been rolled out and now rolls off the conveyor into the mixing machine where the carbon black will be added. What emerges from the mixer looks a lot more like the traditional rubber that we would recognise.

So, we've got the rubber for the blades, but there's more to wipers than just that. The next step is to build the arms that will hold them and carry them back and forth across your windscreen. They're made using this sheet steel. It's fed into this press which is very accurate and wastes very little metal from the roll. This press can produce over 900,000 arms everyday. The shaped arms are still attached to the band of raw steel. They are left like this for now as it makes them easier to transport through the next few stages of the production process.

The wiper's job is to clear your windscreen. This means they'll be exposed to all of the worst weather, so they'll need excellent corrosion protection. This device is used to paint the arms. The particles are electro-statically charged using 80,000 volts. This helps attract the paint to the metallic surfaces. The water curtain keeps the painting area isolated by filtering excess paint out of the air.

The arms are sent to dry and the process of turning the fresh rubber into wiper blades can now begin. First the raw material is squeezed into the extruder. This forces it into the right shape to make a continual band that, when cut in half, will form two separate wiper blades. Now, like this they would be very long wiper blades, but this is only the beginning of the process. Next the long band is passed through a hot, salt bath which toughens up the rubber. And now the band is cut to size. This factory produces wiper blades for a wide range of vehicles, all of which need different model windscreen wipers. With the blades cut to the right length, the two parts can now be separated. A spinning guillotine splits the band in half perfectly. And the result is a fresh wiper blade.

To make sure the rubber quality is high, samples are taken from each batch and run through rigorous tests. Here, dirty water is sprayed on a mock windscreen and the wipers ability to clear it is monitored.

This nozzle is blasting snow at the car in a wind tunnel at minus 30 degrees Celsius. Each wiper has to survive 24 hours in these conditions, non-stop. If they can't clear the snow or they jam for any reason, they fail. The engineers use this time to assess the design of the wiper blade arm as well as the unit's performance. When they're satisfied the blades and arms are up to the task, the engineers will start assembling them.

This is an important but repetitive part of the job in the wiper factory. Her task is to feed fresh blades into the machine so they can be joined together with the arms. The robotic carrier will slide the holders over the blades and clamp them into place.

The new units are then sent past a camera linked to a quality control computer. Human eyes are also used to check each new set of wipers that emerges from the machine. As well as removing all the dirt and muck that are sprayed onto a windscreen, the wiper must also cope with bigger objects like dead leaves in the Autumn. Then there's excessive dirt from a muddy excursion. Manufacturers recommend that you clean your wipers now and then as this will prolong their life span. And a squirt of window wash will keep your view clear whatever the weather.

Did you know?

Did you know that defective windscreen wipers could cost you up to £1,000 and 2 points on your driver's licence ?

加长豪华轿车

扫描二维码，观看中文视频。

我们都见过这种情景。年轻人开派对时，从很长的美国车车顶伸出头来。豪华轿车常与超级明星和显要人物联系在一起，敏锐的企业家看到商机，一个新的产业就此诞生。

从传统的凯迪拉克到运动型多功能车，都可以从普通汽车变成加长的豪华轿车。位于加利福尼亚州的这家工厂，就是专门做加长汽车的。

要把一辆车变成豪华轿车，首先要把它拆散。许多标准零件必须更换，这样才能承受大大增加的负载。当车只剩下空壳，机械师可以开始切割了。并不存在专门拉长的工具，改造过程很简单。他们就是把汽车切成两半。

为了确保汽车，例如这辆凯迪拉克，维持结构的完整性，汽车必须沿框架切割。一旦切成两半后，机械师只需在原车的前一半和后一半之间焊上添加的部分。汽车可以拉长将近3米。这类豪华轿车是针对有钱顾客的。派对豪华车更登峰造极。在美国，汽车可以延长到13米，但它们在许多欧洲城市出行困难。

对于一般的豪华派对轿车，这家公司将它延长5米，但必须正确重建才能通过道路交通安全法规。这些机械师使用18根钢材重新构造框架。

机械师用一根简单的绳子把零件对齐。如果车的框架没有正确安装，开起来就走不了直线。接下来机械师可以安装新底盘了。将它焊接到框架上，然后安装内饰。竖直的侧栏也安上了，显而易见，它们必须精确放置。至此机械师已经重建了豪华轿车的骨架。接下来要涂上新漆并用手工打蜡。

豪华轿车是令人印象深刻的工程产品，但尺寸远远大于初始设计。更重的结构，加上额外的乘员，意味着零件需要提高级别，从刹车和悬挂开始。更大的新刹车盘装好后，机械师打开盒子里的新弹簧。它们比原来的标准弹簧能够承受高得多的负载。豪华轿车越大，就越需要结实的悬挂系统。

窗户也一样，需要3名男子才抬得动。为了满足扩展的外观，窗户最好是一整块玻璃做出来的，但有时候办不到。

木地板也被铺上了，因为它们比地毯更容易清洗。然后还有座位，原装的椅子将被卖掉。

为了让大家在车里面尽情欢乐，通常采用酒吧式座位。长椅子是为每辆新车定制的。座椅由手工安装，即使有3个人帮忙，仍然是一件棘手的活。

派对豪华轿车沿着日落的道路行驶时，怎能没有关键的吧台呢？它已经被整齐地安装到地板上预先切出的位置，并布好了线。现代豪华轿车充满了奢华的电子设备，比如电视，DVD播放器和立体声以供客人娱乐。有些车型甚至配备了迪斯科灯光，营造真正的派对气氛。然

你知道吗？

世界上速度最快的加长豪华轿车是由一款法拉利360摩德纳制成的。它能以大约每小时274千米的速度行驶。

而，所有这些电子设备意味着电池更大的消耗。机械师安装两个额外的发电机，来满足增加的电力需求。

有一种能量对所有汽车都是不变的。不管汽车有多长，油箱总是需要加满的。这款经过改装的运动型多功能车，每行驶 100 千米需要 39 升燃料，是个真正的油老虎。

最初的普通运动型多功能车，已经被改造成道路上最长的车型之一。虽然你再也不能用这辆 4×4 车越野了，但无疑能开着它在好莱坞上下游走。

这就是加长的豪华轿车！

1. 这个工厂专门把普通轿车改装成加长轿车
2. 沿框架把车切成两半
3. 在前后两半之间焊接添加部分
4. 使用 18 根钢材重新构造框架
5. 侧栏必须精确放置到位
6. 提高零件级别更换大刹车盘

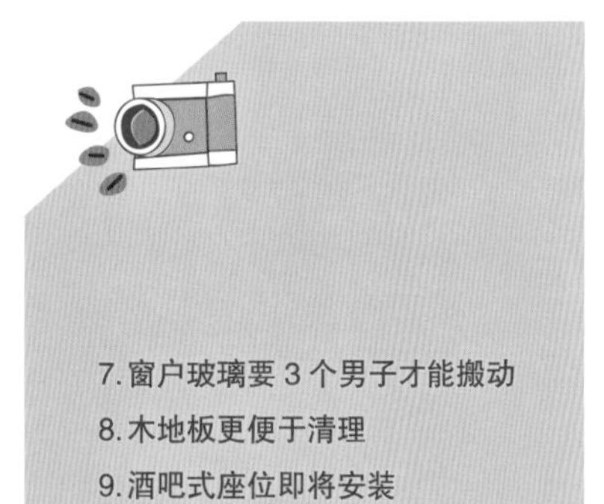

7. 窗户玻璃要 3 个男子才能搬动

8. 木地板更便于清理

9. 酒吧式座位即将安装

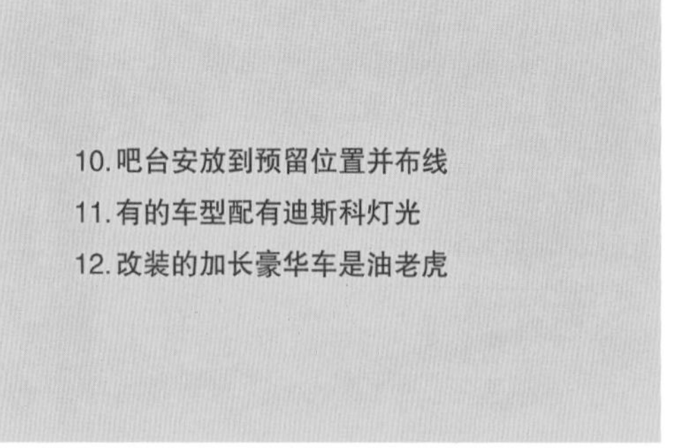

10. 吧台安放到预留位置并布线

11. 有的车型配有迪斯科灯光

12. 改装的加长豪华车是油老虎

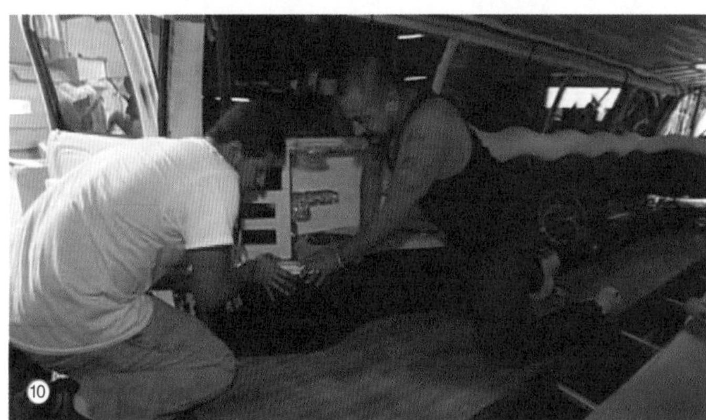

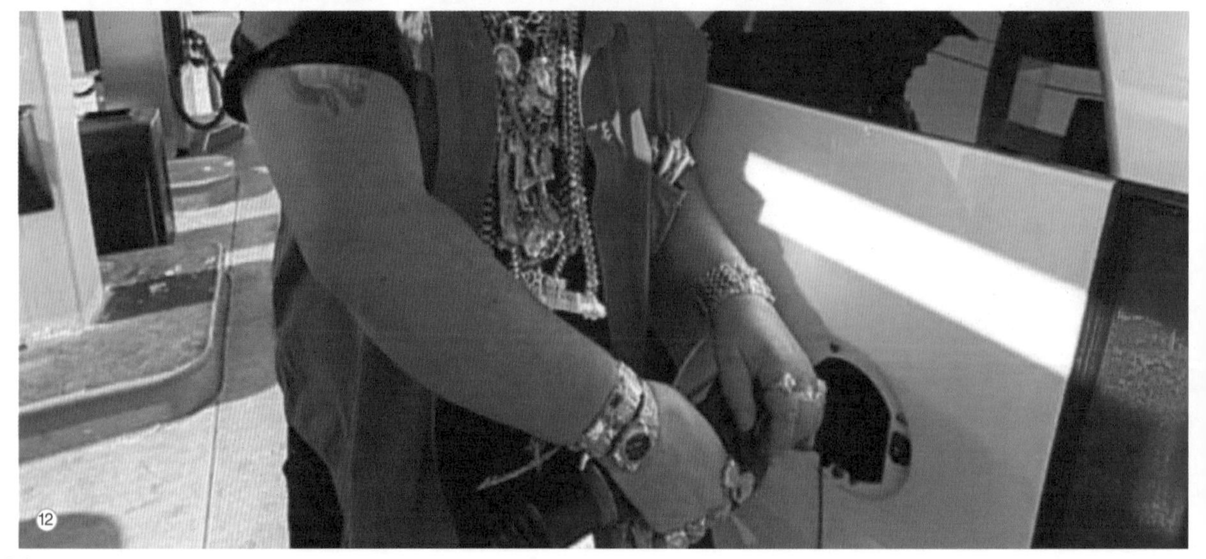

Stretched Limousines

But first, we've all seen them. Young people partying out of the roof of very long American cars. Limousines are often associated with superstars and celebrities, but when savvy entrepreneurs saw an opportunity and a new industry was born.

Everything from the traditional Cadillac to an SUV can be transformed from an ordinary car into an extra long limo And here in California is a factory that specialises in doing just that.

To convert a car into a limousine, first you need to strip it. Many standard parts have to be replaced so they can withstand far higher workloads. Once they're left with the empty shell the mechanics can start cutting. There's no dedicated stretching tool, the process is very simple. They just slice the car straight in half.

To ensure the car, or in this case a Cadillac keeps its structural integrity, they have to cut along the framework. Once they're got two halves, the mechanics simply weld an extension between the original front and back. The car can now to be stretched by almost 3 meters. This sort of limousine would be aimed at the wealthy customers. Party limos are more extreme. In America cars can be stretched up to 13 meters, but they would have difficulty navigating around many European cities.

For the proper party limo, the company will stretch the limo an extra 5 meters, but it has to be properly rebuilt to pass road safety laws. These mechanics will use 18 large steel bars to reconstruct the frame.

Using a simple piece of string helps the mechanic to get everything lined up properly. If the car's frame wasn't true, it wouldn't drive in a straight line. The next mechanic can assemble the new floor. This will be welded onto the frame so the interior can be fitted out. Vertical sidebars are installed too but they have to sit perfectly for obvious reasons. So far the mechanics have rebuilt the new limo's skeleton. The panels come next with a fresh coat of paint and hand rubbed wax.

Limos are impressive pieces of engineering, but they're far bigger than originally intended. The heavier structure, plus the additional passengers means the components will need to be upgraded. Starting with brakes and the suspension. With new larger discs fitted the mechanic opens up boxes of brand new springs. They are capable of supporting far higher loads than the original system that is supplied with the car as standard. The bigger the limo the tougher the suspension needs to be.

And the same goes for the windows which take 3 men to carry. To achieve that extended look, the windows ideally are made out of one piece of glass, but sometimes that just isn't possible.

Wooden floors are fitted because they're easier to clean than carpet. And then there's the seating. The original chairs will be sold.

To get lots of people inside this car having a blast, bar-style seating is the order of the day. Big benches are custom made for each new vehicle. They're then installed by hand which is a tricky business even with 3 guys helping out.

And what party limo could roll along Sunset Strip without the all important bar. It's fitted neatly into the pre-cut space in the floor and wired up. The modern limousine is full of luxury gadgets such as TV's, DVD players and stereo for the guests' entertainment. Some models even feature disco lighting for that real party atmosphere. However, all these gadgets mean a bigger drain on the battery. The mechanics install two extra dynamos to take care of the increased power demands.

But there's one power demand that is a constant with all cars. No matter how long they are that petrol tank always needs to be filled. This SUV with all the conversions takes 39 litres of fuel to do just 100 kilometers. It's a real gas guzzler.

So what started out as your ordinary sports utility vehicle, has been transformed into one of the longest cars on the road today. And although you wouldn't be able to take this 4 X 4 off-road any more, it can certainly handle the highs and lows of downtown Hollywood.

The stretched limousine.

Did you know?

The world's fastest stretched Limousine was made out of a Ferrari 360 Modena. It can travel at speeds of almost 274 kph.

旅行拖车

扫描二维码，观看中文视频。

没什么比拥有自己的空间更舒适了，尤其是移动的空间。并非每个人都喜欢它，但对许多度假者来说，旅行拖车的诱惑在于，当你对一个地点感到无聊时，只需要收拾行李，挂上拖车，就能向新的地点出发。

早在1907年英国就有旅行拖车俱乐部。有些人可能回忆起雨天的假期就心怀恐惧。但今天和过去一样备受欢迎，英国对移动度假屋的喜爱仅次于美国。

不只是英国人喜爱轮子上的假期。这家德国工厂里，他们仍在生产一种成功的蛋形拖车，这个设计始于20世纪50年代。

拖车建造从底盘开始。如果旅行拖车是移动的，并在公共假日汇入道路上的洪流，就需要安装车轮。胶合板制作的底部用螺栓固定在一起，为轮轴做准备。车轴经过镀锌防止生锈，是专为这种拖车发明的特别轻型设计。

一切用螺栓固定到位，将底盘翻转过来，为下个阶段做准备。胶合板涂上一层厚厚的胶水，然后贴上金属层强化地板。两个工人只有1.5小时协同工作，备好底盘部分，确保生产顺利运行。地板就绪后可以装轮子了，首先需要轮罩，将它们粘到底盘上，车轮用螺栓固定在下面。

在普通道路上，旅行拖车能安全达到每小时100千米的速度。而德国赛车手哈利•穆勒决定创造房车的陆上速度世界纪录。在2004年11月，他用这款小拖车达到每小时230千米速度，打破了此前的纪录。

但是，这些旅行拖车是为紧凑舒适而制作的，并非追求速度。实用旅行拖车的诀窍，是尽可能充分利用空间。床铺下有方便的存储柜，座位下也一样，所有这些都精心安顿在一起。没有色彩华丽的材料覆盖和铺垫，就不能算完成的拖车。

已经做好的床铺存储单元，安装到预制底盘上，用螺栓固定到位。

旅行拖车还有其他暗藏的有用小机关。以这个桌子为例。一分钟前它是个床架，下一分钟变成地道的台桌，非常适合传统的拖车野餐。

当然，拖车的乐趣不限于夏季。现代采暖系统远远胜过以往，这些轮子上的蛋形度假屋，大都装有自己的暖气。暖气可以轻松加热15立方米的空间。对小小的房车绰绰有余。所有部件安装到一起后，旅行拖车开始显露出传统的形状。

对于这个独特的型号，它个性化的蛋形是被识别和诱人的重要元素。这种拖车最初设计于20世纪50年代，在东德非常流行。不过，拖车的吸引力在80年代末有所下降，1990年完全停止生产。但它的形状仍保留在德国拖车爱好者群体的记忆中，于是生产恢复了。

你知道吗？

旅行拖车可以很酷。著名的“旅行车主”包括杰伊·凯、杰米·奥利弗、罗比·威廉姆斯，甚至还有超模凯特·莫斯。

现代建造材料帮助它的设计升级，但本身形状保持不变。聚苯乙烯顶棚用螺栓固定牢，使得旅行拖车保温，但它不够经久耐用。为了把坏天气挡在车外，最后将一层薄铝板铺设到位。它重量轻而强度大，能更好抵御恶劣环境。使用定制的铺平设备，工人让铝层变得平滑。塑料包边给人一种现代感，还能带来不同色彩。

最后剩下的是安装冰箱，然后把房车开往各处停放。没有什么比得上在自己的移动旅馆里露营更惬意。这就是现代的蛋形旅行拖车。

1. 蛋形拖车是一种成功的产品
2. 将车轴固定在底部
3. 将底盘翻转过来
4. 将轮胎罩粘到底盘上
5. 车轮用螺栓固定在下面
6. 有限空间布局要紧凑舒适

7. 床铺下的存储格安装到底盘上

8. 紧凑的水池和灶台

9. 给带轮子的小屋装上暖气

10. 最后将薄铝板铺设到顶棚上

11. 让铝层变得平滑

12. 在移动旅馆快乐的露营

Caravans

But first… there's nothing quite like the comfort of your own space, especially when the space is mobile. It isn't everyone's favourite, but for many holiday-makers the lure of caravanning is that when you're bored of one location, all you have to is pack up, hook your caravan up and head off for new horizons.

Caravan clubs have existed in the UK since as far back as 1907. Some may recall rain-drenched holidays with dread. But it's as popular today as ever before. We're second only to the States in our love of mobile holiday homes.

But it's not just the Brits who love their on-road holidays. At this factory in Germany they're still turning out a successful egg-shaped design that was first built in the 50's.

Construction starts with the chassis. If the caravan is going to be mobile and cause havoc on the roads around bank holidays, it needs wheels. The ply wood base is screwed together and prepared for the wheel axle. The axle has been coated with zinc to make it resistant to rust and is an especially light design invented just for this caravan.

With everything bolted firmly into place, the chassis base is flipped over ready for the next stage. A thick layer of glue is spread across the plywood and the metallic laminate floor added in. The two men working together have just one and a half hours to prepare this section of the chassis to keep production running smoothly. With the flooring prepared, the wheels can be fitted but first they need arches. These are glued into place on the chassis and the wheels bolted on underneath.

On normal roads it can safely manage 100 kilometres per hour. However German racing driver Harry Mueller was determined to set the world caravanning landspeed record. In November 2004 he smashed the previous record by reaching 230 kilometres an hour with this little caravan in tow.

But these caravans are built for compact comfort rather than speed. The trick to a practical caravan is making the most out of the available space. Beds have handy compartments beneath them for storage and so do the seats, all of which are carefully put together. And what caravan would be complete without brightly coloured material to cover the cushions?

The completed bed storage units can be fitted to the pre-built chassis and bolted into place.

But caravans have other useful tricks hidden up their sleeves. Take this table for instance. One minute it's a bed base, the next it's a perfect platform, ideal for that traditional caravan meal in the middle of a field.

Of course the delights of caravanning aren't limited to summer. Modern heating systems are far better than they have ever been and most of these egg-shaped holiday homes on wheels come with their own radiator. The heater can easily warm 15 cubic meters of space. More than enough for this little caravan. As all the parts come together, the caravan begins to take on its traditional shape.

For this particular model its characteristic egg-shape is a big part of its identity and appeal. Originally designed back in the 50's it was very popular in East Germany particularly. However, caravanning's appeal declined somewhat in the late 80's and in 1990 production was stopped entirely. But something about the shape stayed in the mind of the German caravanning community and production was revived.

Modern construction materials have helped to bring the design up to date, but the shape itself has remained the same. The polystyrene roof is screwed into place and firmly secured. This helps insulate the caravan, but it's not very durable. To keep the weather out a thin layer of aluminium is the last piece to be put into place. Light weight but strong its far better protection from the elements. Using a custom made ironing device the workers smooth the aluminium down. Plastic edging give it a modern feel and come in a variety of colours too.

All that remains is for the fridge to be filled and the caravan to be driven to a field somewhere and parked up. There's nothing quite like camping in your own mobile hotel. The modern egg-shaped caravan.

Did you know?

Caravans can be cool. Celebrity "caravanners" include Jay Kay, Jamie Oliver, Robbie Williams and even supermodel Kate Moss.

液化石油气汽车改装

扫描二维码，观看中文视频。

汽油的价格在飙升。局面变得如此糟糕，甚至一些偷盗者加油后不付钱就扬长而去。还需要考虑对环境的影响。那么，有什么替代办法呢？去年超过15000名英国司机对自己的汽车进行了改装，使用更环保更便宜的液化石油气燃料。德国汽车制造商以生产油老虎著称，但态度正在起变化。对于司机来说，液态石油气和普通汽油相比有许多好处。

在德国这家小汽修厂里，改装过程从添加了一个新油罐开始。它安装在放备胎的位置。技师为车体开一个洞，让燃料从油罐到发动机的线路从这里通过。他小心地把所有的粗糙边缘打磨光滑，以免损坏其他零件。新的油罐安装到位了，开口对着刚切开的孔洞。带有安全阀的油气管线装配到油罐开口。和易燃品打交道，必须坚持高标准。油气将沿车底的管道向下流动，最终到达这里。此处安装了另一个阀门。这个阀门将液化石油气系统对接到现有的汽油系统，必须连接紧固。

每升液化石油气，比汽油为你跑的里程少，但液化石油气价格也大约只有汽油的一半，所以好处是显而易见的。还有类似于标准汽油滤清器的气体过滤器和压力调节器，以确保运行正常。一切就绪后机械师就能把改装好的车开出去试驾了。

全球有超过900万液化石油气汽车，加油站也正在转变，供应这种更便宜的燃料。在英国，有超过1300个站点提供这种环保的选项。灌装过程很像给标准的汽油箱加油。唯一区别在于管嘴是拧到燃料箱上的，以确保不会意外溅出。

与轿车改装类似，公共汽车也可以改装。燃料罐是特制的，但和普通油箱大体一样，因为液化石油气是液体，而不是气体，并不需要在高压下储存。

带着充满可燃气的备用油箱到处跑，听起来好像很危险，但过去30年间大量的测试，已经开发出更好的储存设备，即使在严重冲击下也能确保安全。尽管科学家们在极端条件下测试这些燃料罐，它们的安全性能也不比普通汽油箱差，甚至暴露在1000摄氏度的火烧试验中。因为没有高压，发生爆炸的风险也很低。

还有个实用的长处。便宜的燃料是明显的优点之一，而拥有两个不同的油箱，使你跑的路程更远，加油次数更少。现在机械师正在进行液化石油气完美的测试工作，包括老板给它加油门，或者应该说“加气”时。

当液化石油气用完时，轻按开关便能返回汽油动力模式。因此液化石油气使驾驶员有了两个燃料箱可供选择。他们只是需要另找地方放置备胎。

你知道吗？

液化石油气动力汽车的陆地速度世界纪录由德国亚琛施耐泽公司保持。2007年，他们的汽车达到每小时318.1千米。

1. 汽油价格飙升促使汽车油气改装
2. 取下备胎为气罐腾出位置
3. 车体钻孔边缘要打磨光滑

4. 气罐开口对着钻好的洞放平
5. 油气管线装配到气罐开口
6. 油气将沿车底管道向下流动

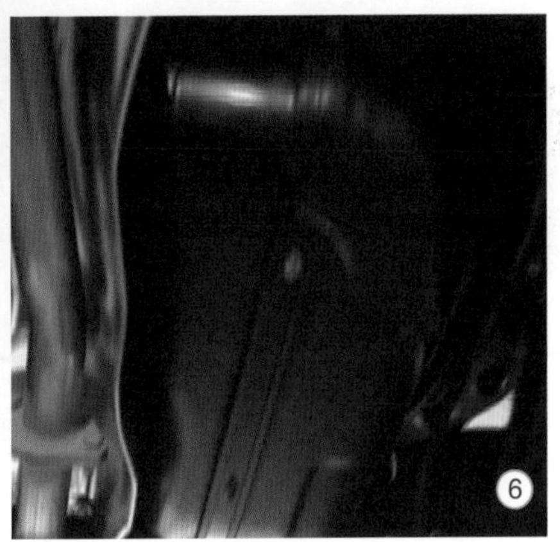

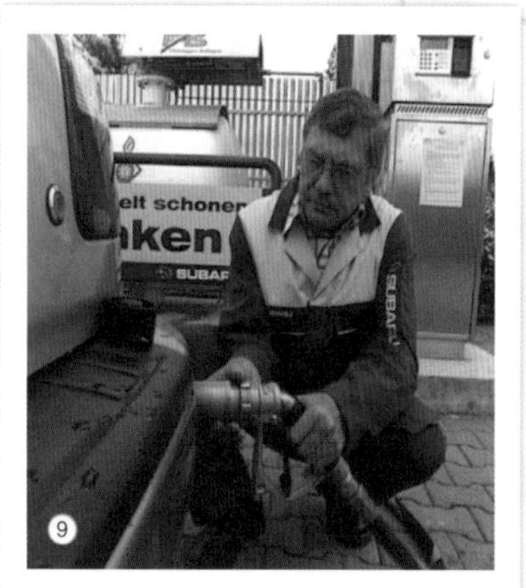

7. 液化石油气和汽油两个系统对接处
8. 安好气体过滤器和压力调节器
9. 液化气管嘴需拧到燃料箱上
10. 公共汽车也可安装液化气罐
11. 车辆起火无高压不会爆炸
12. 有两个不同油箱的车能跑更远

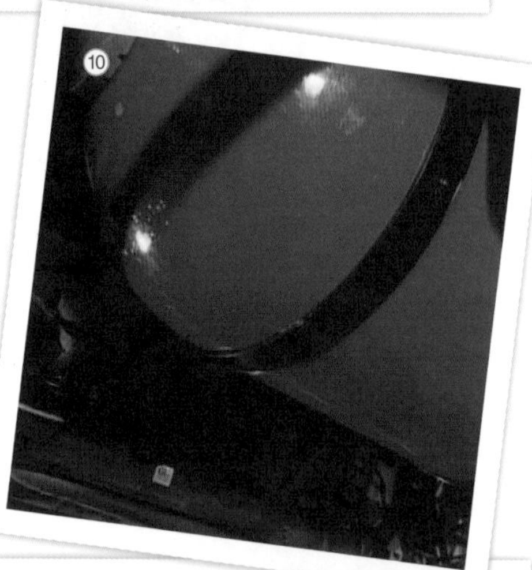

LPG Car Conversions

The price of petrol is rising. It's getting so bad that some petrol pump bandits drive off without paying for it. Then there are the environmental effects to consider. So what's the alternative? Well last year more than 15,000 UK drivers converted their cars to run on cheaper greener LPG fuel. Car producers in Germany are known for their gas guzzlers, but attitudes are changing. Liquid Petrol Gas gives the driver several benefits over normal petrol.

At this small German garage, the conversion process begins with the addition of a new tank. It's installed in the spare tyre well. The mechanic cuts a hole in the body work. This is where the fuel line will travel from the tank to the engine. He takes care to smooth off any rough edges so nothing is damaged. The new tank can now be fitted into place with its opening through the freshly cut hole. A gas line is fitted into the mouth of the tank which includes a safety valve. Standards must be kept high when dealing with flammable fuels. The gas will flow down a pipe along the bottom of the car, it will eventually arrive here where another valve is being fitted. This valve connects the LPG system to the existing petrol system and must be securely attached.

LPG gives you fewer miles per litre than petrol, but LPG is also around half the price so the benefits are obvious. There's also a gas filter similar to a standard petrol filter and a pressure regulator to make sure nothing goes wrong. With everything in place the mechanics can take the converted car out for a test drive.

With over 9 million LPG cars worldwide, petrol stations are also converting to offer this cheaper fuel. In the UK there are over 1,300 forecourts providing this environmentally friendly alternative. The filling process is much like feeding a standard petrol tank. The only difference being the nozzle is screwed into the fuel slot so nothing is split by accident.

扫描二维码，观看英文视频。

Like the car conversion, buses can also be transformed. The tanks are specially made like ordinary fuel tanks and because LPG is a liquid, not a gas, it doesn't need to be stored under high pressure.

Driving around with a spare tank full of flammable gas might sound like a risky proposition, but extensive testing over the last 30 years has helped develop better storage facilities which can survive massive impacts quite safely. Even when the scientists tested these tanks in extreme conditions they didn't prove any more dangerous than your ordinary petrol tank, even when exposed to this test fire burning at over 1,000 degrees Celsius. And as the tanks aren't highly pressurised the risk of explosions is also lower.

There's one other useful benefit. Cheaper fuel is an obvious one, but by keeping two different fuel tanks you can now travel further with fewer fuel stops. Here the mechanics are testing that the LPG system is working perfectly. Even when the boss gives it some gas or should I say liquid.

As the level drops towards empty, a flick of the switch returns the car to petrol power. So the LPG option gives the driver two tanks to choose from. They'll just have to find somewhere else to store the spare tyre.

Did you know?

The world land speed record for an LPG powered car is held by German firm AC Schnitzer. In 2007, their vehicle reached 318.1 kph.

修复老爷车

扫描二维码，观看中文视频。

如果你能坐在这种漂亮的老爷车方向盘后面，那是一种幸运。捷豹厂家建议你不要在养鸡大棚内长时间保存捷豹汽车而不清洗它。但如果你这么做了，那么，你手中拥有的只是一辆锈迹斑斑的破车，但却可以请老爷车修复公司来帮忙。

当这种漂亮汽车刚生产出来时，价值 6000 英镑。修复它的成本会远远比这更高，但修复的原因是喜爱，而不是钱。

捷豹是英国车，但世界领先的专业修复捷豹车公司之一设在德国。我们这辆代表英国汽车历史的生锈车来到这里，准备开始她的康复之旅，先来个淋浴。

当她被清洗完毕后，工程师可以评估后面的工作了。尽管“老妇”从未透露年龄，这位“美女”建造于 1975 年，所以会有很多事情要做。

任何无法修复的零件都可以重建，但通常做法是尽量修复原来的引擎。另一方面生锈的金属必须除掉。20 世纪 70 年代末期的防锈措施不是很先进，车身的一大段都完全锈烂了。工程师必须决定哪些可以留下，并以此为基础重建车体。

这辆“美女”汽车还要涂上一层防锈漆，用来阻止再次发生同样的问题。随着车的结构逐渐恢复，工程师可以进行外观处理了。她已经达到了 32 岁高龄，仅靠整容远远不足挽回魅力。

把整辆车剥开，任何问题或缺陷都将消除。修复的目的是再次给车一种“刚下线”的模样，所以一切必须是完美的。裸露的金属显露无遗，油漆工为汽车框架打一层白色的底漆，准备好涂上新的颜色。

唯一可选的颜色当然是英国赛车绿。工程师使用一台机器，它储存了每一种汽车曾经使用过的颜色记录。告诉人们应该添加什么涂料才能做出各种颜色，从雪佛莱红到那种著名的赛车绿。

新的表层由穿着全身防护服的油漆工喷涂到位。当油漆再次上好后，要到烤箱快速走一遭。在 70 摄氏度下待 1 小时把漆烘干，变成一层坚硬的保护外壳。

与此同时，在车库里的另一处，裁缝专家正重新装潢座位。当年的皮椅需要更换，所以按照原来套子的尺寸，制作出时尚的新座椅。金属线用来加强座位，非常巧妙地缝在下面，以免让乘客感到不适。最后，座椅面料被重新覆盖到原来的框架上，装订到位。

根据工作量的大小和汽车的类型，恢复一辆车可能要花上好几年。修好这种经典车型经常需要寻找或重新制造零件，尤其是发动机。

凭着高品质的英国工艺和现代化的德国技术相结合，这台经典的 4.2 升 V-6 发动机获得了新生。它的重量超过 350 千克，所以工程师需要吊装，仔细放在原有底座上，用螺丝固定。

你知道吗？

布加迪 1931 年的 41 型皇家运动跑车保持着世界上最昂贵的老爷车纪录。它于 1987 年在拍卖会上以 550 万英镑的价格成交。

一个工程师小组紧固所有零件，确保汽车启动后没有任何摇晃松动。

随着发动机修复，内饰也不能马虎。这儿使用的是70年代传统色彩——巧克力深棕色的新地毯。

所有镀铬的零件都要再次电镀，让它们熠熠生辉。首先将老的铬层除去。用硫酸清洗陈旧缺损的表面，露出下面的铜。然后用砂纸对黄铜表层进行打磨。它看上去很亮丽，但容易失去光泽，这就是需要镀铬的原因。几种不同的镀层能保护金属并恢复它的光彩。铜能防锈，镍是银色的基底，铬最后形成一层华贵的外表。

现在把所有部分安装到一起。完成内部装修，便可以安装新的座椅。

该制作新的车顶了。与现代轿车不同，这辆老捷豹使用乙烯树脂车顶。豪华的皮革车顶经受不了一场英国的大雨。

接下来，崭新的镀铬零件开始安装。工程师们要格外小心，刚做好的车身上只要出现最微小的划痕，他们就必须从头再来。

如果全体元件没有各就各位，包括徽标、车灯配件和原装饰品，那就不算真正的经典车。对于修复商来说，这些部件价格昂贵，而且很难寻找。你别想在易趣网上买到。但是，如果修复商运气好，厂家可能使用相似车型的配件。这样便有了更多机会找到替换零件。

一旦大功告成，他检查机油，把车身擦亮，然后松口气来欣赏自己的杰作了。

前个车主显然不知道他在谷仓里藏了一件宝物。虽然修复她的花销远远超过了购买她的费用，但每个地方的修复商都会同意，这位“夫人”不应该被弃如敝屣。

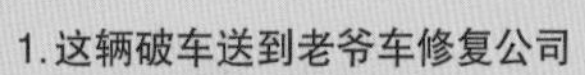
1. 这辆破车送到老爷车修复公司
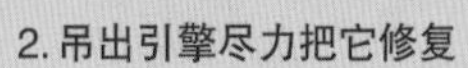
2. 吊出引擎尽力把它修复
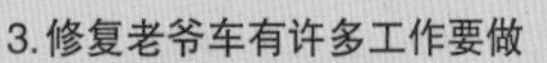
3. 修复老爷车有许多工作要做
4. 车身一大段完全烂透
5. 车的结构逐渐恢复
6. 外观处理环节为汽车框架喷漆

7.给车喷上赛车绿

8.经典发动机吊装回原底座上

9.用螺丝紧固所有零件

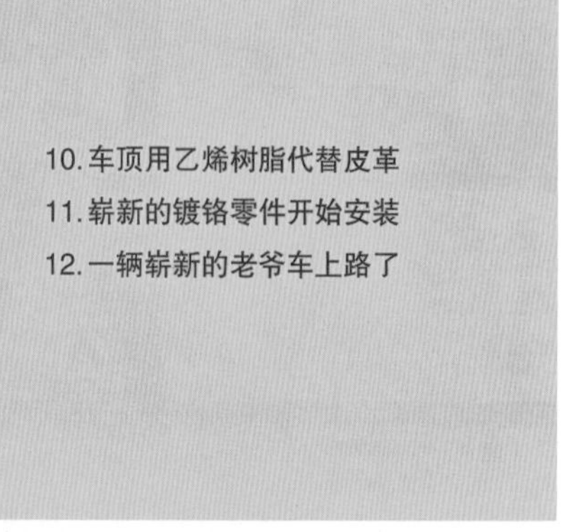

10.车顶用乙烯树脂代替皮革

11.崭新的镀铬零件开始安装

12.一辆崭新的老爷车上路了

Vintage Car Restoration

But first, if you find yourself behind the wheels of one of these beauty's, well lucky you. But the people at Jaguar don't recommend you store it for years in a chicken shed without washing it. However, if that's what you've done and you've now got a rusty wreck on your hands, you can always employ the help of a vintage car restorer.

When they were first made a beauty like this cost £6,000. Putting her back together will cost far more, but restoration is for love not money.

Jaguars are British, but one of the world's leading experts in restoring them is based in Germany. Our rusting piece of British motoring history arrives here ready to start her journey back to health, with a shower.

Once she's been cleaned, the engineers can assess the work ahead. Although a lady never reveals her age, this beauty was built in 1975, so it's likely there's a lot to be done.

Anything beyond repair can be rebuilt, but its general practice to try and restore the original engine. Rusted metal on the other hand, must go. Rust proofing in the late 70's wasn't very good and large sections of this car are quite literally crumbling. The engineers have to work out what they can save and use it as the basis to rebuild the body.

The lady also receives a full coat of rust-proof paint to try and stop the same problems from re-occurring. With the structure being restored, the engineers can turn to the lady's looks. Having reached the ripe old of 32, it's going to take more than a face lift to restore her charms.

The entire car is stripped so that any problems or defects can be ironed out. The aim of restoration is to give a car its "fresh off the production line" look once more, so everything must be perfect. With only bare metal showing, the painter primes the structure with a white undercoat, ready for her new colours.

And the only colour that would do of course, is British racing green. The engineers use a machine which has records of every car colour ever used. It tells them exactly which shades to add to create everything from Chevvy Red to that famous green.

The fresh coat is air brushed into place by painters in full protective clothing. When she's fully dressed once more, she will take a quick trip to the oven. 1 hour at 70 degrees Celsius dries the lacquer to a hard protective shell.

Meanwhile in an entirely different part of the garage, an expert seamstress is re-upholstering the seats. The original leather work needs replacing so she fashions new seats and uses the original covers to cut them down to the right size. Wire ribbing is used to strengthen them, but it's very cleverly sewn in underneath so as not to hurt the passengers delicate seating. Finally the coverings are reattached to the original framework and stapled into place.

Depending on the amount of work and the type of car, restoration can take years. Often there are parts to find or completely build from scratch to restore a classic like this, particularly the engine.

However, owing to the combination of quality English workmanship, and modern German technology this classic 4.2 litre V-6 engine has been given a new lease of life. It weighs over 350 kilos so the engineers hoist it into position. They'll then carefully lower it onto its original mounts and screw it into place. An army of engineers then tightens everything to ensure nothing wobbles when she's started up again.

With the engine coming along nicely the interior also gets some attention. New carpets are being laid in that traditional 70's colour, dark chocolate brown.

And to put the sparkle back into her eyes, all the chrome gets replated. First, the old chrome is removed. A bath in sulphuric acid strips away the old, pitted metal revealing copper underneath. That's then sanded down to reveal the base layer which is brass. This shines beautifully but tarnishes easily which is where the chrome comes in. Several different layers are now applied to protect the metal and restore her shine. Copper for rust-proofing, nickel for a silvery foundation and finally chrome to bring back that stunning finish.

All the pieces are coming together. With the interior completed, the new seats can now be fitted.

Then it's time for the new roof covering. Unlike modern cars, this old jaguar had a vinyl roof. A luxury leather covering wouldn't stand a chance against the English rain.

Next, the freshly chromed metal work is refitted. The engineers have to be particularly careful at this point because if there's even the tiniest of scratches to the fresh body work, they would have to start all over again.

Now a classic wouldn't be right without all the elements in place including the badges, light fittings and all the original ornaments. For the classic restorer these parts are expensive and very hard to find. You can't just pop onto eBay for spares. However if the restorer is lucky, the manufacturer may have used similar fittings on a range of vehicles. This improves his chances of finding replacements parts.

Once he's sure everything's in place, he checks the oil, gives it a quick polish, and sits back to admire his handiwork.

The previous owner clearly had no idea what kind of treasure he had hidden in his barn. Although the investment to restore her far exceeds the price that he could get, restorers every where would agree that this is one lady who should never have been left on the shelf.

Did you know?

The Bugatti 1931 Type 41 Royal Sports Coupe hold the record for the most expensive vintage car in the world. It was sold at auction in 1987 for £5.5 million.

实验室玻璃仪器

链锯

制革

人造革

拾零篇

防弹背心

手铐

欧元硬币

阿斯米尔花市

实验室玻璃仪器

扫描二维码，观看中文视频。

有没有想过如何制作一个试管？简单的实验室器材和更复杂的设备，都是由工艺师和精密机器在这样的工厂里完成的。玻璃器皿在科学领域扮演着重要的角色。在这间屋子里，他们用手工和玻璃打交道，为科学行业制造工具。

玻璃耐久，耐热、耐化学腐蚀，还容易清洗，因此是理想的试验室器材——从简单的试管到复杂的蒸馏器。这些蒸馏器把化合物分离成单一元素。但首先，它们需要一部分一部分地被制造出来。

首先将两个螺旋管互相插套起来。这个喷灯使用天然气和氧气，产生的热可高达 1850 摄氏度。蒸馏器的主外管由一个大玻璃管制成。工匠把它套在一个类似陶工旋盘的机器上。开口处要狭窄，才能和附着的小管子相匹配。

工匠用一块耐热板操纵熔化的玻璃，与此同时，向管子内吹入稳定气流以避免玻璃塌陷，用一根棒子来增减和调整玻璃形成开口。当尺寸恰好时，就可以把两个玻璃管接到一起。这是技术含量极高的手艺。要获取吹玻璃工匠的资质，至少需要三年的时间。此后还需要多年经验的积累，才做得了如此复杂的器件。

显而易见，玻璃吹制需要大量的燃气。这个工厂的燃气账单一定非常可观，因为它一天的用量，相当于一个普通家庭一年的消耗。最后，玻璃吹制师再接上另一个大玻璃管。当它们被高温熔合到一起后，主外管就此完成。这意味着其中的螺旋管可以放入了。螺旋管被平稳地放置到位。凭借细致的调整和精湛技术，玻璃吹制师在同事刚刚做好的主外管上添加进口和出口。

每一道工序中，玻璃被重新加热，而持续的高温会给产品带来损害。做成一个蒸馏器需要大量的辛苦工作和一双非常稳健的手。当蒸馏器做好后，很快就会在实验室分离化合物。在工厂的另一端，试管、烧瓶等较为简单的实验室玻璃器皿，则由自动化生产线完成。自动化生产技术由玻璃工吹制技术发展而来，能以很高速度使玻璃成形。为保险起见，自动化生产的最终成品仍需人工进行检验。

在这里，实验烧瓶正经历收尾工序。烧瓶被逐个装上机器加热，以便机械手对瓶颈部位塑形。仅仅几秒后，玻璃就冷却定型了。

每个做好的烧瓶，会和其他几百个一起进入炉子。在那里，烧瓶被加热后迅速冷却，让玻璃结构变得平滑而坚固，这个过程被称为回火。这是前后的对比。右边是回火后的烧瓶。它透明而结实，能经受住科学实验室的反复使用，甚至在学校化学试验课上被大量演示。

工人对每个烧瓶进行擦拭，也是检查烧瓶是否还有瑕疵的机会。此后烧瓶被装箱，即将在科学世界里扮演它们虽然渺小但至关重要的角色。

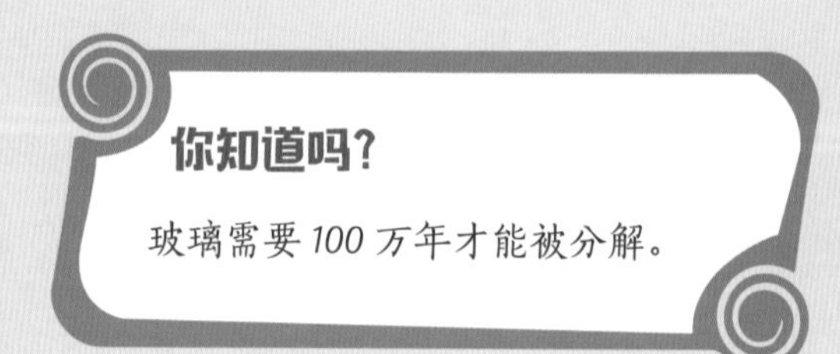

1. 工匠用手工为科学制作仪器
2. 将两个螺旋管插套起来
3. 主外管套在旋盘机器上
4. 主外管与附着的小管相匹配
5. 小管与主外管窄口熔在一起
6. 玻璃仪器制作需要精湛技艺

7. 蒸馏器外管进入下个工序
8. 在外主管上添加进口和出口
9. 玻璃工匠需要 3 年时间获得资质
10. 自动化技术源自人工吹制
11. 烧瓶回火后透明而结实
12. 擦拭光滑的烧瓶检查后装箱

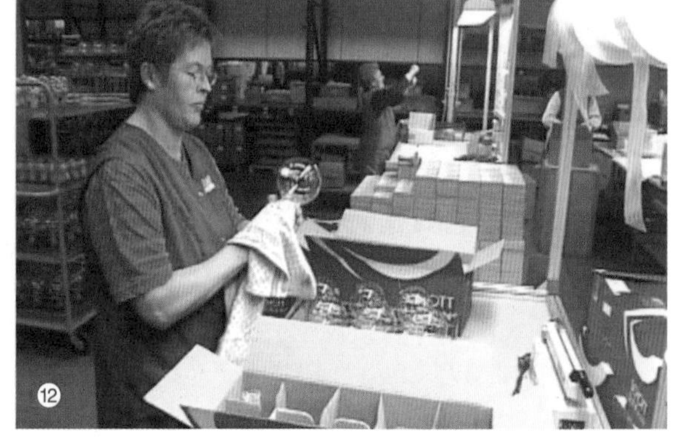

Laboratory Glassware

Ever wondered how you make a test tube? The simple bit of lab kit and the more complex equipment is made by skilled men and complicated machines in factories like this one. Glassware has played an important role in the world of science and in this room they work the glass by hand to build some of the tools for the scientific trade.

Glass is durable, heat-resistant and chemical-resistant and easy to clean so it's ideal for lab equipment – from simple test tubes to intricate distillers. These distillers separate compounds into individual elements. But first they have to be made piece by piece.

It begins with two spiral tubes that are inserted inside one another. This torch uses natural gas and oxygen and creates heat of up to 1,850 degrees Celsius. The main casing of the distiller is made from a large tube. This worker mounts it on his equivalent of a potter's wheel. The opening has to be narrowed so it fits the smaller tube it will be attached too.

He manipulates the molten glass with a heat proof paddle and blows a steady stream of air inside the tube to stop it collapsing. Using a rod, he adds and removes glass to form the opening. When he's got it to the right size he can join the two tubes together. This is very technical stuff. It takes at least three years to get certified as a glass blower, but many more years of experience to graduate to complicated pieces like this one.

As you can see glass blowing requires large quantities of gas and this factories gas bill must be fairly hefty as they use as much gas in one day as the average family would consume in a year. Finally the glass blower joins another large tube. When they're all fused together the casing is finished, which means that the inner spirals can be inserted. They slide smoothly into place. Using delicate adjustments and his expertise, the blower constructs entrance and exit holes in the body his colleague has just made.

At each step the glass has to be fired again and the constant heat takes its toll. To make the distiller it's taken a lot of hard work and two very steady pairs of hands. And now it is complete it will soon be in the lab separating compounds. Over at the other side of the factory. Simpler laboratory glassware, such as test tubes and flasks, is made by robots. They use the techniques developed by glassblowers to shape the glass at high speed. Just to be safe, human eyes still check the final products.

Here, the finishing touches are being applied to this laboratory flask. Each flask is loaded individually and heated so the robots can shape the neck. In a matter of seconds the glass is cooled and it's set in shape.

Once each flask is finished, it is sent off along with hundreds of others to a furnace. In there it will be fired and then uickly cooled down which evens and toughens the glass structure, this process is called tempering. And here's the before and after. The tempered glass on the right is now clear and is tough enough to withstand repeated use in science labs, or even in school chemistry experiments.

Workers give each item a good polish which also gives them a chance to check for any flaws or damage. They are then boxed up and ready to play their small but vital part in a world of science.

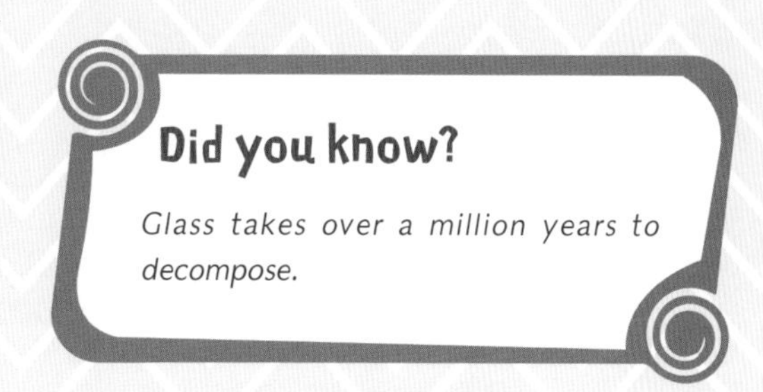

链锯

砍伐大树意味着几个小时挥舞斧头的沉重劳动。现在伐木只需要高功率的手持链锯就行。早在1926年，德国有个名为斯蒂尔的公司发明和制造了第一个链锯，但需要两个人开动。直到20世纪50年代，新的链锯设计能让一个人独立操作。

现代链锯最重要的部分是导板。切割链以每秒20米到25米的速度沿导板运动。这些星形像自行车上的齿轮。它引导链条围绕导板，持续循环旋转刀片锯倒树木。每个星形用螺栓固定到位，整个部件涂上保护层。苍白的帘幕看起来像一个固体表面，实际上是持续流动的液态油漆。为了让链锯轻便，适合一个人使用，许多部件由塑料制成。以降低链锯的整体重量。颗粒材料通过迷宫一样的管道进入注塑机，塑料在这里被加热并注塑成型。冷却变硬后，就能把成品取出。

不同的组件装配时，工人同时开始建造驱动链锯的发动机。像任何现代化生产线一样机器人被用来迅速完成高精度作业。机器做着重复的工作比如给启动马达的绳子打结。马达本身和汽车马达的设计相类似。这是一个单缸四冲程发动机，但只有大约10马力，比一般的4门轿车少得多。曲柄组件做好后连接到气缸顶部，完成的发动机安装到预制的塑料外壳里。制造链锯使用250多个不同的零件，当它们沿着生产线组装起来时更熟悉的外形显现了。

最后要添加的是起动马达和拉绳。马达是如何驱动旋转的链条进行切割呢？链锯内部的引擎利用离心力工作。当马达旋转越来越快，内部弹簧伸长推动表面与外齿轮接触。外齿轮旋转并依次带动链条转动。

现代的链锯生产不仅考虑伐树的最大效率，厂商还需要考虑废气排放，在许多国家链锯必须遵守严格的排放限制。普通手提式链锯工作时发出大约100分贝噪声。相比之下人类交谈的声音大约是60分贝。每一台新的链锯必须贴上标注告知买家噪声有多大。成品的性能也要进行检查。尽管尺寸比较小，但链锯非常强劲，能轻松切开包有金属层的木材。在日益流行的锯树比赛中选手使用更强大的“竞赛”级链锯。有次比赛中一台链锯使用竞赛摩托车的发动机，功率高达62马力。

随着动力部分组装完成，工人在包装和发售前检查每台马达。盖上印章表示经过检验，然后装箱。但箱子里只有锯条盒而没有锯条，这是有原因的。每个马达能使用不同种类的链锯条。客户购买马达后，锯条需要单另购买。完工的成品进行包装，用玻璃纸裹起来发送到世界各地。就像蒙蒂·派松曾经唱的那样，现代伐木工可以整夜睡觉，整天辛苦，手拿轻盈而强大的链锯砍倒树木。

你知道吗？

有个德国男子与妻子离婚时，用链锯将自己的凉亭锯成两半，放在卡车上拉走，另一半留给妻子。

1. 链锯让砍伐大树变得轻松
2. 现代链锯最重要的部分是导板
3. 星形齿轮引导链条绕导板循环旋转
4. 每个星形齿轮用螺栓固定到位
5. 许多部件由塑料制成
6. 塑料加热并注塑成型

7. 马达和汽车马达的设计相类似
8. 曲柄组件连接到气缸顶部
9. 发动机安放到塑料外壳里
10. 链锯沿着生产线被组装起来
11. 噪声测试新的链锯
12. 选手使用竞赛级链锯比赛

Chain Saws

Felling tall trees used to mean hours of back-breaking, axe-wielding hard labour. Now, all that's required to topple timber is a high powered, hand-held chain saw. Back in 1926 a company called Stihl in Germany produced some of the first chain saws ever invented but they needed two people to operate them. It wasn't until the 50's that new chain saws were designed that could be operated by just one person.

One of the most important parts of the modern saw is the guide bar. The cutting chain travels around this bar at between 20 and 25 meters a second. These stars act like the cog on your bicycle. They guide the chain around the bar creating the continual loop of spinning metal that saws through the tree. Once each star has been bolted into place, the whole unit is coated for protection. This pale curtain may look like a solid surface, but it's actually a continual flow of liquid varnish. To keep the chain saw light and easy for one person to use, many of the components are made out of plastic. This helps reduce the overall weight. The granulated material is fed through this maze of tubing into injection moulding machines. Here it will be superheated and forced into shape. When the plastic cools and hardens, the finished pieces can be removed.

While the different components are being assembled, workers can start building the engine that will drive the chain saw. Like any modern production line, robots are also used for precision jobs that need to be done quickly. Machines do repetitive jobs like tying knots in the rope used to start the motor. The motor itself is a similar design to the one you would find in a car. It's a single cylinder 4-stroke engine but it only generates about 10 horse power, much less than the average 4-door saloon. The crank assembly is built and then attached to the top of the cylinder, and the completed unit can then fitted into the pre-formed plastic housing. More than 250 different parts are used to build a chain saw and as they're assembled along the line a more familiar shape emerges.

One of the final pieces to be added is the starter motor and pull rope, but how will these motors power the spinning chain that does the cutting? The engine inside a chain saw works using centrifugal force. As the motor spins faster and faster, springs inside expand, pushing surfaces into contact with the outer cog wheel. This spins, and in turn spins the chain as well.

Now making a modern chain saw isn't just about cutting down trees as efficiently as possible. Emissions are also a concern for the manufacturers and there are strict limits in many countries which chain saws must adhere to. The average hand held chain saw works at around 100 decibels. A human conversation is roughly 60 decibels by comparison. Each new chain saw must carry stickers that inform the buyer just how loud their new tool is going to be. The performance of a finished model is also checked. Despite its relatively small size, this chain saw packs a punch, and is powerful enough to cut through wood wrapped in a layer of metal with ease. In increasingly popular tree-sawing contests, competitors are using ever more powerful 'competition' chain saws. At one event a saw powered with a racing motorbike engine generated a whopping 62 horse power.

With the assembly of the power unit complete, the workers will inspect each finished motor before preparing it to be packed up and sent to suppliers. A stamp is applied to show it's been inspected and it's boxed up. But there's no blade, only a blade cover, and there's good reason for this. Each motor unit can work with a variety of different saw blades. When a customer buys a motor, the saw blade is bought separately. The finished units are packed, wrapped in cellophane and loaded up to be delivered all over the world. So as Monty Python once put it, the modern lumberjack can sleep all night, and work all day ... cutting down trees with a lightweight, yet powerful hand-held chain saw.

Did you know?

While divorcing his wife, a German man chainsawed his summer house in half, put it on a truck and drove it away, leaving his wife the other half.

制革

菲斯作为中世纪摩洛哥的中心，具有生产金属制品、地毯和皮革的悠久传统。这里现在仍在使用古老的方法，把牛、绵羊和山羊的皮变成高品质的皮革。

首先需要从动物皮上除去毛发。动物皮被涂上一层腐蚀性的石灰膏，使毛发松动。另一名工人粗略地把毛刮下来。

你鞋上的皮革可能来自高科技的生产线，但这里的皮革，几百代人一直用同样的方式生产。

一旦毛发去除，皮被浸泡在石灰里一个星期。工艺在这些白色的石质容器内进行，它们数百个排列成蜂巢一样的结构。皮吸收了液体，在处理之前需要清洗。这项工作在木桶里完成。

下面的过程不适合敏感的鼻子：洗过的皮在鸽粪溶液中浸泡 3 天。富含氨的溶液使皮变柔软，并能够在下一个步骤中吸收其他成分。在褐色大桶里，皮被浸泡在鞣革剂里长达 3 周。丹宁酸来自磨碎的树皮，使动物皮变成皮革而不会腐烂，同时保持它的平滑度并且易于加工。皮革厂有着悠久的传统，但对于在这里工作的人却不利健康，比如站在高浓度液体中会让皮肤过敏。

鞣制后的皮是赭石色。它们要么这样处理，要么在旁边的大桶里染色。这里一个工人正在使用从罂粟花中提取的红色素。

菲斯有一种标志性的皮革制品：黄山羊皮革。它被用来制造传统的拖鞋和手袋。这些山羊皮已经被有机鞣革液处理过，具有浅颜色。然后将染料涂在皮上。染料由撒哈拉沙漠的罗望子制成。这种皮革会被当地工匠用来制造各种物品，然后销售给游客或出口到世界各地。

把皮革铺在房顶上，染了黄色的一面朝下，强烈的摩洛哥太阳就不会把颜色晒脱掉。干燥后给皮革上油，然后带到市场出售给皮革制造商。一张皮革的价格可以高达 6 英镑。听起来可能不值得讨价还价，但在摩洛哥，这大概是一天的平均工资。

在这里，皮革已经以同样的方式生产和销售了一千多年，至今仍然有着很大的需求。

你知道吗？

一家工厂最近开始回收制革厂使用的铬，此举使塞布河的污染减少了 90%。

1. 菲斯因生产金属制品、地毯和皮革而闻名
2. 现在仍用传统方法生产皮革
3. 给动物皮涂层石灰膏后把毛刮下

4. 皮子被浸泡在石灰水里 1 星期
5. 皮子投进木桶进行清洗
6. 洗过的皮子在鸽粪溶液中浸泡 3 天

7. 皮子被泡在鞣革剂里 3 周
8. 石缸里的染色剂是从罂粟花中提取的
9. 菲斯标志性皮革黄山羊皮用有机鞣革液处理
10. 给黄山羊皮涂上罗望子制成的染料
11. 将小羊皮涂有颜色的一面朝下铺张曝晒
12. 皮革带到市场出售给制造商

Leather Tanners

The medieval centre of Fez in Morocco has a long tradition of producing metalwork, rugs and leather. Here age-old methods are used to turn cow, sheep and goat skin into high-quality leather.

First the hair needs to be removed from the animal skins. The skins are covered in a corrosive lime paste which loosens the hair. Another worker roughly scraps it off.

The leather for your shoes might have come from a high tech production lines, but here leather has been made the same way for hundreds of generations.

Once the hair has been removed the skins soaked in a lime bath for a week. That happens in these white stone vessel, one of hundreds arranged in a honeycomb-like structure. The skins have soaked up the liquid so before they can be processed they need to be washed. That's done in these wooden barrels.

The following procedure is not for the faint-nosed: the washed skins are soaked in a solution of pigeon dung for 3 days. The ammonia-rich liquid makes the skin supple and able to absorb the substance in the next step. In the brown vats, the skins are soaked for up to 3 weeks in tanning agents. The chemical tannin, gained from ground bark, turns the skins into leather that won't rot, keep its consistency and is easy to process. The tanneries have a long tradition, but for the men who work here there are health hazards, such as skin irritations from the highly concentrated fluid they wade in.

After tanning the skins are an ochre colour. They are either processed like this one or dyed in the neighbouring vats. Here a worker is using a red pigment extracted from poppy flower.

There's one signature leather product from Fez: yellow goat leather. It's used to make traditional slippers and bags. These goat skins have been tanned in an organic tanning fluid which gives the skins a light colour. Then a dye is spread on. It's made from tamarind from the Sahara. This leather will be used by local craftsmen to make an array of goods that will be sold to tourists or exported around the world.

The leathers laid out on the rooftops with the yellow side down so the intense Moroccan sun doesn't bleach it. When it's dried the leather is oiled and then taken to the market to be sold to leather manufacturers for up to £6 a piece. That might not sound much to haggle over but in Morocco it's about an average's day's pay.

Leather has been produced and sold here the same way for over a thousand years and it's still in high demand.

Did you know?

A plant has recently begun recycling the chromium which is used by the tanneries and reduced pollution in the Sebou River by 90%.

人造革

扫描二维码，观看中文视频。

这些鞋看起来好像全是由皮革制成的，但有些实际上是仿造品。在20世纪30年代，制造商更容易获得聚氯乙烯原料，便开始选择廉价的替代品，人造革就此诞生了。

人造革生产从简单的成分开始。这种液体被称为增塑剂，有助于保持材料柔软。稳定剂为保持良好形状，香料用来模仿真皮的气味。当然还有主要成分：聚氯乙烯或称PVC。所有这些都放在一起搅拌30分钟，达到均匀一致。

在工厂的另一边，它们需要与染料混合。人造革最受欢迎的颜色之一是黑色，要制作这种颜色，只需要把水和大量的工业烟灰——就像老烟囱里扫出来的东西混合在一起。其他颜色都存储在这些大桶里。为了得到理想的颜色效果，计算机控制系统把不同颜料混合在一起。这和你在油漆店看到的系统类似。一旦所有的颜色添加完毕，便被搅拌在一起。

PVC混合物用轮桶推过来，将染料倒进去。继续搅拌30分钟，这种糊膏就可以变成合成皮革了。

这张纸用硅酮覆盖，不会与人造革黏连，便于稍后将它们分离。一名工人把糊膏涂到纸上。然后经过一条0.3毫米的狭缝被压平。在200摄氏度下，PVC糊膏变成一层固体平板。为了给它提供额外的强度，需要增加第二层PVC，但这层的混合物配方中有膨松剂。当两层一起烘烤时，第二层就像烤箱里的蛋糕一样发起。除了提高强度，还能赋予材料更自然的感觉。

目前材料厚1.3毫米，但还不够结实。涂上胶水后，又可以增加一层材料。这是棉花和聚酯的混合物，使人造革不会被轻易撕裂。材料再次通过烘箱，在200摄氏度下，三层紧密黏合在一起。在显微镜下，可以看到截面底部含棉和聚酯的纤维层。材料现在可以从涂有硅酮的纸张上剥离了。工厂能重复使用这些硅酮纸多达十余次。

人造革差不多完工了，但目前的状态看起来像橡胶。下一工序是为它造出有纹理的表面。当材料通过下一组的辊子时，被加上一层人造树脂，使材料具有光泽，向皮革的模样又靠近一步。这个房间里有超过300个带纹理的辊子，包括水牛、甚至鸵鸟皮的纹理。人造革被设计师选择的辊子压过，获得一种自然效果。

看起来像皮革，摸起来像皮革。甚至闻起来也像皮革，你能说出它们的区别吗？

你知道吗？

在聚氯乙烯出现之前，有一部分密纹唱片是由虫胶制成的，虫胶是一种从南亚紫胶甲虫身上获得的材料。

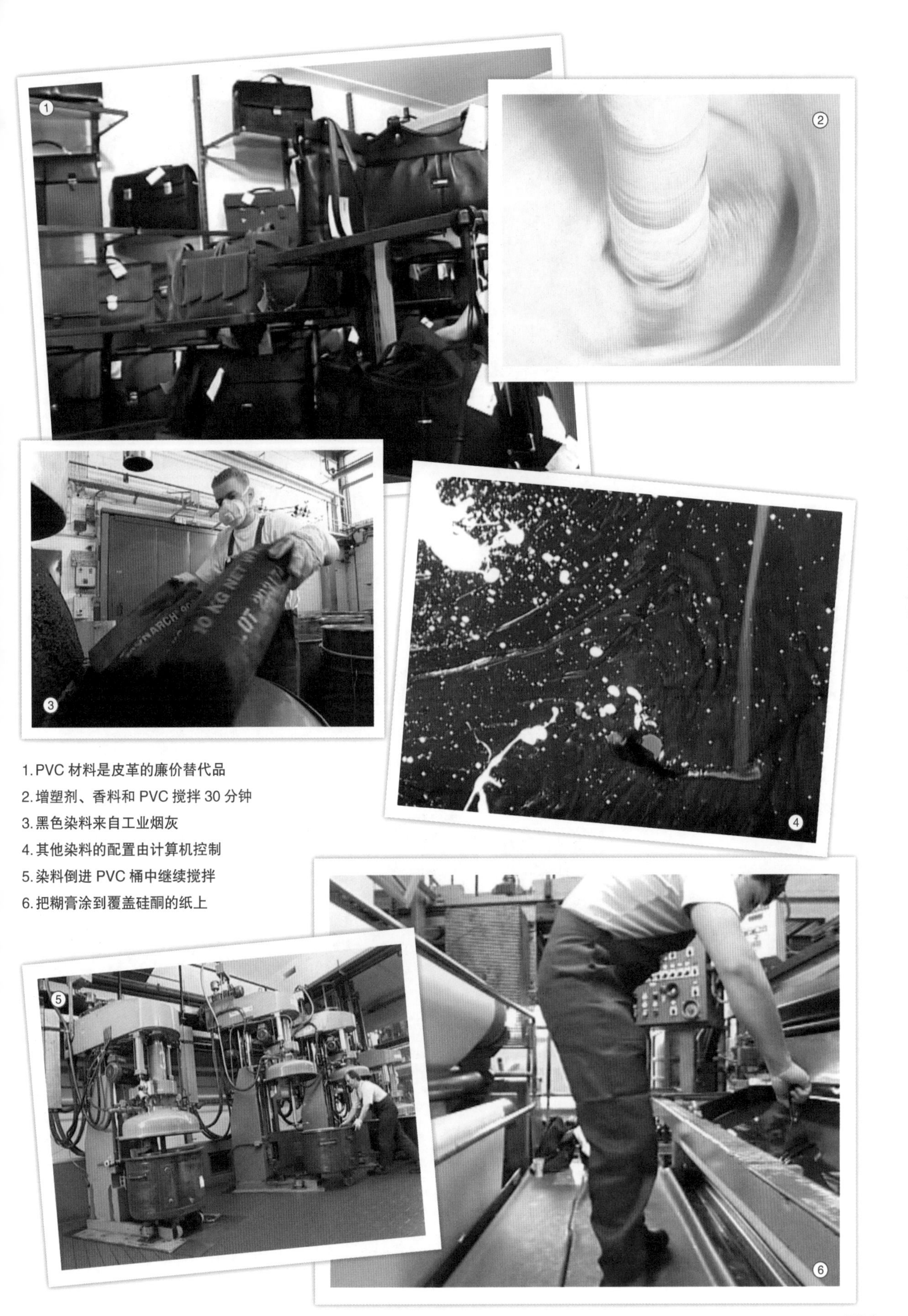

1. PVC 材料是皮革的廉价替代品
2. 增塑剂、香料和 PVC 搅拌 30 分钟
3. 黑色染料来自工业烟灰
4. 其他染料的配置由计算机控制
5. 染料倒进 PVC 桶中继续搅拌
6. 把糊膏涂到覆盖硅酮的纸上

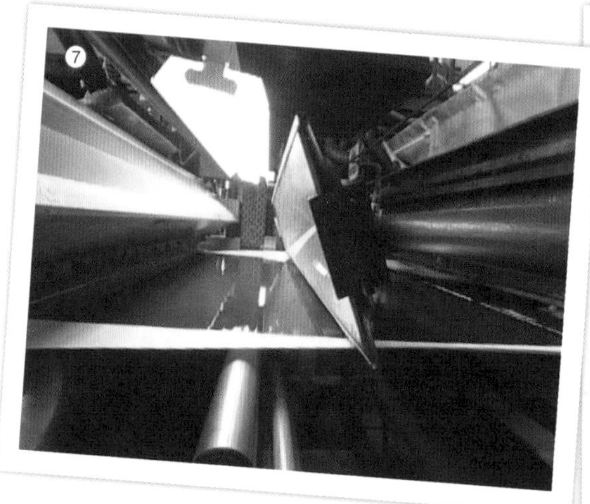

7. 糊膏被压平后进入加热区
8. 显微镜下观察纤维层
9. 涂上胶水加上棉花和聚酯
10. 通过烘箱三层紧密黏合
11. 这里有带不同纹理的辊子
12. 辊子压过人造革带上纹理

Synthetic Leather

These might look like they're all made from leather but some of them are actually made from an imitation. In the 1930's, PVC became more widely available to manufactures they started to make a cheaper alternative to the real thing and synthetic leather was born.

It begins with simply ingredients. A liquid called plasticizer which will help keep the material supple. A stabilizer to keep it in good shape. A fragrance for the authentic leather aroma. And of course the main ingredient: polyvinyl chloride or PVC. All of these are stirred together for thirty minutes for thirty minutes to ensure a smooth consistency.

On the other side of the factory they need to mix up a dye. One of the most popular colour for synthetic leather is black and to make it they simply mix water with a vast quantity of industrial sot just like the kind you would sweep out of an old chimney. All the other colours are stored in these vats. To get the exact shade they're after a computer controlled system mixes the dyes together. It's the same kind of system as you'd find in a paint shop. Once all the colours have been added they are all stirred together.

The PVC mix is wheeled over and the dye is poured in. It's blended for a further 30 minutes, and then the paste is ready to become synthetic leather.

This paper has been covered with silicon so that it won't bond with the synthetic leather and they can separate it later on. A worker spreads the paste over the paper. And then it's evened out as it's pressed through a point 3mm slit. At 200 degrees Celsius the PVC paste transforms into a solid sheet. To give it extra strength they add a second layer of the PVC formula but this time there've mixed in a rising agent. When the two layers are baked the second layer rises just like a cake in the oven. As well as strengthening the material it also gives it a more natural feel.

The material is now 1.3 mm thick but it's not strong enough yet. Glue is spread on so they can give it yet another layer. It's a mix of cotton and polyester and it stops the material from tearing easily. Once more the material makes its way through an oven where the three layers are bonded by a 200 degree Celsius heat. In the microscopic cross section the fibrous layer of cotton and polyester can be seen at the bottom. The material can now be detached from the original paper silicone base. They can re-use these bases up to ten times.

The synthetic leather is nearly ready, but in its present state the material looks like rubber. The next stage of the process is to give it a textured finish. As the material passes through the next set of rollers a coat of artificial resin is added. This gives the cloth a sheen that takes it one step closer to the leather look. This room contains over 300 textured rollers including varieties such as water buffalo and even artificial ostrich. The imitation leather's pressed with a roller of the designers choice to give it a natural finish.

Looks like leather, feels like leather. And even smells like leather, could you tell the difference?

Did you know?

Before PVC, LP records were partly made from shellac, a material gained from the South Asian Lac Beetle.

防弹背心

扫描二维码，观看中文视频。

面对这种情况可能是致命的。但如果你穿着最新的高科技防弹衣就不那么危险了。每天在世界各地，很多警察穿着防弹铠甲。由于枪支在街头日益泛滥，警察穿上这种防弹服保护自己。

防弹背心与大多数服装一样，从模板开始制作。可能让你惊讶的是，防弹部分不是用金属而是用一种布做出来的。它叫作芳纶，是一种超高强度的合成纤维，比棉花强 15 倍。

蚕丝是第一种用来缓冲子弹的纺织物，但枪械师造出了更强大的武器，使丝绸失去了阻止弹头的能力。但由于科学的发展和芳纶这类合成材料出现，防弹服装再次成为现实。

这种材料被织成一大张，存储在这些卷轴上。每张布太单薄，不足以挡住子弹，但背心由多层组成。它们协同工作就可以吸收子弹的能量，防止它穿透人的身体。因为事关生死，所以材料要严格测试。对每个卷筒进行取样，如果有一项测试失败，整卷材料都会被拒收。技术人员对每个样品做各种不同测试。如果一个样品达不到规定标准，整个卷轴也会结果相同。这种材料是用来挽救生命的，所以不允许失败。

防弹布的另一个重要试验地是靶场。要让警官以生命相托付，各种新款式都必须证明是防弹的。每项新的设计和创新都在这个靶场经过多次测试。工程师用固定在工作台上的枪，向新防弹背心多个回合射击，并评估它的效力。激光有助于精确瞄准。当一切都排列停当后，他们躲进枪后的安全室，准备，射击。

但要了解它们是如何阻挡子弹的，我们首先要看看背心是怎么做成的。就像普通裁缝使用布料，芳纶需要被切割成形。我们知道防弹背心需要多层，所以多张防弹材料被重叠起来。然后用面料把各层覆盖。这有助于密封，将其中空气排出。

用真空密封固定住芳纶材料进行切割，以确保所有材料层具有相同的形状。铡刀把材料切割成形。虽然芳纶能抵挡子弹，但它很容易被切开。防弹能力来自材料的编织和多层的组合。子弹大小的金属不能从线之间穿过。而尖针能穿透，这对缝制防弹衣很重要。多层芳纶缝合在一起，能提高它们的复合强度。

随着越来越多的女警官在一线打击犯罪，防弹背心必须适应女性体形。裁缝在这里做出的褶皱设计，可以更舒适地符合女警官体形。

紧密缝合后，就把防弹层放置在塑料保护外层中，并密封到位。防弹背心现在准备进行最终检验。一颗子弹正中靶心，结果令人难忘。

防弹服的秘密，在于它吸收和分散冲击的能力。不让子弹穿过。子弹的力量可以穿透前几层，但芳纶层通过吸收能量让子弹从突破到减速，然后停止。

现代防弹服根据使用地点不同而差别很

你知道吗？

世界上第一件防弹外套于 2007 年以 1000 英镑的价格出售。它可以阻止子弹、刀甚至注射针头的袭击。

大。防弹背心可以为保护士兵、警察，甚至保镖而定制。轻的布料更舒适，而且不显眼，但它们带来另一个问题。

虽然防弹背心吸收了初始冲击，穿着者的身体仍会吸收一些子弹的力量。更轻的防弹背心意味着受内伤的风险更大。为解决这个问题，使用橡皮泥进行了测试。它的弹性和人体相当。这里测量到的伤害会告诉技术人员，穿着者能否经受子弹的冲击存活下来。40 毫米是允许的最大深度。在 36 毫米处，穿着者可能遭受肋骨骨折，或内脏受损，但仍会活下来。

在更严酷的环境，比如战区，防弹铠甲是用陶瓷做出来的。它很笨重，也不时尚……但和保命相比，这是个很小的代价。宁可牺牲生命而追求华美，防弹背心显然不是这样。

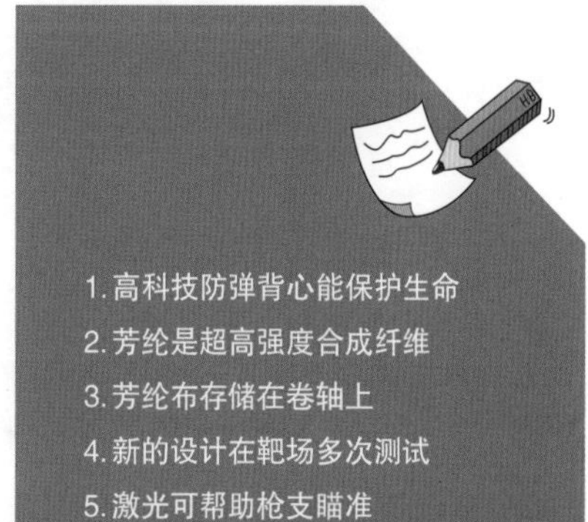

1. 高科技防弹背心能保护生命
2. 芳纶是超高强度合成纤维
3. 芳纶布存储在卷轴上
4. 新的设计在靶场多次测试
5. 激光可帮助枪支瞄准
6. 多层芳纶布真空密封后切割

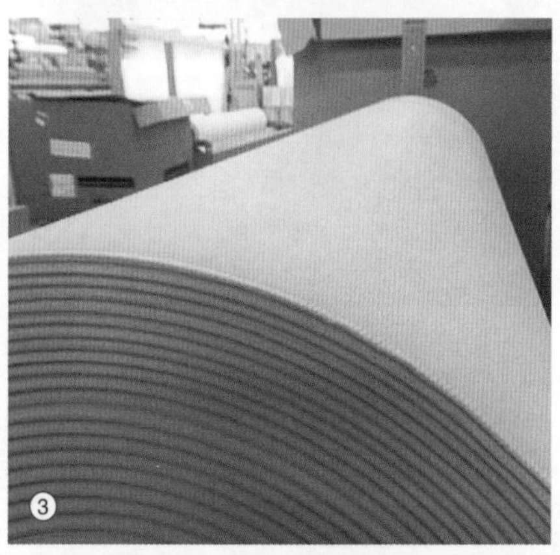

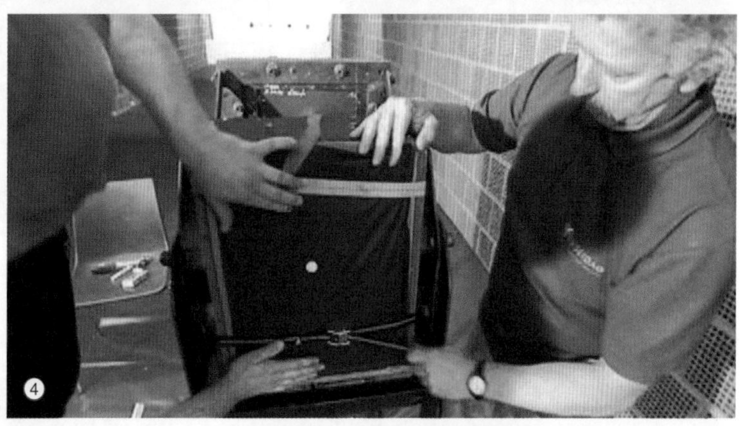

7. 多层芳纶布缝合在一起
8. 防弹层放进保护层里密封
9. 穿防弹衣身体仍会吸收一些能量
10. 橡皮泥的弹性和人体相当，用于试验
11. 伤口深 36 毫米仍可存活
12. 在战区用陶瓷做防弹铠甲

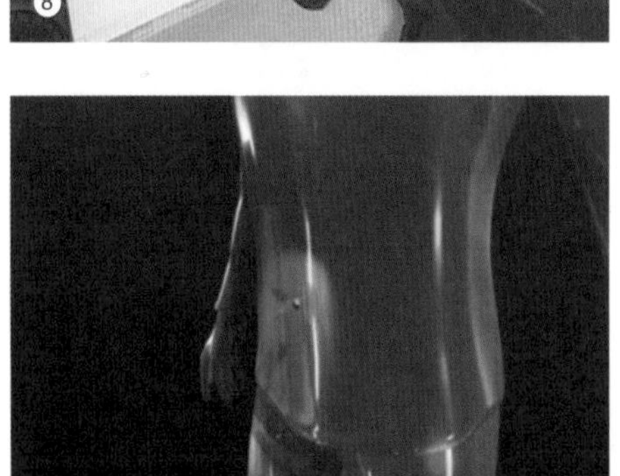

Bulletproof Vests

But first, facing a situation like this could be fatal. But not if you're wearing the latest high tech bullet proof body armour. Everyday all over the world, many police officers wear bullet-proof armour. As guns become more prevalent on the streets, officers are using this form of ballistic clothing so they can protect themselves. Bullet proof vests start out, like most types of clothing, with a template. But you may be surprised to learn that the bullet proof part isn't made out of metal but a type of cloth. It's called aramid and it's a super-tough synthetic fibre that's 15 times stronger than cotton.

Silk was the first textile used to slow a bullet down, but as gun-smiths produced more powerful weaponry, silk lost its ability to stop hot lead. But thanks to science and synthetic materials like the aramid, bullet proof clothing is again a reality.

The material is woven into vast sheets and stored on these reels. Now each sheet is too thin to stop a bullet, but the vest is made up of multiple layers. Working together they absorb the energy of the bullet, stopping it from penetrating the wearer's body. As it could mean the difference between life and death, the material is stringently tested. A sample is taken from each roll, and if it fails a single test, the entire roll is rejected. Technicians will run a variety of different tests on each sample. If one piece doesn't meet the required standards the same could be true for the whole roll. This material is supposed to save lives, so failure is not an option.

Another important testing ground for the bullet proof cloth is the firing range. If an officer is to trust one with his life, each new style must be proven to be bullet-resistant. Every new design and innovation is targeted repeatedly at this firing range. Using a gun strapped to a bench, the engineers can fire test-rounds at the new jackets and assess their effectiveness. A laser sighting helps to get the perfect aim. And when they're satisfied everything's lined up properly, they take shelter in the safety room behind the gun… prepare… and fire.

But to understand how they stop a bullet, first we need to see how these vests are put together. Like the cloth used by an ordinary tailor, the aramid needs to be cut to shape. However we know the jacket needs multiple layers, so several sheets of the bullet proof material are laid one on top of the other. A cover is then placed over the layers. This will help create a seal so that all the air can be removed.

The aramid material is held in its vacuum-sealed package when it's being cut, ensuring all the layers are shaped the same way. A guillotine trims the sheets into shape. Although the aramid can resist bullets, it can be cut quite easily. The bullet proof quality comes from a combination of the material's weave and the multiple layers. A thick piece of metal like a bullet can't penetrate between individual threads. However, a thin needle can, which is important for putting a jacket together. The multiple layers are sewn up which improves their combined strength.

With more and more female officers on the frontline fighting crime, vests must also be made to fit the feminine form. The folds the tailor is including here are designed to accommodate the female officer's shape more comfortably.

And once they are sewn up tightly, the ballistic layers are placed into the protective plastic outer layer, and sealed into place. The vest is now ready for the ultimate test.A bullet is fired dead centre with impressive results.

The secret to this bullet proof clothing is its power to absorb and disperse the shock of the impact without letting the bullet pass through. The bullet's power may penetrate the first few layers, but by absorbing its energy, those layers cause the bullet to mushroom, slow then stop.

Modern bullet proof armour varies greatly depending on where it will be used. Jackets can be tailored to protect soldiers, police and even bodyguards. Lighter cloth varieties are more comfortable, and less noticeable, however they do create a different problem.

Although the jacket absorbs the initial impact, the wearer's body also absorbs some of the bullet's force. Lighter jackets mean greater risks of internal injury. To deal with this problem, tests are conducted using plasticine. Its elasticity is comparable to the human body. Measuring the damage here will tell technicians whether the wearer would have survived this bullet or not. 40 millimetres is the maximum depth allowed. At 36 the wearer may suffer broken ribs, or damaged organs, but would probably still live.

For the tougher environments such as war zones, armour is made out of ceramics. It's bulkier and not very fashionable … but that's a small price to pay. While haute couture may be to die for, but the bullet proof jacket certainly isn't.

Did you know?

The world's first bulletproof coat went on sale for £1,000 in 2007. It stop bullets, knives and even hypodermic needle attacks.

手铐

扫描二维码，观看中文视频。

无论在哪里，撬门入室绝对会引起执法者的注意。在德国，公众对试图盗窃他人财产时被抓住的任何人都持谴责态度。德国警官曾使用这样的旧式手铐。然而，它们有重大的设计问题。首先，它们可能让犯罪嫌疑人受伤，这是一个严重的缺陷。而对警察暴力执法的指控是所有警务人员都希望避免的。其次，如果戴手铐时上下倒置了，开锁会很麻烦。最后，犯罪嫌疑人只需要一个简单的回形针就可以打开它们。世界上没有一个警官会对这点感到高兴。

所以很明显，老式手铐达不到现代警务的要求，但究竟有什么解决办法呢？

是的，需要一个完全崭新的设计。有尖锐棱角的钢臂必须用更好受的束缚来取代，连接的铰链需要改良为复杂的固定中央闩栓。以下是如何将理论设计变为真正原型的过程。这是一台 CNC 铣床或叫电脑数控铣床。它有一个内存芯片用来存储产品设计，引导钻头雕刻出精确的设计形状。这里用的是铝块。这个系统在批量生产中不划算，但它用来做原型是非常有用的，随后原型可以进行各种测试。

铣头以每分钟超过 2 万转的速度切割金属时，空气喷嘴将金属碎片吹走。当铣床完成工作后，手铐原型的一半已经清晰可见，但与旁边的一个完成版相比，很明显还有一些工作要做。除了升级的锁定机制，新手铐还有更舒适的、新设计的束缚功能。

还需要一个模具制作新设计的橡胶覆盖层，以保护犯罪嫌疑人的手腕。他们再次使用电脑数控铣床做出一个模型，和有橡胶覆盖层的成品手铐臂完全一样。将它放进橙色盒子里，在周围倒入硅凝胶。把盒子放入真空室，其中凝胶受到空气脉冲的轰击，让其中的所有气泡迸裂，使凝胶完全包住手铐臂模型。这对下一阶段很重要。

凝胶被放置凝固，一旦硬度足够，就被切成两半。模型手铐臂被拿走形成一个模具。稍小的金属手铐臂放入其中，准备加上橡胶覆盖层。安上一根管子后，模具被放回真空装置里，其中已经安着一个装有液体橡胶的容器。倒进液体橡胶，充满金属手铐臂和硅胶模具之间的空隙。通过填充这一缝隙，设计师制造出一个安全的橡胶屏障，它可以保护犯罪嫌疑人不会因为裸露的金属而受伤。当手铐臂做好后，内部锁定系统的制造也可以完成了。

添加各种额外的附件来解决老问题。这些小突起可用来“锁定”手铐臂。这样就不会太紧。老式的手铐能很轻易用一个回形针打开，但使用三扣锁定杆，让新手铐打开要难得多。执法人员面临的另一个问题也得到了解决。锁孔从两侧都能打开，意味着手铐再也不会上下倒置了。所以，一个完全的改观，意味着警员使用这种新款的手铐更容易，被这双高科技手铐拘押的嫌犯也更安全。

你知道吗？

2007 年，305 名伦敦警察排成一条长龙，创下被戴上手铐人数最多的“另类世界纪录”。

1.旧式手铐易伤手腕且存在倒置问题
2.嫌疑人用回形针就能开锁
3.新设计解决了旧手铐的问题

4.把铝块放进铣床
5.手铐全真模型的一半清晰可见
6.有橡胶层的手铐臂模型放入盒内

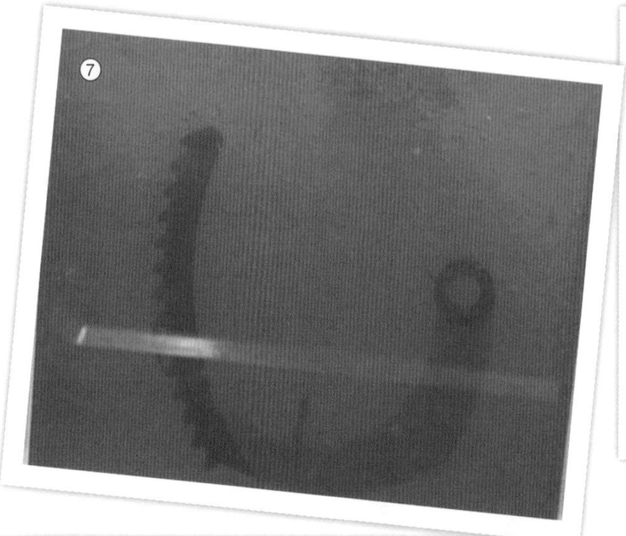

7.真空使凝胶包住手铐臂模型

8.把手铐臂型放进模具内

9.液体橡胶填充硅胶模具空隙

10.手铐臂橡胶减少了手臂受伤

11.三扣锁定杆不易打开

12.新设计的锁孔在两侧均能打开

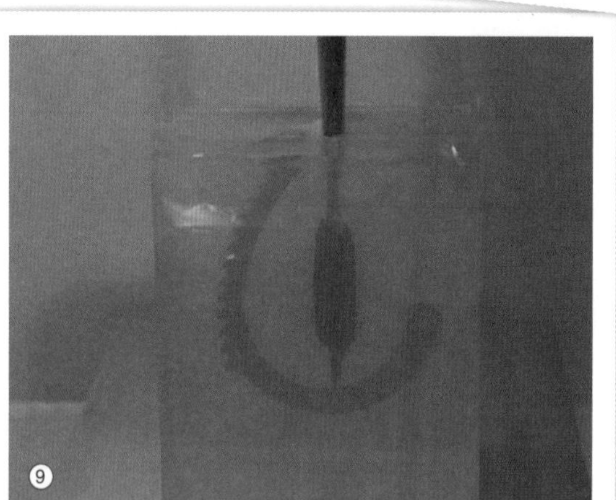

Hand Cuffs

Breaking and entering is a sure-fire way of catching the attention of the local lawman, wherever you are. Here in Germany, they take a very dim view of anyone caught trying to steal someone else's property. German officers used to use old-style hand cuffs like these. However, they suffered from significant design problems. One serious problem was that they could injure the suspect, and allegations of police brutality are something all police officers are keen to avoid. Secondly, if the cuffs were put on upside down, they were awkward to unlock. And thirdly a suspect only needed a simple paperclip to open them. And there isn't an officer anywhere in the world who would be happy about that.

So it's clear the old style cuffs weren't up to the requirements of the modern policing, but what was the solution?

Well, a complete redesign was called for. The sharp steel arms had to be replaced with more comfortable restraints, and the connecting chain improved with a sophisticated rigid central bar. Here's how that theoretical design became a real prototype … This is a CNC Mill or Computer Numerical Control mill. It has a memory-chip that stores the design. Drill heads are then directed to carve out that exact design shape. In this case, into an aluminium block. This system wouldn't be cost –effective in mass production, but it is very useful to make a prototype which can then be put to various tests.

An air nozzle blasts away the metal fragments as the heads spin at over 20,000 revolutions per minute cutting through the solid metal. When the mill has finished its job, one half of the handcuff prototype is clearly visible, but next to a finished version, it's clear there's still some work to be done. As well as an upgraded locking mechanism, the new cuffs will also feature more comfortable, redesigned restraints. A mould is needed to make the new rubber covering that will protect a suspect's wrists.

They use the CNC mill once more to create an exact model of a finished rubber-coated arm. This is placed in the orange box and a silicon gel is poured around it. The box is placed into a vacuum chamber where the gel is bombarded with pulses of air. This bursts any bubbles in the gel causing it to envelop the arm completely. This is important for the next stage.

The gel is left to solidify and once it's hard enough, it's sliced in half. The model arm is removed, creating a mould. The slightly smaller metal arm of a new cuff is inserted, ready for its rubber coating. A pipe is attached and the casing is returned to the vacuum device which has been loaded with a container of liquid rubber. It's poured in, filling the space between the metal arm and the silicon mould. By filling this gap, the designers create a safe, rubber barrier which protects suspects from the kind of injuries that could be caused by the bare metal. With the arm finished, the construction of the internal locking system can be completed.

A variety of extras have been added to address the old problems. These nubbins can be used to "lock" the cuff arms. This way they can't be over-tightened. Old fashioned cuffs could be easily undone with a paper clip, but the use of 3 locking levers has made the new model much harder to pick. And another problem that law enforcers faced has also been cleared up. The locks are accessible from both sides, meaning the cuffs can never be put on upside down. So, a complete make-over means this new style of manacle is easier for police officers to use, and safer for any suspects restrained in a pair of the latest high-tech police handcuffs.

In 2007, London police took the World Alternative Record for the most people handcuffed together, by joining 305 officers in one long line.

欧元硬币

扫描二维码，观看中文视频。

向太阳进发？如果你去欧洲，就需要欧元。它具有防伪设计，由外圈的环和中间填充的复杂多层金属组成。

这些高科技硬币的生命旅程，从巨大的金属废料场开始，堆积如山的回收铜被用作原料。庞大的液压机把金属压成巨型方块。每个重达8吨。它们被送到铸造车间，铜在这里熔化。炉子里出来的液态铜有1200摄氏度，发出绿色的光焰。但是这仅仅是第一步。铜已经变为可以加工的形式了，但仍有很多事情要做。

首先金属必须冷却。从机器里出来的两大块铜锭共重70吨，可以制作约200万枚硬币。它们被切成5米长的整块，然后进行热轧。金属热轧需加热至大约900摄氏度。这真有点热。然后金属在一系列辊子间来回传动并被碾压。金属受热后远比冷却时更容易处理，通过辊压，工厂生产出很长的铜板。这种形态比一大块铜更适合做硬币。

下一步是清洁。加热和冷却让铜粘上一层肮脏的颜色，需要刮洗干净。

铜被再次送回给碾压机进一步延展。当它卷起来时，像一卷闪闪发光的卫生纸，有将近400米长。

铜材只是1欧元和2欧元硬币背后的一小部分制造技术。为了买饮料、巧克力或者是火车票，机器必须知道你是否投入了正确的金额，因此硬币要进行严格的计量和测试。

最重要的测试是硬币有多大磁性，这很有高科技含量。不同的金属有不同的磁性图谱。2欧元的特征取决于填充中孔的材料。它由3层金属压在一起。顶层和底层是铜，中间是镍。为了制作填充料，不同金属被送入机器，并压在一起不可分离。

我们现在可以用这种夹层金属冲压出硬币来了。这台压力机每分钟击打铜板250次，为大个儿2欧元硬币切出中间的填充料。如果仔细观察，你会看到在上层的铜合金，中间的镍，以及下层的铜合金这种三明治结构。

与此同时，工厂另一边的另一台机器把硬币外圈冲压出来。这些外圈将填进“铜镍三明治”。被去掉的中间部分返回到炉内重新开始生产过程。

外圈有点脏，但把它们放进混有滚珠的酸性溶液里，就能让脏东西被磨掉而闪闪发光，可以变成钱了。

有了外圈，也做好了填充物，但还缺少点东西。我们做出来的都是空白金属。需要一个模板，让它们变成硬币。

雕刻师的工作，就是给硬币做一个对应的橡皮图章，把币值永久地印到硬币上。用工业压力机把一个图案压到冲模上，压力机施加160吨压力，所以必须非常小心。

你知道吗？

世界上最值钱的硬币是1933年的双鹰20美元金币。它在2002年以480万英镑的惊人价格被拍卖。

冲压模具在870摄氏度的炉子里硬化。没有这道工序，模具上的图案会很快消失。而一旦经过硬化，每个模具可以印出多达200000枚硬币。

用清洗剂除去表面所有污渍。在投入使用前进行检查和抛光。模具被装入机器，直到此时，我们还没有完整的硬币用来冲压，但所有的部分即将汇拢到一起了。

外圈放置到位，填充物落入孔中。两部分在冲压模具下被敲击到一起。既把币值压到硬币上，又使两部分充分结合。

从管道里出来的是一个全新的2欧元硬币，可以花费了。

这样，通过把那些普通的铜和镍压在一起，我们最终得到了精巧雅致的高科技2欧元硬币。

1. 欧元硬币由多层金属组成
2. 熔化的铜被送到铸造车间
3. 通过辊压生产出很长的铜板
4. 铜板经刮洗碾压后卷起来
5. 机器测试硬币的磁性
6. 2欧元硬币外层是铜，中间的填充材料是镍

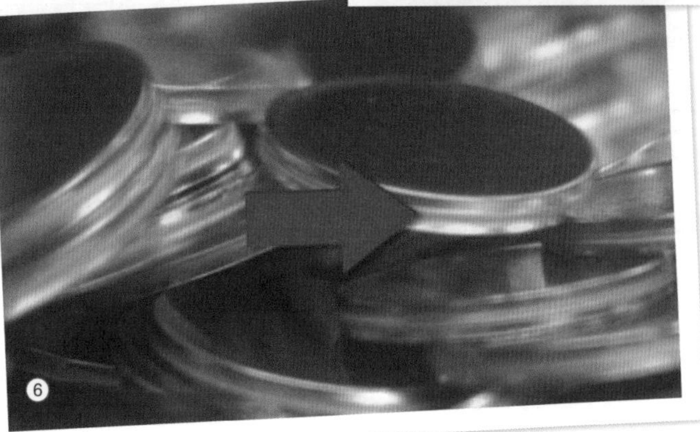

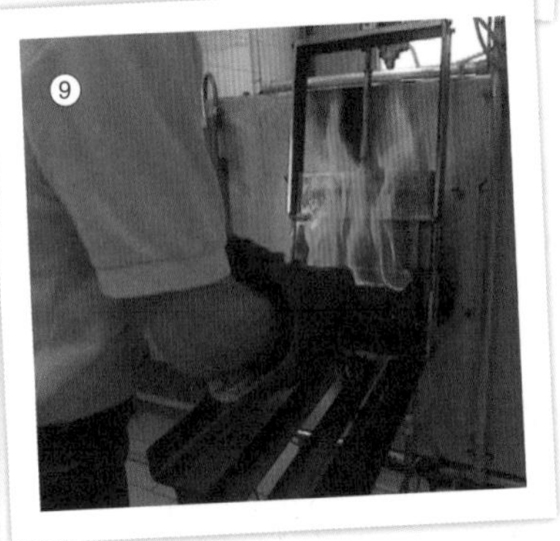

7. 外圈放进混有滚珠的酸性溶液中清洗
8. 雕刻师要为硬币做一个冲压模具
9. 冲压模具送入炉内硬化
10. 用清洁剂除去模具表面所有污渍
11. 外圈放置到位填充物落入孔中
12. 测量 2 欧元硬币厚度

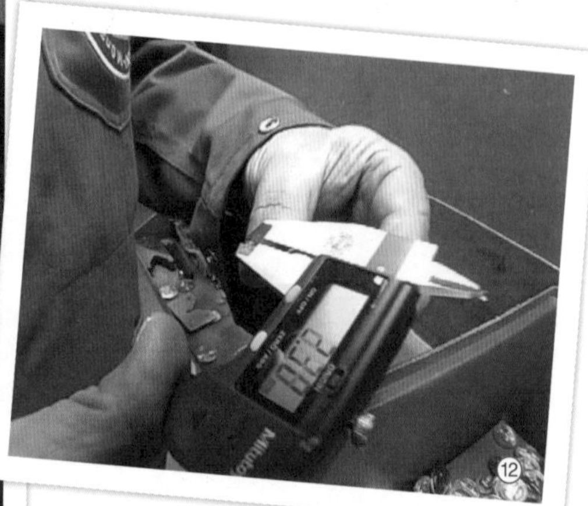

Euro Coins

Heading for the sun? Well if you're going to Europe, you'll need Euros. They were designed to prohibit counterfeiting and using an external ring and a sophisticated sandwich of metal for the filling.

Life for these high-tech coins starts out here at this enormous scrap metal junkyard, where Mountains of recycled copper are being prepared for use. They're squeezed by an immense hydraulic ram which makes giant cubes of the stuff. weighing in at 8 tons each. They're then sent to the foundry where the copper is melted down. Molten copper leaves the furnace at 1,200 degrees Celsius emitting brilliant green flames. But this is only the first step. We've got the copper into a workable form, but there's still plenty more to do.

First the metal must be cooled. The two enormous slabs that emerge from the machine weigh 70 tons between them and you could press about 2 million coins from them in all. They're chopped up into 5 meter long chunks, then its time for some hot-rolling. To hot-roll the metal, it's heated to about 900 degrees Celsius. That's the hot bit. Then the metal is passed back and forth between a series of rollers to spread it out. Hot metal is far easier to manipulate than cold, so by pressing it through the rollers, the factory ends up with a really long sheet of copper. This form is far better for making coins than a big lump.

The next stage is to clean the metal. The heating and cooling have given the copper a dirty colour so now it's scraped clean.

And once again, it's sent back to the presses to be rolled out some more. By the time they roll it up like a very shiny roll of toilet paper It's almost 400 meters long.

Copper is just one small part of the technology behind 1 and 2 Euro coins. To buy a drink, some chocolate or even a train ticket, the machine must know you've put the right money into it so the coins are measured and tested rigorously.

The most important test is how magnetic the coin is and that's the high tech bit. Different metals have different magnetic fingerprints. The 2 Euro's special characteristic is the filling for its hole. This is made up of three layers of metal compressed together. There's copper on the top and bottom, and the middle is nickel. To make this filling different metals are fed into the machine and compressed together so hard they can't be separated.

From this sandwich of metal, we can now start to stamp out some coins. This press hammers the sheet 250 times a minute as it cuts the fillings for the big 2 Euro coins. If you look closely, you'll see the nickel sandwich with copper alloy on top, nickel in the middle and copper alloy on the bottom. Meanwhile, in another part of the factory, another machine is stamping out the outer rings. These big hoops will be filled with the copper, nickel sandwich. The middles that have been removed will be returned to the furnace to start the process all over again.

The rings are a little dirty but a quick bath in acid mixed with ball bearings grinds away the filth leaving them sparkling and ready to be turned into money.

So we've got our outer rings, and we've already made up the fillings, but there's still something missing. All of our pieces are blank bits of metal. A template is needed to make turn into coins.

The engravers job is to make the coin equivalent of a rubber stamp. This permanently marks the value onto the coin. Using an industrial press, he crushes a pattern onto the stamp block, it exerts 160 tons of pressure so he's got to be very careful.

The stamps are hardened in this furnace at 870 degrees Celsius. Without this treatment, the motif would disappear very quickly. But once it's been hardened, each stamp can engrave as many 200,000 coins.

The surfaces are scraped clean using fine glass to remove any imperfections. Then they're checked and polished before being put to work. The stamp is fitted into the machine, however, at this point we don't have any complete coins to stamp, but everything's about to come together.

The ring is put into place, and the filling dropped into the hole. The two are then placed under the stamp which hammers them together. This both impresses the value onto the coin and compresses the two pieces together. What emerges from the pipe is a brand new 2 Euro coin, ready to be spent.

So by crushing together some ordinary copper and nickel we end up with the elegant sophistication of the high-tech 2 Euro coin.

Did you know?

The most valuable coin in the world is the 1933 gold Double Eagle $20 piece. It was auctioned in 2002 for an amazing £4.8million.

阿斯米尔花市

扫描二维码，观看中文视频。

这不仅是一朵情人节的花。这是一朵玫瑰，一宗大生意。鲜花买卖是一项巨大的全球产业，荷兰的阿斯米尔市场是世界最大的供应商。它有200个足球场大，成为世界上最大的商业地产。

每天早晨，来自欧洲各地的数百辆卡车、货车和面包车到达这里。他们运来的易损货物是20多万支花和盆栽植物，它们被卸载后储存到冷库中。

为便于能追踪谁在买卖，各批次的每件货都要求有详细信息。然后再装上这些拖车送到拍卖厅。文书工作完成后，数百名工人把拖车按秩序排成整齐的队列。电车系统确保阿斯米尔市场能顺利进行大规模交易。从拍卖开始的一刻起，直到最后的花在夜间售出，电车一直在庞大的仓库中奔忙。这里有16.2千米轨道，它们的工作是把拖车带到一个地方——拍卖厅。

一年中有超过17亿株玫瑰，6亿株郁金香及数百万种其他花卉和盆栽植物在这里出售。质量控制对阿斯米尔的成功至关重要。这么多花木待价而沽，购买者必须能迅速评估。阿斯米尔的专家对每个销售者选取样本和评定级别。从颜色和气味到植物的耐久性都要掂量。买方依赖这些重要信息做出决定。一旦他们掌握了底细，就可以开始采购。

装花的车厢开过来，显示器把植物的信息告诉买主，种者何人，价格多少。但不同于普通拍卖，价格是逐渐降低的，而不是升高。像任何拍卖一样，在阿斯米尔竞拍的窍门是时机。如果你行动过早就可能出价太高。如果您长久等待更低的价格，鲜花将会销售一空。

心理脆弱的人不适合干这工作。有时会犯错，但这里的大部分买家都是经验丰富的老手。所有的竞标信息回馈到主销售员那儿。他在花木离开拍卖厅前，把买家的信息分发到位。员工将相关信息粘贴到拖车上，使花木能找到正确的货车进行装运。

另一个团队直接用电动车把花木送到买主的面包车上。

根据成交后贴在每批货上的凭单，司机能找到正确的买主。买主现在要做的，是核查送来的与购买的是否相符。如有损坏或出错可以退回，一切正常的话，花木就可以装车并运送到他们的商店。每天数百辆卡车离开这个交易站，驶往欧洲各地的花店。

而对于有些旅途漫长的花木，为确保运输中不受损坏，在发货前会包上玻璃纸保护层。

阿斯米尔吸引了全球各地的买家，其中有些花的最终归宿是非常高贵的花瓶，在半个地球之外的国家。这些货物晚间装上飞机，清晨便能到达遥远的莫斯科甚至东京。

所以，当你下次买一束花的时候，它很有可能经过了漫长旅程，从阿斯米尔花市来到你的身边。

你知道吗？

郁金香在17世纪非常珍贵，曾引发过一场“郁金香狂潮”。一株郁金香的售价可达到30万英镑。

1. 阿斯米尔花市有 200 个足球场大
2. 早上从欧洲运来 20 多万株花木
3. 每件货物都要有详细信息
4. 16.2 千米的轨道上拖车排成列
5. 专家定级是买家的重要依据
6. 显示器上有植物的各种信息

7. 价格从高起拍出手时机很重要
8. 主销售员分发买家信息
9. 电车把花木送到买主车上
10. 司机按货单为买主运送花木
11. 买主验收购买的花卉
12. 花木于清晨飞达莫斯科

Aalsmeer flower market

This is not just a flower for Valentine's day. This is a rose and it's big business. Buying and selling flowers is a massive global industry and Aalsmeer market in Holland is the biggest supplier in the world. It's the size of 200 football pitches, making it the biggest commercial property anywhere in the world.

Every morning hundreds of trucks, lorries and vans arrive from all over Europe. Their fragile cargo consists of over 20 million cut flowers and potted plants. Once they've been unloaded they're stored in refrigerated warehouses.

To keep track of who's buying and selling, detailed information is assigned to the individual lots in each shipment. They're then loaded onto these carriages to be taken to the auction rooms. With the paperwork complete, hundreds of workers organise the carriages into orderly queues and join them up. This trolley system is what helps Aalsmeer's huge operation to run so smoothly. From the moment the auctions begin till the last flowers are sold at night, these carts will trundle through the vast warehouses. There's over 16.2 kilometres of track in here their job is to bring the carriages to one place, the auction room.

In the course of a year over 1.7 billion roses, more than 600 million tulips and millions of other assorted flowers and pot plants are sold here. Quality control is vital to Aalsmeer's success. With so many plants on offer, the buyers must be able to assess the produce quickly. Experts at Aalsmeer take samples from each seller and grade them. Everything from colour and smell, to the plants' frailty is assessed. The buyers depend on this vital information for making their decisions. Once they're armed with the facts, they can start shopping.

The flowers are wheeled past on the carriages and a display tells the buyers what the plants are, who grew them and their value. But unlike normal auctions, bids are made as the price goes down, NOT up. Like any auction, the trick at the Aalsmeer is timing. If you bid too early you may pay too much. If you wait too long for a lower price, the flowers might sell out.

It's not for the faint-hearted. Mistakes can be made, but most of the buyers here are experienced operators. All the bidding information is fed to the master salesman. He will then assign the buyers' information to the plants as they leave the hall. The staff attach the information to the carriages so the plants can be delivered to the right lorries for transportation.

扫描二维码，观看英文视频。

Using motorised trolleys, a different team can now deliver the flowers directly to the buyers vans.

Using the stickers that have been placed with each lot after they were bought, the drivers are able to find the right buyer. All the buyers have to do now is to check their delivery tallies with what they've paid for. Any plants that are damaged or not right can be sent back, but if all's well it can be loaded up and delivered to their stores. Hundreds of lorries leave this depot everyday to stock florists all over Europe.

For some flowers however, the journey is far longer. To ensure they aren't damaged on their travels, they're wrapped in a protective layer of cellophane before they're shipped off.

Aalsmeer attracts buyers from all over the globe, and some of these flowers are destined for some very exclusive vases in countries half way around the world. These shipments are loaded onto planes in the evening, and by morning can be as far away as Moscow or even Tokyo.

So, the next time you're buying a bouquet, it's highly likely that it may have made to the long journey to you via the Aalsmeer flower market.

Did you know?

Tulips were very valuable in the 17 century, starting a craze called "Tulipmania". A tulip could sell for the equivalent of £300,000.